Dissimilar Metal Welding

Dissimilar Metal Welding

Special Issue Editors

Pierpaolo Carlone
Antonello Astarita

MDPI • Basel • Beijing • Wuhan • Barcelona • Belgrade

MDPI

Special Issue Editors

Pierpaolo Carlone
University of Salerno
Italy

Antonello Astarita
University of Naples "Federico II"
Italy

Editorial Office
MDPI
St. Alban-Anlage 66
4052 Basel, Switzerland

This is a reprint of articles from the Special Issue published online in the open access journal *Metals* (ISSN 2075-4701) from 2018 to 2019 (available at: https://www.mdpi.com/journal/metals/special_issues/dissimilar_metal_welding).

For citation purposes, cite each article independently as indicated on the article page online and as indicated below:

LastName, A.A.; LastName, B.B.; LastName, C.C. Article Title. *Journal Name* **Year**, *Article Number, Page Range.*

ISBN 978-3-03921-954-4 (Pbk)
ISBN 978-3-03921-955-1 (PDF)

Contents

About the Special Issue Editors

Pierpaolo Carlone has been Associate Professor in Manufacturing Technologies and Systems at the Department of Industrial Engineering of the University of Salerno since 2018. Formerly (since 2007), he was Assistant Professor at the Department of Mechanical Engineering at the same university. He received his Master's Degree (summa cum laude) in Mechanical Engineering at the University of Salerno in 2003 and his Ph.D. in Mechanical Engineering from the same University in 2007. He has been a member of the Doctoral Council of Mechanical Engineering since 2007 and of the Doctoral Council of Industrial Engineering since 2011. He has been a member of the Italian Association of Mechanical Technology (AITeM) and of the European Scientific Association for Material Forming (ESAFORM) since 2004. In 2014, he was elected to the Board of Directors of the ESAFORM. He has been member of the ASM International since 2016.His research interests are mainly focused on advanced and light-weight materials processing, in particular composites manufacturing and solid state similar and dissimilar metal welding. He is author/co-author of approximately 100 papers published in reputaable scientific journals and proceedings of international conferences, one scientific book, and some book chapters.

Antonello Astarita is Assistant Professor at the University of Naples "Federico II". He achieved received his Master's Degree in Mechanical Engineering in 2008 and his Ph.D. in Manufacturing Engineering in 2013, both at the University of Naples. He was visiting scientist at the University of Manchester and visiting Professor at the University of Cadiz, and has also worked as external consultant for several companies. His research activities are in the field of material science, and in particular in the innovative processes related to the production of titanium and aluminium parts. Formerly, he focused his studies on the solid joining of metals for aeronautical applications and on the precision forming of titanium parts. In recent years, he has focused his research interests on additive manufacturing techniques, such as electron beam melting, selective laser melting, and cold dynamic spray deposition. He is co-author of more than 50 journal papers on these topics.He received the "ASM-IMM Visiting Lectureship Award" in 2016 from ASM International and the "ESAFORM Scientific Prize" in 2018 from the European Scientific Association for Forming Materials. Dr. Astarita is member of the Editorial Board of *Surface Engineering and Journal of Materials Engineering and Performances*; he was also in the scientific committee of several international conferences. He teaches manufacturing processes at the University of Naples and advanced materials and manufacturing at the University of Salerno. He is also member of ASM International, ESAFORM, and the Italian Association for Manufacturing (AITEM).

metals **MDPI**

Editorial

Dissimilar Metal Welding

Pierpaolo Carlone [1],* and Antonello Astarita [2],*

[1] Department of Industrial Engineering, University of Salerno, 84084 Fisciano (SA), Italy
[2] Department of Chemical, Materials and Production Engineering, University of Naples "Federico II", 80138 Naples, Italy
* Correspondence: pcarlone@unisa.it (P.C.); antonello.astarita@unina.it (A.A.);
  Tel.: +39-089-964310 (P.C.); +39-081-7682364 (A.A.)

Received: 30 October 2019; Accepted: 7 November 2019; Published: 9 November 2019

1. Introduction and Scope

The combination of distinct materials provides intriguing opportunities in modern industry applications, whereas the driving concept is to design parts with the right material in the right place. Consequently, a great deal of attention has been directed towards dissimilar welding and joining technologies. In the automotive sector, for instance, the concept of "tailored blanks", introduced in the last decade, has further highlighted the necessity to combine dissimilar materials. As far as the aeronautic field is concerned, most structures are built combining very different materials and alloys in order to match lightweight and structural performance requirements.

In this framework, the application of fusion welding techniques, namely tungsten inert gas or laser welding, is quite challenging due to the difference in physical properties, in particular the difference in melting points between adjoining materials. On the other hand, solid state welding methods, for example, friction stir welding and linear friction welding processes, have already proved to be capable of manufacturing sound Al-Cu [1], Al-Ti [2], Al-Mg [3,4] joints. Recently, promising results have also been obtained using hybrid methods. The main focus of this special issue is to discuss some recent advances in the field of dissimilar metal joining. Selected applications of major welding and joining processes have been highlighted. Special attention has been given to mechanisms behind the joining of dissimilar metals for special purpose applications, investigating the adoption of traditional experimental approaches in addition to computational modelling, for deeper information gathering. In the following section an overview of the selected articles is provided.

2. Contributions

This special issue of *Metals* covers sixteen articles [5–20] focused on dissimilar metal joining techniques. Some of the published reports have confirmed the increasing interest in solid state welding processes, in particular friction based welding [6–10] and electromagnetic pulse welding [11,12], due to benefits related to the properties and achievable microstructure, and to energy and environmental considerations. Other papers dealt with fusion welding techniques, mainly laser based [13–16], among others [5,17], and brazing processes [18,19]. Most of the applications are related to the automotive and aerospace sector, nevertheless dissimilar joints, characterized by improved fracture resistance, were indicated as an indispensable part of nuclear power plants for connecting the safe end (austenitic stainless steel 316L) to the pipe-nozzle (ferrite low-alloy steel A508) [20].

More specifically, a thorough review article proposed by Patel et al., has shed light on the potential of friction stir welding (FSW) in dissimilar welding of distinct aluminium alloys [6], commenting on microstructure, mechanical properties, corrosion and fatigue behaviour. The authors discussed in detail, aspects related to the processing parameters and setup, in terms of placement of the adjoining materials and tool offsets. Furthermore, pros and cons given by the application of bobbin and stationary shoulder tools were evidenced as well. Li et al. demonstrated the capabilities of FSW to mitigate some limiting factors associated with Al/steel fusion welding, attributable to the formation of brittle Al/Fe intermetallic compounds (i.e., AlFe3, AlFe, Al2Fe, Al3Fe, Al5Fe2, and Al6Fe), welding distortion, cavities, and cracks, providing some intriguing opportunities for the automotive industry [7]. In particular the effect of revolutionary pitch on interface microstructure and mechanical behaviour of friction stir lap welds of AA6082-T6 to galvanized DP800 dual-phase steel sheets was investigated. The experimental results were commented on, taking into account numerical calculations provided by an iso-strain-based linear mixture law of the stir zone [7]. The automotive sector has witnessed the emerging trend of incorporating Cu-based materials in electrical components. The solid state joining of dissimilar Cu alloys, and of Cu alloys with Al alloys, is the focus of articles [8,9], respectively. In the former, Sun et al. successfully welded dissimilar CuNiCrSi and CuCrZr in a butt joint configuration using FSW. The microstructure and mechanical properties were investigated, highlighting the absence of the typical heat affected zone [8]. The transformation of coarse grains in the base metal (BM) into fine equiaxed grains in the nugget zone (NZ) was observed. In the latter, Eslami et al. pointed out that an adjustment of the cross-section is required to realise electrical conductors free of resistive losses [9]. In [10], Zhou et al. carried out friction stir spot welding-brazing of aluminium alloy and a hot-dip aluminized titanium alloy, using a Zn interlayer to extend the extremely narrow joining area, generally addressed as the main drawback of FSSW process. The formation of the brazing zone between the Al alloy and Al coating on Ti6Al4V alloy was successfully introduced by the addition of a Zn interlayer. A dramatic enhancement of the fracture load was proved using this hybrid technique.

Magnetic pulse welding (MPW) is an eminent impact welding process which utilizes the high-speed collision between two metallic surfaces in order to promote the creation of metallurgical connections. Bellmann et al. [11] discussed the influence of temperature in dissimilar MPW, assuming aluminium alloy EN AW-6060 for the outer tube and C45 steel for the inner rod [11]. Their experiments showed that jetting in a strong material flow was not mandatory for a successful MPW process. A cloud of particles ejected during the impact, with lower velocities, can in turn enable welding. Faes et al. investigated electromagnetic pulse welding process to join copper to steel tubular elements, comprehensively discussing the role of stand-off distance and discharge energy [12].

As far as dissimilar fusion welding processes are concerned, relevant efforts have been directed toward laser based methods [13–16]. Dual-beam laser welding has been investigated for dissimilar welding of steel/Al [13]. Cui et al. studied the effect of the major process parameters, including the dual-beam power ratio and dual-beam distance on steel/Al joint features, in terms of weld shape, interface microstructure, tensile resistance and fracture behaviour. Intermetallic compound (IMC) layer formation (needle-like θ-Fe4Al13 phases) was also highlighted [13]. The article by Pereira et al. [14] deals with dissimilar metal laser welding between DP1000 Steel and AA1050, by employing a pulsed Nd: YAG laser. Welding parameters such as laser beam power, laser beam diameter, pulse duration and welding speed were optimized for the obtainment of a better set of weld joints, even for highly dissimilar materials. On the similar note, Xue et al. [15] investigated the interfacial features of a dissimilar Ti6Al4V/AA6060 lap joint produced by pulsed Nd:YAG laser beam welding. The potential phases, TiAl, TiAl2, and TiAl3, were observed near the Ti/Al interface. The phase change was situated mainly in the Al-rich melted zone. By using an orthogonal experimental design method, the sensitivity order of the selected key process parameters on peak shear strength were: overlap, duration, laser beam diameter and power. Jarwitz et al. also focused on laser beam welding of different set of materials, in order to clarify the influence of the oscillation parameters on the weld seam geometry, and the implications on the electrical resistance of the joints [16].

Xue et al. [18] inspected the microstructure and properties of a Cu/304 stainless steel dissimilar metal joint brazed with a low silver Ag16.5CuZnSn-xGa-yCe braze filler after aging treatment. The addition of Ce reduced the intergranular penetration depth of the filler metal into the stainless steel during the aging process by 48.8%. The Ag16.5CuZnSn-2Ga-0.15Ce brazed joint showed optimum performance compared to the other joints. Yu et al. proposed the method of welding/brazing to realise a high quality welding of dissimilar metals, using 5A06 aluminium and galvanized steel welding using laser beam as the main heat source, and a trailing arc in an assisting role [19]. Under suitable welding parameters, a sound welding seam was obtained. The highest tensile strength was observed to be 163 MPa, which was nearly 74% 5A06 aluminium alloy when the fracture occurred at the weld seam. Near the aluminium welding brazing seam, two different IMC formations appeared [19].

3. Conclusions and Outlook

A varying range of dissimilar welding processes and configurations have been discussed. Evidently, the major focus in these investigations was to overcome the challenges posed by dissimilar metal joining and to achieve sound joints with mechanical and metallurgical property changes. The usage of solid state and hybrid/mixed techniques have yielded interesting results in terms of joint performance. Nevertheless, there are still many challenges to address, related to both material and processing aspects.

Acknowledgments: Editors would like to extend their sincere thanks to all reviewers for their invaluable efforts in the improvement of the quality of this special issue.

Conflicts of Interest: The author declares no conflict of interest.

References

1. Carlone, P.; Astarita, A.; Palazzo, G.S.; Paradiso, V.; Squillace, A. Microstructural aspects in Al–Cu dissimilar joining by FSW. *Int. J. Adv. Manuf. Technol.* **2015**, *79*, 1109–1116. [CrossRef]
2. Jeong-Won, C.; Huihong, L.; Hidetoshi, F. Dissimilar friction stir welding of pure Ti and pure Al. *Mater. Sci. Eng. A* **2018**, *730*, 168–176.
3. Boccarusso, L.; Astarita, A.; Carlone, P.; Scherillo, F.; Rubino, F.; Squillace, A. Dissimilar friction stir lap welding of AA 6082-Mg AZ31: Force analysis and microstructure evolution. *J. Manuf. Processes* **2019**, *44*, 376–388. [CrossRef]
4. Mehta, K.P.; Carlone, P.; Astarita, A.; Scherillo, F.; Rubino, F.; Vora, P. Conventional and cooling assisted friction stir welding of AA6061 and AZ31B alloys. *Mater. Sci. Eng. A* **2019**, *759*, 252–261. [CrossRef]
5. Shin, S.; Park, D.-J.; Yu, J.; Rhee, S. Resistance Spot Welding of Aluminum Alloy and Carbon Steel with Spooling Process Tapes. *Metals* **2019**, *9*, 410. [CrossRef]
6. Patel, V.; Li, W.; Wang, G.; Wang, F.; Vairis, A.; Niu, P. Review—Friction Stir Welding of Dissimilar Aluminum Alloy Combinations: State-of-the-Art. *Metals* **2019**, *9*, 270. [CrossRef]
7. Li, S.; Chen, Y.; Kang, J.; Amirkhiz, B.S.; Nadeau, F. Effect of Revolutionary Pitch on Interface Microstructure and Mechanical Behavior of Friction Stir Lap Welds of AA6082-T6 to Galvanized DP800. *Metals* **2018**, *8*, 925. [CrossRef]
8. Sun, Y.; He, D.; Xue, F.; Lai, R.; He, G. Microstructure and Mechanical Characterization of a Dissimilar Friction-Stir-Welded CuCrZr/CuNiCrSi Butt Joint. *Metals* **2018**, *8*, 325. [CrossRef]
9. Eslami, N.; Harms, A.; Deringer, J.; Fricke, A.; Böhm, S. Dissimilar Friction Stir Butt Welding of Aluminum and Copper with Cross-Section Adjustment for Current-Carrying Components. *Metals* **2018**, *8*, 661. [CrossRef]
10. Zhou, X.; Chen, Y.; Li, S.; Huang, Y.; Hao, K.; Peng, P. Friction Stir Spot Welding-Brazing of Al and Hot-Dip Aluminized Ti Alloy with Zn Interlayer. *Metals* **2018**, *8*, 922. [CrossRef]
11. Bellmann, J.; Lueg-Althoff, J.; Schulze, S.; Hahn, M.; Gies, S.; Beyer, E.; Tekkaya, A.E. Thermal Effects in Dissimilar Magnetic Pulse Welding. *Metals* **2019**, *9*, 348. [CrossRef]
12. Faes, K.; Kwee, I.; de Waele, W. Electromagnetic Pulse Welding of Tubular Products: Influence of Process Parameters and Workpiece Geometry on the Joint Characteristics and Investigation of Suitable Support Systems for the Target Tube. *Metals* **2019**, *9*, 514. [CrossRef]
13. Cui, L.; Chen, H.; Chen, B.; He, D. Welding of Dissimilar Steel/Al Joints Using Dual-Beam Lasers with Side-by-Side Configuration. *Metals* **2018**, *8*, 1017. [CrossRef]
14. Pereira, A.B.; Cabrinha, A.; Rocha, F.; Marques, P.; Fernandes, F.A.; Alves de Sousa, R.J. Dissimilar Metals Laser Welding between DP1000 Steel and Aluminum Alloy 1050. *Metals* **2019**, *9*, 102. [CrossRef]
15. Xue, X.; Pereira, A.; Vincze, G.; Wu, X.; Liao, J. Interfacial Characteristics of Dissimilar Ti6Al4V/AA6060 Lap Joint by Pulsed Nd:YAG Laser Welding. *Metals* **2019**, *9*, 71. [CrossRef]
16. Jarwitz, M.; Fetzer, F.; Weber, R.; Graf, T. Weld Seam Geometry and Electrical Resistance of Laser-Welded, Aluminum-Copper Dissimilar Joints Produced with Spatial Beam Oscillation. *Metals* **2018**, *8*, 510. [CrossRef]

17. Dokme, F.; Kulekci, M.K.; Esme, U. Microstructural and Mechanical Characterization of Dissimilar Metal Welding of Inconel 625 and AISI 316L. *Metals* **2018**, *8*, 797. [CrossRef]
18. Xue, P.; Zou, Y.; He, P.; Pei, Y.; Sun, H.; Ma, C.; Luo, J. Development of Low Silver AgCuZnSn Filler Metal for Cu/Steel Dissimilar Metal Joining. *Metals* **2019**, *9*, 198. [CrossRef]
19. Yu, X.; Fan, D.; Huang, J.; Li, C.; Kang, Y. Arc-Assisted Laser Welding Brazing of Aluminum to Steel. *Metals* **2019**, *9*, 397. [CrossRef]
20. Yang, J.; Wang, L. Optimizing the Local Strength Mismatch of a Dissimilar Metal Welded Joint in a Nuclear Power Plant. *Metals* **2018**, *8*, 494. [CrossRef]

metals

MDPI

Review

Friction Stir Welding of Dissimilar Aluminum Alloy Combinations: State-of-the-Art

Vivek Patel [1,2], Wenya Li [1,*], Guoqing Wang [3], Feifan Wang [3], Achilles Vairis [1,4] and Pengliang Niu [1]

[1]   State Key Laboratory of Solidification Processing, Shaanxi Key Laboratory of Friction Welding Technologies, School of Materials Science and Engineering, Northwestern Polytechnical University, Xi'an 710072, China; profvvp@yahoo.com or vivek.patel@sot.pdpu.ac.in (V.P.); vairis@staff.teicrete.gr (A.V.); penglniu@126.com (P.N.)

[2]   Department of Mechanical Engineering, School of Technology, Pandit Deendayal Petroleum University, Gandhinagar 382007, India

[3]   China Academy of Launch Vehicle Technology, Beijing Institute of Astronautical Systems Engineering, Beijing 100076, China; vivsforyou@gmail.com (G.W.); wangifw@hotmail.com (F.W.)

[4]   Mechanical Engineering Department, TEI of Crete, Heraklion 71004, Greece

*   Correspondence: liwy@nwpu.edu.cn; Tel.: +86-29-8849-5226; Fax: +86-29-8849-2642

Received: 17 January 2019; Accepted: 11 February 2019; Published: 26 February 2019

Abstract: Friction stir welding (FSW) has enjoyed great success in joining aluminum alloys. As lightweight structures are designed in higher numbers, it is only natural that FSW is being explored to join dissimilar aluminum alloys. The use of different aluminum alloy combinations in applications offers the combined benefit of cost and performance in the same component. This review focuses on the application of FSW in dissimilar aluminum alloy combinations in order to disseminate research this topic. The review details published works on FSWed dissimilar aluminum alloys. The detailed summary of literature lists welding parameters for the different aluminum alloy combinations. Furthermore, auxiliary welding parameters such as positioning of the alloy, tool rotation speed, welding speed and tool geometry are discussed. Microstructural features together with joint mechanical properties, like hardness and tensile strength measurements, are presented. At the end, new directions for the joining of dissimilar aluminum alloy combinations should guide further research to extend as well as to improve the process, which is expected to raise further interest on the topic.

Keywords: aluminum; dissimilar; friction stir welding; FSW; hardness; microstructure; tensile

---

1. Introduction

Friction stir welding (FSW) is a solid state welding process which was invented at The Welding Institute (TWI) in UK in 1991 [1]. FSW is regarded as an environmentally friendly and energy efficient joining technique providing one of the best alternatives to fusion welding in order to produce a good combination of microstructure and properties in the joints. FSW has already proved its superiority in joining aluminum (Al) alloys as well as magnesium (Mg) alloys over fusion welding processes because of its solid-state nature. FSW uses a non-consumable rotating tool which has a shoulder and a pin (or more formally probe) at its end which plunges into the base material (BM) and advances in the welding direction [2], as shown in Figure 1. During the process, the shoulder touches the top surface of the BM and the pin moves yielded material around it. As a result of this action, heat is generated by frictional and plastic deformation of the BM by advancing the rotating tool. The shoulder of the tool has a forging action as it restricts the expulsion of plasticized material from the BM, while the pin extrudes material and produces a material flow between the advancing side (AS) and the retreating side (RS) of the joint. FSW has shown great potential in welding Al alloys for structural applications. More recently, Ma et al. [3] published a critical review paper on recent developments in FSW of Al alloys. Al alloys have remained the prime selection for structural material in aerospace, shipbuilding and automotive industries for their excellent strength to weight ratio. In order to improve performance while controlling the cost of Al alloys in these industries, there is an increasing demand to weld dissimilar Al joints with FSW. Because of the different physical and chemical properties in dissimilar Al alloy combinations, challenges such as solidification cracking, porosity, formation of intermetallic and so forth, are present. Therefore, the FSW of dissimilar Al alloy combinations has gained attention over the recent years, demonstrating the potential of the process to join these. The present review aims to discuss and analyze the available literature on FSWed dissimilar Al alloy combinations so far.

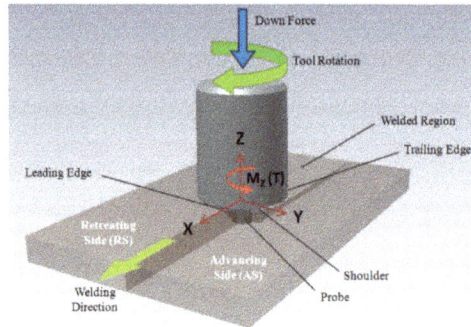

Figure 1. Schematic of friction stir welding, reproduced from [4], with permission from Elsevier, 2014.

2. General Progress in FSW of Dissimilar Al-Al Combinations

There are review papers available on FSW of same Al alloy joints, which discuss various aspects of the process such as tool design, process parameters, heat generation, microstructure and mechanical properties [4–11]. The number of research papers on FSW of dissimilar Al alloy joints published to date is shown in Figure 2 (search on 15 December 2017 found 68 papers from Web of Science). The vast majority of the publications has been in the past 5 years, reaching a peak on 2018. In addition, Magalhães et al. [12] studied research and the extent of industrial application of FSW of similar and dissimilar material joints as shown in Figure 3. The similar material joints of Al alloys are being studied to a far larger extent compared to other alloys and the same trend is observed in the dissimilar material combinations. This trend observed literature clearly identifies the interest on the FSW of dissimilar Al alloy joints, which is expected to increase over the coming years.

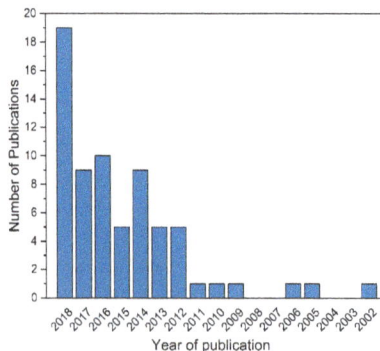

Figure 2. Journal papers published on FSW of dissimilar Al alloy joints.

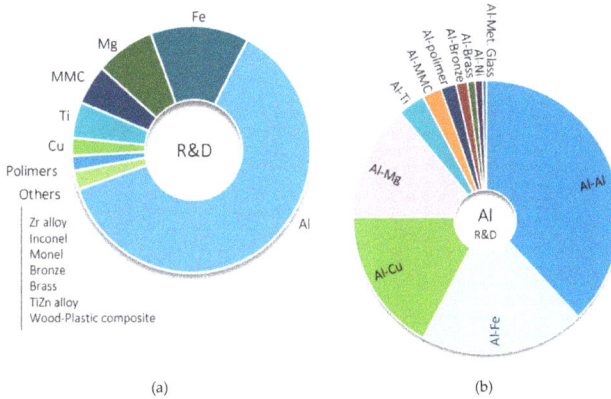

Figure 3. Papers on FSW: (**a**) same material joints and (**b**) dissimilar materials joints, reproduced from [12], with permission from Taylor & Francis, 2017.

All papers from the top 10 ranked journals published on FSW, classified as Q1 by Scimago Journal & Country Rank.

Summary of Published Works

In order to identify the key findings on various aspects a summary of existing literature follows (Table 1). For the FSW of dissimilar Al alloy combinations there are the preliminary welding parameters such as the BM placement, the tool rotational speed and welding speed. The placement of the BM affects material flow, while rotational and welding speeds control heat input on both sides of the joint during welding. All of these parameters have been investigated for the different material combinations (see Table 1). In addition, the effects of welding parameters on the mechanical properties that is, the hardness and the joint strength have been investigated. As it can be seen a number of studies have been performed on the effect of the placement of BM (i.e., whether a particular material is placed on the AS or the RS side) on the material flow and the resulting microstructure in the SZ and the mechanical properties of the weld. Other papers have focused on the effect of tool geometry that is, shoulder diameter to pin diameter ratio and pin profile (cylindrical, conical, polygonal) on the microstructure and mechanical properties of the weld.

Table 1. Summary of FSW of dissimilar Al alloy joints studied in literature.

No.	Author (s)	Thick (mm)	Alloy Positioning		Welding Parameters		Objective of Study
			AS	RS	Rotation Speed (rpm)	Welding Speed (mm/min)	
1	Niu, et al. [13]	6.35	2024	5083	600	150	Strain hardening behavior and mechanism
2	Niu, et al. [13]	6.35	7075	2024	600	150	Strain hardening behavior and mechanism
3	Hasan, et al. [14]	6	Both	both	900	150	Effect of pin flute radius and alloy positioning
4	Ge, et al. [15]	7075-T6 & 2024-T3 Lap joint: 7075-upper; 2024-lower	3	NA	NA	600	30, 60, 90, 120	Effect of pin length and welding speed
5	Kalemba-Rec, et al. [16]	6	Both	Both	280, 355, 450, 560	140	Influence of tool rotation speed, pin geometry and alloy positioning
6	Safarbali, et al. [17]	4	2024	7075	1140	32	Effect of post-weld treatment
7	Palanivel, et al. [18]	6	6351	5083	800, 1000, 1200	45, 60, 75	Optimization of shoulder profile, rotational speed and welding speed
8	Hamilton, et al. [19]	6	Both	Both	355	112	Phase transformation maps
9	Gupta, et al. [20]	5083-O & AA6063-T6	6	NR	NR	700, 900, 1100	40, 60, 80	Optimization of tool geometry, rotational speed and welding speed
10	Huang, et al. [21]	5052&AlMg₂Si	8	Al-Mg₂Si	5052	1000	80	Microstructure and mechanical properties
11	Moradi, et al. [22]	2024-T351& 6061-T6	6	2024	6061	800	31.5	Texture evolution
12	Prasanth and Raj [23]	6061-T6 & 6351-T6	6.35	NR	NR	600, 900, 1200	30, 60, 90	Optimization of rotational speed, welding speed and axial force
13	Azeez and Akinlabi [24]	6082-T6 & 7075-T6	10	7075	6082	950, 1000	80, 100	Double-sided weld
14	Azeez, et al. [25]	6082-T6 & 7075-T6	10	7075	6082	950, 1000	80, 100	Single-sided weld
15	Peng, et al. [26]	6061-T651 & 5A06-H112	5	6061	5A06	600, 900, 1200	100, 150	Nanoindentation hardness and fracture behavior
16	Das and Toppo [27]	6101-T6 & 6351-T6	12	6101	6351	900, 1100, 1300	16	Influence of rotational speed on temperature and impact strength
17	Sarsilmaz [28]	2024-T3 & 6063-T6	8	2024	6063	900, 1120, 1400	125, 160, 200	Microstructure, tensile and fatigue behavior
18	Kookil, et al. [29]	2219-T87 & 2195-T8	7.2	Both	Both	400, 600, 800	120, 180, 240, 300	Effect of rotational speed and welding speed
19	Hamilton, et al. [30]	2017A-T451 & 7075-T651	6	Both	Both	355	112	Positron lifetime annihilation spectroscopy
20	Kopyscianski, et al. [31]	2017A-T451 & Cast AlSi9Mg	6	2017A	AlSi9Mg	355	112	Microstructural study
21	Ghaffarpour, et al. [32]	5083-H12 & 6061-T6	1.5	6061	5083	700, 1800, 2500	25, 30, 212.5, 400	Optimization of rotational speed, welding speed and tool dimensions

Table 1. *Cont.*

No.	Author (s)	Alloy Combinations	Thick (mm)	Welding Parameters				Objective of Study
				Alloy Positioning		Rotation Speed (rpm)	Welding Speed (mm/min)	
				AS	RS			
22	Bijanrostami, et al. [33]	6061-T6 & 7075-T6	5	6061	7075	1000, 1375, 1750, 2125, 2500	50, 125, 200, 275, 350	Underwater FSW: optimizations of rotational and welding speeds on tensile properties
23	Kasman, et al. [34]	5083-H111& 6082-T6	5	NR	NR	400, 500, 630, 800	40, 50, 63, 80	Effect of probe shape, rotational speed, welding speed.
24	Palanivel, et al. [35]	5083-H111 & 6351-T6	6	6351	5083	800-1200	45-85	Macrostructure examination at different rotational and welding speeds
25	Doley and Kore [36]	5052 & 6061	1, 1.5	6061	5052	1500	63, 98	Study of welding speed
26	Saravanan, et al. [37]	2024-T6 & 7075-T6	5	2024	7075	1200	12	Effect of shoulder diameter to probe diameter
27	Yan, et al. [38]	Al-Mg-Si & Al-Zn-Mg	15	Both	Both	800	180	Effect of alloy positioning on fatigue property
28	Yan, et al. [39]	Al-Mg-Si & Al-Zn-Mg	15	Both	Both	800	180	Study of Fatigue behavior
29	Hamilton, et al. [40]	2017A-T451 & 7075-T651	6	Both	Both	355	112	Numerical simulation
30	Zapata, et al. [41]	2024-T3 & 6061-T6	4.8	2024	6061	500, 650, 840	45, 65	Effect of rotational and welding speeds on residual stress
31	Sun, et al. [42]	UFG 1050 & 6061-T6	2	Both	Both	800	400, 600, 800, 1000	Microstructure and mechanical properties at different welding speeds
32	Texier, et al. [43]	2024-T3 & 2198-T3	3.18	2198	2024	NR	NR	Heterogeneities in microstructure and tensile properties at the shoulder-affected regions
33	Rodriguez, et al. [44]	6061-T6 & 7050-T7451	5	7050	6061	270, 340, 310	114	Fatigue behavior
34	Yoon, et al. [45]	6111-T4 & 5023-T4 Lap joint	1	NA	NA	1500 1000	100 700	Mechanism of onion ring formation
35	Rodriguez, et al. [46]	6061-T6 & 7050-T7451	5	7050	6061	270, 340, 310	114	Microstructure and mechanical properties
36	Ilangovan, et al. [47]	5086-O & 6061-T6	6	6061	5086	1100	22	Effect of probe profiles
37	Reza-E-Rabby, et al. [48]	2050-T4 & 6061-T651	20	Both	Both	150 300 300	101 203 406	Effect of probe features
38	Donattus, et al. [49]	5083-O & 6082-T6	NR	5083	6082	400	400	Anodizing behavior
39	Karam, et al. [50]	A319 & A413 cast	10	A413	A319	630, 800, 1000	20, 40, 63	Influence of rotational and welding speed
40	Ipekoglu and Cam [51]	7075-O & 6061-O 7075-T6 & 6061-T6	3.17	6061	7075	1000 1500	150 400	Effect of initial temper conditions and postweld heat treatment
41	Cole, et al. [52]	6061-T6 & 7075-T6	4.6	Both	Both	700-1450	100	Effect of temperature
42	Song, et al. [53]	2024-T3 & AA7075-T6 Lap joint	5	NA	NA	1500	50, 150, 225, 300	Effect of alloy positioning and welding speed on defects and mechanical properties

Table 1. *Cont.*

No.	Author (s)	Alloy Combinations	Thick (mm)	Alloy Positioning		Welding Parameters		Objective of Study
				AS	RS	Rotation Speed (rpm)	Welding Speed (mm/min)	
43	Jannet and Mathews [54]	5083-O & 6061-T6	6	6061	5083	600, 750, 900	60	Effect of rotational speed
44	Palanivel, et al. [55]	6351-T6 & 5083-H111	6	6351	5083	950	36, 63, 90	Effect of welding speed
45	Jonckheere, et al. [56]	2014-T6 & 6061-T6	4.7	Both	Both	500, 1500	90	Effect of alloy positioning and tool offset on temperature and hardness
46	Palanivel, et al. [57]	6351-T6 & 5083-H111	6	6351	5083	600-1300	36-90	Optimization of process parameters (probe shapes, rotational and welding speeds, axial force) for UTS
47	Ghosh, et al. [58]	A356 & 6061-T6	3	6061	A356	1000	70-240	Effect of welding speed
48	Velotti, et al. [59]	2198-T351 & 7075-T6 Lap joint	3 & 1.9	NA	NA	830	40	Stress corrosion cracking investigation
49	Koilraj, et al. [60]	2219-T87 & 5083-H321	6	2219	5083	400-800	15-60	Optimization of process parameters (probe shapes, rotational and welding speeds, shoulder to probe diameter ratio) for UTS
50	Dinaharan, et al. [61]	6061 cast &6061 rolled	6	Both	Both	800, 1000, 1200, 1400	50	Effect of rotational speed and alloy positioning
51	Palanivel, et al. [62]	6351-T6 & 5083-H111	6	6351	5083	600, 950, 1300	60	Effect of rotational speed and probe profile
52	Song, et al. [63]	5052-H34 & 5023-T4	~1.5	5052	5023	1500	100-700	Liquation cracking study
53	Ghosh, et al. [64]	A356 & 6061-T6	3	6061	A356	1000, 1400	80, 240	Effect of rotational and welding speed
54	Kim, et al. [65]	5052-H34 & 5023-T4	1.5 & 1.6	Both	Both	1000, 1500	100, 200, 300, 400	Effect of alloy positioning
55	Prime, et al. [66]	7050-T7451 & 2024-T351	25.4	2024	7050	NR	50.8	Residual stress study
56	Miles, et al. [67]	5182-O & 5754-O 5182-O & 6022-T4 5754-O & 6022-T4	~2	NR	NR	500, 1000, 1500	130, 240, 400	Formability study
57	Ouyang and Kovacevic [68]	6061-T6 & 2024-T3	12.7	NR	NR	637	133	Material flow study

3. Welding Parameters

3.1. Positioning of Alloy

The placement of the alloy affects material flow as it strongly influences material stirring and mixing. This can be crucial in the final joint microstructure when the BM combination selected have significant differences in mechanical properties [69,70]. As the material flow during FSW is quite complex on its own, the placement of materials becomes an important parameter in welding, similar to the importance of the rotation and the welding speeds (see Table 1). For example Yan et al. [38] showed this for the Al-Zn-Mg and the Al-Mg-Si combination. There is an interesting material flow resistance behavior at the RS due to the difference in mechanical properties. When the Al-Zn-Mg alloy is placed at the AS, there was limited movement of the Al-Mg-Si alloy material to the AS side because of its stronger ability to flow as shown in Figure 4a. When the Al-Mg-Si was placed at the RS, there was no RS material (Al-Zn-Mg) flow to AS due to the strong resistance to flow by this high strength material as shown in Figure 4b. As it can be seen from Figure 4, the zig-zag line bonding interface formed due to excellent material mixing. The bonding interface may have vortex type in case of poor combination of rotational speed and welding speed and it becomes more prominent for BMs with significant difference in the properties. Niu et al. [71] investigated an AA2024-AA7075 joint and found that the top section of the SZ was composed of the BM of RS, whereas the middle and bottom sections by the BM of AS as shown in Figure 5. Kim et al. [65] also showed that by placing the high strength Al alloy on the AS generates excessive agglomerations and defects due to limited material flow. In essence, the high strength Al should be placed at the RS to minimize the effect of the resistance to material flow.

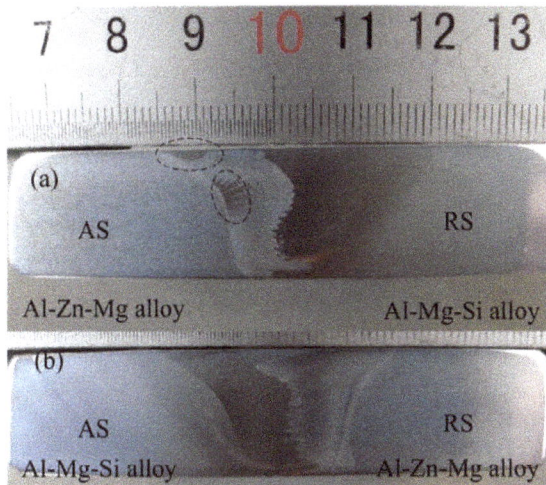

Figure 4. Cross sectional photos of the joints: (**a**) AS: Al-Zn-Mg and (**b**) AS: Al-Mg-Si, reproduced from [38], with permission from Elsevier, 2016.

In the case of the lap joint, the BM placement affects the material flow and leads to the generation of the ubiquitous hook defect. Now the material movement is in an upward direction that is, from the bottom sheet to the top sheet, creating hook defects of various sizes. As expected, in addition to the rotation and welding speed, the placement of the BM affects the hook size as well [53,72–74]. As it can be seen from Figure 6, the hook height is larger at the RS when the AA2024 is placed at the top, while it decreases when the AA7075 is placed as a top plate.

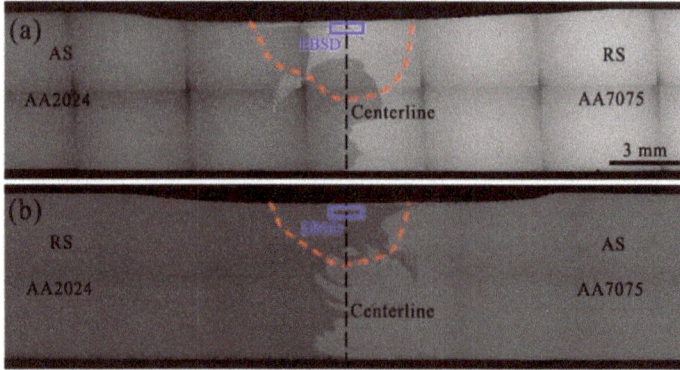

Figure 5. Cross-sectional SEM macrostructure of the AA2024–AA77075 joints: (**a**) AS: AA2024 and (**b**) AS: AA7075, reproduced from [71], with permission from Elsevier, 2019.

Figure 6. Cross sections of lap joints produced at various welding speeds: AA2024 as top plate (**a**) 50, (**b**) 150, (**c**) 225, (**d**) 300 mm/min; AA7075 as top plate (**e**) 50, (**f**) 150, (**g**) 225, (**h**) 300 mm/min, reproduced from [53], with permission from Elsevier, 2014.

3.2. Tool Rotation and Welding Speeds

Tool rotation and welding speeds control heat generation or heat input as they relate to the material plastic flow during FSW. The tool rotation speed affects the intensity of plastic deformation and through this affects material mixing. Kalemba-Rec et al. [16] showed a proportional relationship between material mixing and tool rotation speed for a dissimilar AA7075-AA5083 joint. However, very large rotation speeds lead to numerous imperfections such as poor surface (flash), voids, porosity, tunneling or formation of wormholes because of the excessive heat input [75–77], as shown in Figure 7. Low welding speeds increase heat input and are associated with defects like tunneling [55,58,75,78,79]. It is therefore necessary to select the appropriate combination of tool rotation and welding speed for a defect free joint with a good metallurgical bond and mechanical properties. As it can be seen in Table 1, quite a lot of papers have focused on the optimization of these parameters for different combinations of Al alloys [23,29,32,33,35,36,41,42,50,54,55,58,64,80].

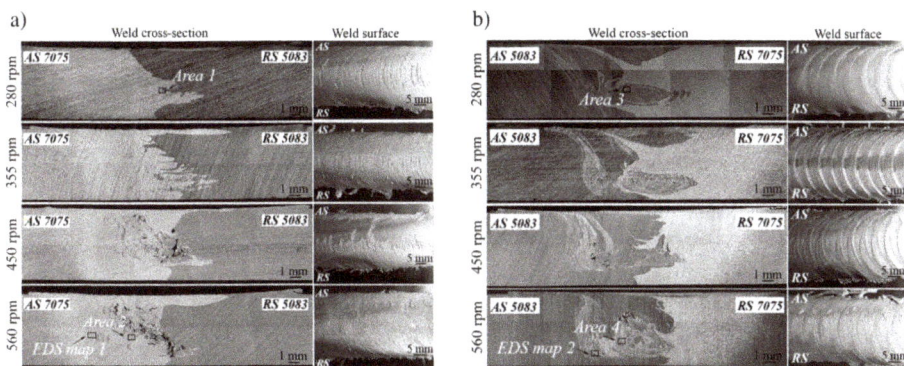

Figure 7. Cross sectional and top surface photos of an (**a**) AS 7075–RS 5083 weld and (**b**) an AS 5083–RS 7075 weld (AS—advancing side, RS—retreating side), whereas marked areas indicate further microstructural analysis; Triflute pin employed [16], with permission from the authors.

3.3. Tool Geometry

The geometry of the shoulder and the pin profile govern heat generation and material flow during welding [81]. The shoulder contributes to a large extent to heat input due to its size. The common shoulder profiles employed are the flat, the concave and the convex. Additional features on the pin such as a spiral or a groove improve frictional behavior as well as material flow. Palanivel et al. [18] reported on the effect of shoulder profiles on the AA5083-AA6351 combination by using three different shoulder features, the partial impeller (PI), the full impeller (FI) and the flat grove (FS) as shown in Figure 8. The full impeller shoulder tool produced the optimum mechanical strength due to the enhanced material flow it produced. The pin profile greatly affects material stirring and mixing. Cylindrical or conical pin profiles which may have features like threads or threads with flats have been used for dissimilar Al alloy combinations as shown in Figure 8. When used without threads a smaller surface is provided to the material, while the threaded and flat features on it increase the contact area while threads guide material flow around the pin in a rotational as well as a translation direction [14,16,47,82]. The polygonal pin profiles produce pulses in the flow during material stirring and mixing, leading to material adhering to the pin [83–86]. This pulsating effect hinders material flow significantly in the case of dissimilar Al alloy combinations. It is therefore recommended to use a cylindrical or a conical pin profile with various features in the dissimilar Al alloy joints for good material flow to produce sound joints.

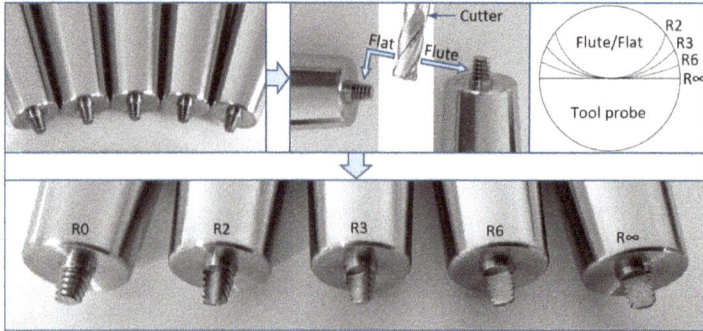

(a)

(b)

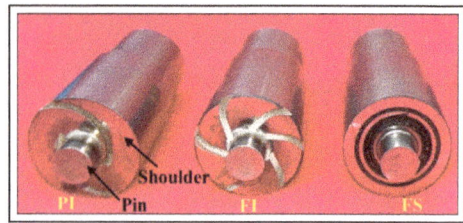

(c)

(d)

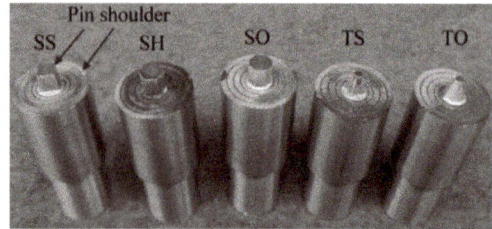

(e)

Figure 8. Tools of different geometries used in different Al alloy combinations. (**a**) AA2024-AA7075, reproduced from [14], with permission from Springer, 2018; (**b**) AA5083-AA7075, reproduced from [16], with permission from Springer, 2018; (**c**) AA5083-AA6351, reproduced from [18], with permission from SAGE, 2018; (**d**) AA5083-AA6082 [34], with permission from the authors; (**e**) AA5083-AA6351 [57], with permission from Springer, 2013.

4. Microstructure Evolution

The typical microstructure of a FSW joint consists of three distinct zones that is, HAZ, TMAZ and SZ [87,88]. These zones form depending on the thermal and mechanical deformation that the tool induces during welding. The SZ undergoes extensive grain refinement, producing fine grain microstructures, while the TMAZ has an elongated grain structure [89,90]. The microstructure evolution depends on the welding parameters (as discussed in the previous section), as the material movement or flow plays a more important role in the case of dissimilar material combinations compared to same material joints. The appropriate selection of all process parameters results in excellent material mixing on the both sides (AS and RS) of the joint and produces a sound weld. Recently, a comprehensive EBSD investigation for the AA5083-AA2024 joint was reported by

Niu et al. [91], as shown in Figure 9. As it can be seen from the EBSD orientation maps (Figure 9a–d), tilted and elongated grains in the TMAZ and fine grains d in the SZ developed due to dynamic recrystallization. Grain boundary orientations also varied in all three zones as shown in Figure 9(e–h). A higher fraction of large (>10°) angular grain boundaries was present in the SZ, while more of low (2–10°) angular grain boundaries were present in HAZ. Also, a more intense texture in the SZ was formed compared to other zones.

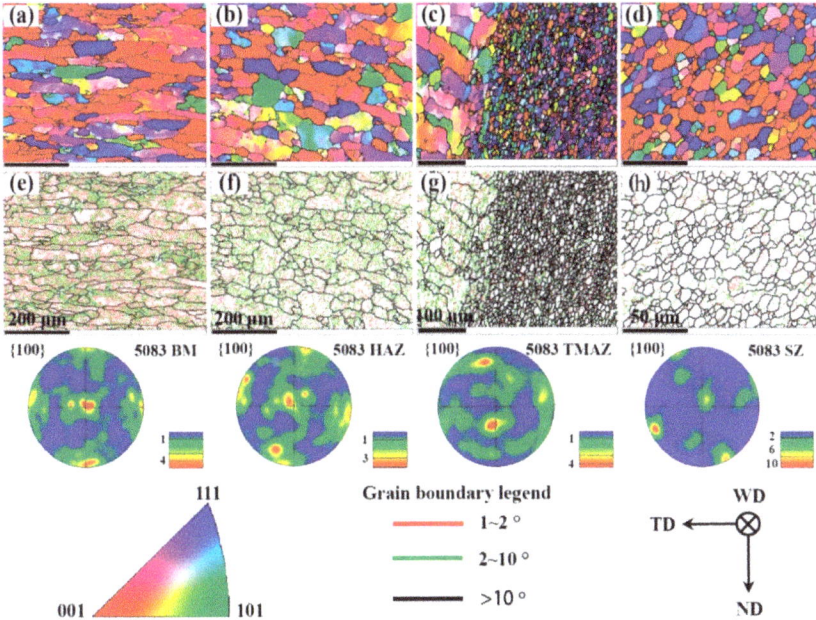

Figure 9. EBSD orientation maps and grain boundaries of the dissimilar AA5083-AA2024 aluminum alloy joint on the AA5083 side: (**a,e**) BM, (**b,f**) HAZ, (**c,g**) TMAZSZ interface and (**d,h**) SZ, reproduced from [91], with permission from Elsevier, 2019.

5. Mechanical Properties

5.1. Hardness

The hardness of the FSW joint is related to the joint strength and its deformation behavior, especially in the case of dissimilar material combinations. The hardness distributions of various different Al alloy combinations are shown in Figure 10. The common highly asymmetrical hardness distribution along the cross-section of dissimilar material joints is due to the different microstructural zones (SZ, TMAZ, HAZ) which develop due to the thermo-mechanical history during welding. Since the maximum temperature is reached at the SZ, precipitates or strengthening particles dissolve partially or completely decreasing hardness in SZ. Whereas the lowest hardness values are found in the HAZ due to the coarsening of precipitates or over aging. Therefore, the HAZ always remains the most common zone or site where failure occurs during tensile deformation. It is also worth noting that SZ has higher hardness values compared to the BM (which may be of low strength) because of the combined effect of grain refinement and the effect of both of the BMs in the SZ. However, it is not always true due to different initial conditions of heat-treatable alloy combinations. Recently, Niu et al. [13] reported an interesting hardness behavior of joints prior to and following fracture, by quantifying hardening with the ratio of HV_f/HV_w, where HV_f and HV_w are the microhardness of the fractured and the as-welded joints, respectively. This ratio was over one in the SZ, TMAZ and HAZ, which confirmed the strain hardened behavior of the joints as shown in Figure 11. In summary, hardness distribution in the dissimilar material joints is closely associated with mechanical behavior such as strain hardening and the fracture origin.

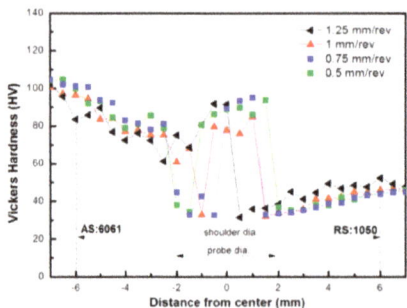

(a) AA6061-AA1050 [42]

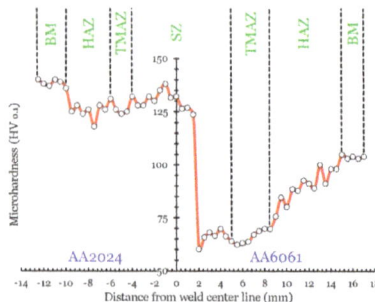

(b) AA2024-AA6061 [22]

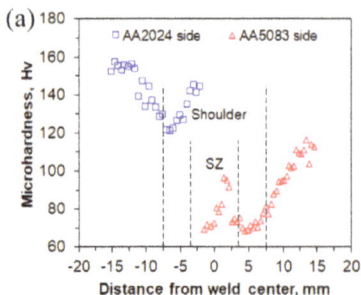

(c) AA2024-AA5083 [91]

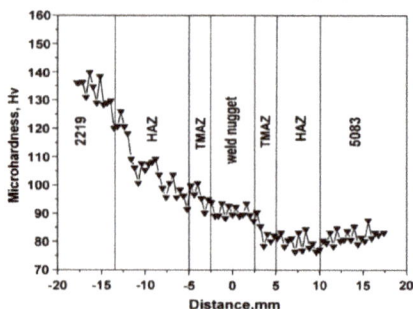

(d) AA2219-AA5083 [60]

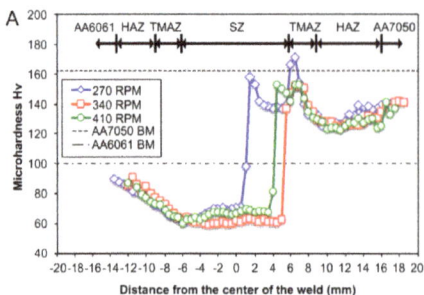

(e) AA6061-AA7075 [46]

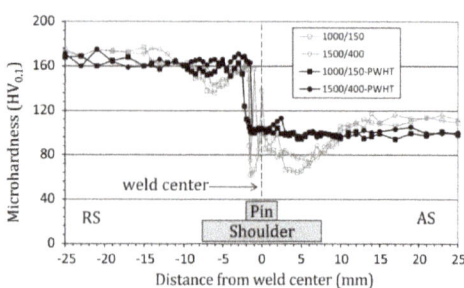

(f) AA6061-AA7075 T6 [51]

Figure 10. Hardness distribution along the cross section of the dissimilar Al combination joints. (**a**) AA6061-AA1050 [42]; with permission from the authors. (**b**) AA2024-AA6061, reproduced from [22], with permission from Elsevier, 2108; (**c**) AA2024-AA5083 reproduced from [91], with permission from Elsevier, 2019; (**d**) AA2219-AA5083 reproduced from [60], with permission from Elsevier, 2012; (**e**) AA6061-AA7075 reproduced from [46], with permission from Elsevier, 2015; (**f**) AA6061-AA7075 T6], reproduced from [51], with permission from Springer, 2014.

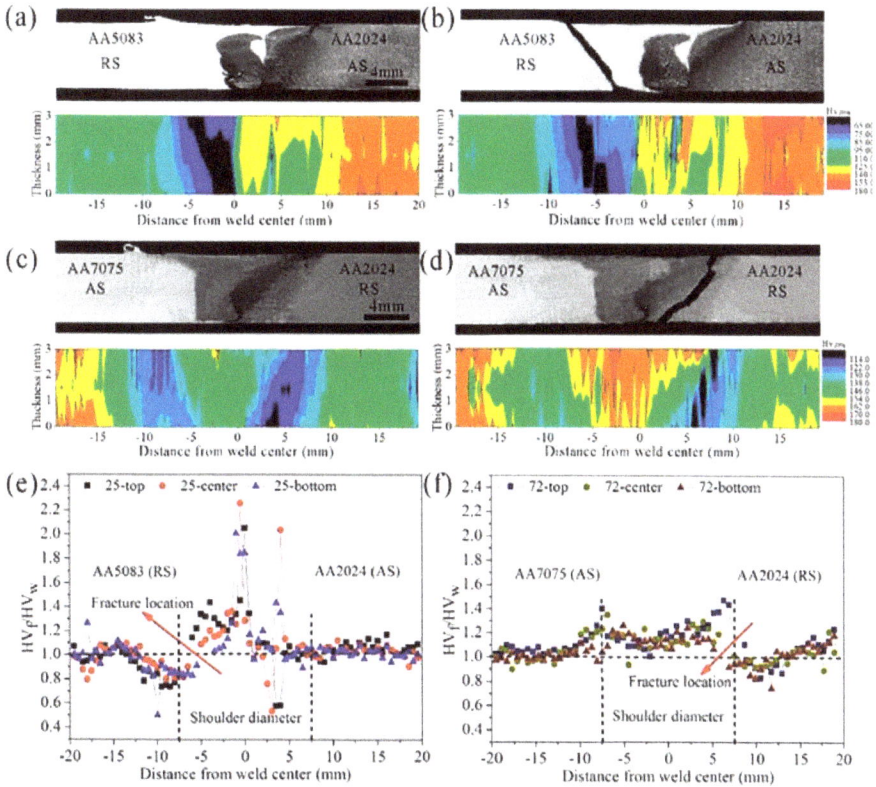

Figure 11. Cross-sectional macrostructures and hardness distributions of the FSWed dissimilar joints: (**a**) 25-joint before fracture, (**b**) 25-joint after fracture, (**c**) 72-joint before fracture, (**d**) 72-joint after fracture; hardening level across the FSWed joints: (**e**) 25-joint and (**f**) 72-joint, reproduced from [13], with permission from Elsevier, 2018. Note: 25-joint means AA2024-AA5083 and 72-joint means AA7075-AA2024 joint.

5.2. Tensile Strength

The number of published papers investigating welding the 5xxx-6xxx series alloys to identify the effect of process parameters (especially the tool rotation speed and welding speed) on the joint strength is shown in detail in Table 1. The joint strength increases with the rotation speed due to the enhanced material mixing effect [18,54,57,62]. The tool rotation speed intensifies plastic deformation and welding speed controls the thermal cycle, residual stresses and rate of production. So, it is essential to select the appropriate combination of these speeds for weld quality or joint strength. Bijanrostami et al. [33] investigated the AA6061-AA7075 joint to identify that maximum joint strength is achieved with a combination of moderate rotation and low welding speed. When high heat input conditions are used (i.e., high rotation and low welding speeds) large grains and lower dislocation densities develop in the SZ. On other side when low heat input condition are selected (i.e., low rotation and high welding speeds) defects are generated. So, grain size strengthening and low dislocation densities are necessary for joint strength. However, the maximum joint strength of an A356-AA6061 joint was achieved with low rotation and welding speed by Ghosh et al. [58,64]. Evidence of fine grain size, fine distribution of Si particles and reduced residual stresses in the SZ were found for low rotation and welding speeds. Together with rotation and welding speeds, the effect of tool geometry like the pin profile or features [14,18,47,48,62], pin shapes [34,57] and shoulder diameter to pin diameter ratio [37,60] on joint strength have been investigated. The pin profile or feature controls material flow and in effect material mixing at the joint interface, the pin shape affects SZ size as well as material movement and the shoulder to pin diameter ratio controls frictional heat generation between the tool and the BM. The conical threaded pin was identified as the best possible configuration for the AA 6061–AA5086 joint due to the production of a uniformly distributed precipitates and the distinct generation of the onion rings as material was mixed appropriately in the SZ, as reported by Ilangovan et al. [47]. In summary, the tensile strength of the dissimilar FSWed Al joints relies on the microstructure evolution during FSW, which in turn depends on the heat input as governed by the welding parameters (as discussed in Section 4).

6. Summary and Outlook

With regards to the research published and the appropriate future work to be performed in the FSW of dissimilar Al alloy combinations, the following comments can be proposed:

6.1. Al Alloy Combinations

Almost all of the investigations conducted concerned BM in the as-rolled condition that is, 2xxx-5xxx, 2xxx-6xxx, 2xxx-7xxx, 5xxx-6xxx, 5xxx-7xxx Al series. It would be interesting to explore dissimilar Al alloy combinations in as-cast conditions and as a combination between as-cast and as-rolled conditions, depending on the application.

6.2. Base Metal Placement

Limited number of papers on the effect of placement is available and still remains inconclusive. Base material placement becomes an issue in the cases where there are significant differences in mechanical properties of the BMs as in the 6xxx-7xxx and the 5xxx-7xxx combinations.

6.3. Tool Offset

There is a very limited number of welding parameters optimization studies to study tool offset. It needs further comprehensive evaluation using microstructure characterization to understand the material flow in the SZ.

6.4. Bobbing Tool and Stationary shoulder Tool

The bobbin tool [92] and the stationary shoulder tool are considered as a strategic variant of FSW, which have distinct benefits over the conventional FSW tool. Stationary shoulder tool offers low heat input during welding and processing [93–95] and would benefit Al alloy dissimilar joints [96].

6.5. Corrosion and Fatigue Behavior

Finally, corrosion and fatigue behavior studies of various combinations of dissimilar Al alloy joints would be beneficial to expand its industrial use.

Author Contributions: Conceptualization, W.L. and V.P.; methodology, V.P. and P.N. resources, W.L.; data curation, G.W. and F.W.; writing—original draft preparation, V.P.; writing—review and editing, V.P. and AV.; supervision, W.L.; project administration, W.L.; funding acquisition, W.L.

Funding: The authors would like to thank for financial support National Natural Science Foundation of China (51574196, U1637601). We would also like to express our gratitude to the Guest Editor, Antonello Astarita (University of Naples Federico II, Italy) for the invitation to contribute this article to METALS.

Conflicts of Interest: The authors declare no conflict of interest.

References

1. Thomas, W.M.; Nicholas, E.D.; Needham, J.C.; Murch, M.G.; Temple-Smith, P.; Dawes, C.J. Friction Stir Butt Welding. International Patent Application No. PCT/GB92/02203; GB Patent Application No. 9125978.8; U.S. Patent Application No. 5,460,317, 6 December 1991.
2. Mishra, R.; Ma, Z.; Mishra, R. Friction stir welding and processing. *Mater. Sci. Eng. R* **2005**, *50*, 1–78. [CrossRef]
3. Ma, Z.Y.; Feng, A.H.; Chen, D.L.; Shen, J. Recent Advances in Friction Stir Welding/Processing of Aluminum Alloys: Microstructural Evolution and Mechanical Properties. *Crit. Rev. Solid State Mater. Sci.* **2017**, *43*, 269–333. [CrossRef]
4. Gibson, B.; Lammlein, D.; Prater, T.; Longhurst, W.; Cox, C.; Ballun, M.; Dharmaraj, K.; Cook, G.; Strauss, A.; Gibson, B. Friction stir welding: Process, automation, and control. *J. Manuf. Processes* **2014**, *16*, 56–73. [CrossRef]
5. Nandan, R.; Debroy, T.; Bhadeshia, H. Recent advances in friction-stir welding—Process, weldment structure and properties. *Prog. Mater Sci.* **2008**, *53*, 980–1023. [CrossRef]
6. Padhy, G.; Wu, C.; Gao, S. Friction stir based welding and processing technologies - processes, parameters, microstructures and applications: A review. *J. Mater. Sci. Technol.* **2018**, *34*, 1–38. [CrossRef]
7. Shah, P.H.; Badheka, V.J. Friction stir welding of aluminium alloys: An overview of experimental findings—Process, variables, development and applications. *Proc. Inst. Mech. Eng. Part L: J. Mater. Des. Applic.* **2017**, *6*, 1464420716689588. [CrossRef]
8. Wahid, M.A.; Khan, Z.A.; Siddiquee, A.N. Review on underwater friction stir welding: A variant of friction stir welding with great potential of improving joint properties. *Trans. Nonferrous Met. Soc. China* **2018**, *28*, 193–219. [CrossRef]
9. Scherillo, F.; Curioni, M.; Aprea, P.; Impero, F.; Squillace, A.; Zhou, X. Study of the Linear Friction Welding Process of Dissimilar Ti-6Al-4V–Stainless Steel Joints AU - Astarita, Antonello. *Materials and Manufacturing Processes* **2016**, *31*, 2115–2122. [CrossRef]
10. Sepe, R.; Armentani, E.; di Lascio, P.; Citarella, R. Crack Growth Behavior of Welded Stiffened Panel. *Procedia Engineering* **2015**, *109*, 473–483. [CrossRef]
11. Astarita, A.; Prisco, U.; Squillace, A.; Villano, P.; Scherillo, F.; Coticelli, F. Theoretical analysis of a friction stir welded panel in comparison with the baseline version. *Weld. Cutt.* **2016**, *15*, 238–240.
12. Magalhães, V.M.; Leitão, C.; Rodrigues, D.M. Friction stir welding industrialisation and research status. *Sci. Technol. Weld. Joining* **2017**, *23*, 400–409. [CrossRef]
13. Niu, P.; Li, W.; Chen, D. Strain hardening behavior and mechanisms of friction stir welded dissimilar joints of aluminum alloys. *Mater. Lett.* **2018**, *231*, 68–71. [CrossRef]
14. Hasan, M.M.; Ishak, M.; Rejab, M.R.M. Effect of pin tool flute radius on the material flow and tensile properties of dissimilar friction stir welded aluminum alloys. *Int. J. Adv. Manuf. Technol.* **2018**, *98*, 2747–2758. [CrossRef]
15. Ge, Z.; Gao, S.; Ji, S.; Yan, D. Effect of pin length and welding speed on lap joint quality of friction stir welded dissimilar aluminum alloys. *Int. J. Adv. Manuf. Technol.* **2018**, *98*, 1461–1469. [CrossRef]
16. Kalemba-Rec, I.; Kopyściański, M.; Miara, D.; Krasnowski, K. Effect of process parameters on mechanical properties of friction stir welded dissimilar 7075-T651 and 5083-H111 aluminum alloys. *Int. J. Adv. Manuf. Technol.* **2018**, *97*, 2767–2779. [CrossRef]
17. Safarbali, B.; Shamanian, M.; Eslami, A. Effect of post-weld heat treatment on joint properties of dissimilar friction stir welded 2024-T4 and 7075-T6 aluminum alloys. *Trans. Nonferrous Met. Soc. China* **2018**, *28*, 1287–1297. [CrossRef]
18. Palanivel, R.; Laubscher, R.; Vigneshwaran, S.; Dinaharan, I. Prediction and optimization of the mechanical properties of dissimilar friction stir welding of aluminum alloys using design of experiments. *Proc. Inst. Mech. Eng. Part B: J. Eng. Manuf.* **2016**, *232*, 1384–1394. [CrossRef]
19. Hamilton, C.; Dymek, S.; Kopyściański, M.; Węglowska, A.; Pietras, A. Numerically Based Phase Transformation Maps for Dissimilar Aluminum Alloys Joined by Friction Stir-Welding. *Metals* **2018**, *8*, 324. [CrossRef]

20. Gupta, S.K.; Pandey, K.; Kumar, R. Multi-objective optimization of friction stir welding process parameters for joining of dissimilar AA5083/AA6063 aluminum alloys using hybrid approach. *Proc. Inst. Mech. Eng.* **2016**, *232*, 343–353. [CrossRef]

21. Huang, B.W.; Qin, Q.D.; Zhang, D.H.; Wu, Y.J.; Su, X.D. Microstructure and Mechanical Properties of Dissimilar Joints of Al-Mg2Si and 5052 Aluminum Alloy by Friction Stir Welding. *J. Mater. Eng. Perform.* **2018**, *27*, 1898–1907. [CrossRef]

22. Moradi, M.M.; Aval, H.J.; Jamaati, R.; Amirkhanlou, S.; Ji, S. Microstructure and texture evolution of friction stir welded dissimilar aluminum alloys: AA2024 and AA6061. *J. Manuf. Processes* **2018**, *32*, 1–10. [CrossRef]

23. Prasanth, R.S.S.; Raj, K.H. Determination of Optimal Process Parameters of Friction Stir Welding to Join Dissimilar Aluminum Alloys Using Artificial Bee Colony Algorithm. *Trans. Indian Inst. Met.* **2017**, *71*, 453–462. [CrossRef]

24. Azeez, S.; Akinlabi, E. Effect of processing parameters on microhardness and microstructure of a double-sided dissimilar friction stir welded aa6082-t6 and aa7075-t6 aluminum alloy. *Mater. Today: Proc.* **2018**, *5*, 18315–18324. [CrossRef]

25. Azeez, S.; Akinlabi, E.; Kailas, S.; Brandi, S. Microstructural properties of a dissimilar friction stir welded thick aluminum aa6082-t6 and aa7075-t6 alloy. *Mater. Today: Proc.* **2018**, *5*, 18297–18306. [CrossRef]

26. Peng, G.; Ma, Y.; Hu, J.; Jiang, W.; Huan, Y.; Chen, Z.; Zhang, T. Nanoindentation Hardness Distribution and Strain Field and Fracture Evolution in Dissimilar Friction Stir-Welded AA 6061-AA 5A06 Aluminum Alloy Joints. *Adv. Mater. Sci. Eng.* **2018**, *2018*, 1–11. [CrossRef]

27. Das, U.; Toppo, V. Effect of Tool Rotational Speed on Temperature and Impact Strength of Friction Stir Welded Joint of Two Dissimilar Aluminum Alloys. *Mater. Today: Proc.* **2018**, *5*, 6170–6175. [CrossRef]

28. Sarsılmaz, F. Relationship between micro-structure and mechanical properties of dissimilar aluminum alloy plates by friction stir welding. *J. Therm. Sci.* **2018**, *22*, 55–66. [CrossRef]

29. No, K.; Yoo, J.-T.; Yoon, J.-H.; Lee, H.-S. Effect of Process Parameters on Friction Stir Welds on AA2219-AA2195 Dissimilar Aluminum Alloys. *Korean J. Mater. Res.* **2017**, *27*, 331–338. [CrossRef]

30. Hamilton, C.; Dymek, S.; Dryzek, E.; Kopyściański, M.; Pietras, A.; Węglowska, A.; Wróbel, M. Application of positron lifetime annihilation spectroscopy for characterization of friction stir welded dissimilar aluminum alloys. *Mater. Charact.* **2017**, *132*, 431–436. [CrossRef]

31. Kopyściański, M.; Dymek, S.; Hamilton, C.; Weglowska, A.; Pietras, A.; Szczepanek, M.; Wojnarowska, M. Microstructure of Friction Stir Welded Dissimilar Wrought 2017A and Cast AlSi9Mg Aluminum Alloys. *Acta Phys. Pol. A* **2017**, *131*, 1390–1394. [CrossRef]

32. Ghaffarpour, M.; Kazemi, M.; Sefat, M.J.M.; Aziz, A.; Dehghani, K. Evaluation of dissimilar joints properties of 5083-H12 and 6061-T6 aluminum alloys produced by tungsten inert gas and friction stir welding. *Proc. Inst. Mech. Eng.* **2015**, *231*, 297–308. [CrossRef]

33. Bijanrostami, K.; Barenji, R.V.; Hashemipour, M. Effect of Traverse and Rotational Speeds on the Tensile Behavior of the Underwater Dissimilar Friction Stir Welded Aluminum Alloys. *J. Mater. Eng. Perform.* **2017**, *26*, 909–920. [CrossRef]

34. Kasman, S.; Kahraman, F.; Emiralioğlu, A. A Case Study for the Welding of Dissimilar EN AW 6082 and EN AW 5083 Aluminum Alloys by Friction Stir Welding. *Metals* **2016**, *7*, 6. [CrossRef]

35. Palanivel, R.; Laubscher, R.F.; Dinaharan, I.; Murugan, N. Developing a Friction-Stir Welding Window for Joining the Dissimilar Aluminum Alloys AA6351 and AA5083. *Mater. Technol.* **2017**, *51*, 5–9. [CrossRef]

36. Doley, J.K.; Kore, S.D. A Study on Friction Stir Welding of Dissimilar Thin Sheets of Aluminum Alloys AA 5052–AA 6061. *J. Manuf. Sci. Eng.* **2016**, *138*, 114502. [CrossRef]

37. Saravanan, V.; Rajakumar, S.; Banerjee, N.; Amuthakkannan, R. Effect of shoulder diameter to pin diameter ratio on microstructure and mechanical properties of dissimilar friction stir welded AA2024-T6 and AA7075-T6 aluminum alloy joints. *Int. J. Adv. Manuf. Technol.* **2016**, *87*, 3637–3645. [CrossRef]

38. Yan, Z.; Liu, X.; Fang, H. Effect of Sheet Configuration on Microstructure and Mechanical Behaviors of Dissimilar Al–Mg–Si/Al–Zn–Mg Aluminum Alloys Friction Stir Welding Joints. *J. Mater. Sci. Technol.* **2016**, *32*, 1378–1385. [CrossRef]

39. Yan, Z.-J.; Liu, X.-S.; Fang, H.-Y. Fatigue Behavior of Dissimilar Al–Mg–Si/Al–Zn–Mg Aluminum Alloys Friction Stir Welding Joints. *Acta Metall. Sinica* **2016**, *29*, 1161–1168. [CrossRef]

40. Hamilton, C.; Kopyściański, M.; Węglowska, A.; Dymek, S.; Pietras, A. A Numerical Simulation for Dissimilar Aluminum Alloys Joined by Friction Stir Welding. *Metall. Mater. Trans. A* **2016**, *47*, 4519–4529. [CrossRef]

41. Zapata, J.; Toro, M.; López, D. Residual stresses in friction stir dissimilar welding of aluminum alloys. *J. Mater. Process. Technol.* **2016**, *229*, 121–127. [CrossRef]

42. Sun, Y.; Tsuji, N.; Fujii, H. Microstructure and Mechanical Properties of Dissimilar Friction Stir Welding between Ultrafine Grained 1050 and 6061-T6 Aluminum Alloys. *Metals* **2016**, *6*, 249. [CrossRef]

43. Texier, D.; Zedan, Y.; Amoros, T.; Feulvarch, E.; Stinville, J.; Bocher, P. Near-surface mechanical heterogeneities in a dissimilar aluminum alloys friction stir welded joint. *Mater. Des.* **2016**, *108*, 217–229. [CrossRef]

44. Rodriguez, R.; Jordon, J.; Allison, P.; Rushing, T.; Garcia, L. Low-cycle fatigue of dissimilar friction stir welded aluminum alloys. *Mater. Sci. Eng. A* **2016**, *654*, 236–248. [CrossRef]

45. Yoon, T.-J.; Yun, J.-G.; Kang, C.-Y. Formation mechanism of typical onion ring structures and void defects in friction stir lap welded dissimilar aluminum alloys. *Mater. Des.* **2016**, *90*, 568–578. [CrossRef]

46. Rodriguez, R.; Jordon, J.; Allison, P.; Rushing, T.; Garcia, L. Microstructure and mechanical properties of dissimilar friction stir welding of 6061-to-7050 aluminum alloys. *Mater. Des.* **2015**, *83*, 60–65. [CrossRef]

47. Ilangovan, M.; Boopathy, S.R.; Balasubramanian, V. Effect of tool pin profile on microstructure and tensile properties of friction stir welded dissimilar AA 6061–AA 5086 aluminium alloy joints. *Defence Technol.* **2015**, *11*, 174–184. [CrossRef]

48. Reza-E-Rabby, M.; Tang, W.; Reynolds, A.P. Effect of tool pin features on process response variables during friction stir welding of dissimilar aluminum alloys. *Sci. Technol. Weld. Joining* **2015**, *20*, 425–432. [CrossRef]

49. Donatus, U.; Thompson, G.E.; Zhou, X. Anodizing Behavior of Friction Stir Welded Dissimilar Aluminum Alloys. *J. Electrochem. Soc.* **2015**, *162*, C657–C665. [CrossRef]

50. Karam, A.; Mahmoud, T.S.; Zakaria, H.M.; Khalifa, T.A. Friction Stir Welding of Dissimilar A319 and A413 Cast Aluminum Alloys. *Arab J. Sci. Eng.* **2014**, *39*, 6363–6373. [CrossRef]

51. Ipekoğlu, G.; Çam, G. Effects of Initial Temper Condition and Postweld Heat Treatment on the Properties of Dissimilar Friction-Stir-Welded Joints between AA7075 and AA6061 Aluminum Alloys. *Metall. Mater. Trans. A* **2014**, *45*, 3074–3087. [CrossRef]

52. Cole, E.G.; Fehrenbacher, A.; Duffie, N.A.; Zinn, M.R.; Pfefferkorn, F.E.; Ferrier, N.J. Weld temperature effects during friction stir welding of dissimilar aluminum alloys 6061-t6 and 7075-t6. *Int. J. Adv. Manuf. Technol.* **2013**, *71*, 643–652. [CrossRef]

53. Song, Y.; Yang, X.; Cui, L.; Hou, X.; Shen, Z.; Xu, Y. Defect features and mechanical properties of friction stir lap welded dissimilar AA2024–AA7075 aluminum alloy sheets. *Mater. Des.* **2014**, *55*, 9–18. [CrossRef]

54. Jannet, S.; Mathews, P.K. Effect of Welding Parameters on Mechanical and Microstructural Properties of Dissimilar Aluminum Alloy Joints Produced by Friction Stir Welding. *Appl. Mech. Mater.* **2014**, *592*, 250–254. [CrossRef]

55. Palanivel, R.; Mathews, P.K.; Dinaharan, I.; Murugan, N. Mechanical and metallurgical properties of dissimilar friction stir welded AA5083-H111 and AA6351-T6 aluminum alloys. *Trans. Nonferrous Met. Soc. China* **2014**, *24*, 58–65. [CrossRef]

56. Jonckheere, C.; De Meester, B.; Denquin, A.; Simar, A. Torque, temperature and hardening precipitation evolution in dissimilar friction stir welds between 6061-T6 and 2014-T6 aluminum alloys. *J. Mater. Process. Technol.* **2013**, *213*, 826–837. [CrossRef]

57. Palanivel, R.; Mathews, P.K.; Murugan, N. Optimization of process parameters to maximize ultimate tensile strength of friction stir welded dissimilar aluminum alloys using response surface methodology. *J. Cent. South Univ.* **2013**, *20*, 2929–2938. [CrossRef]

58. Ghosh, M.; Husain, M.M.; Kumar, K.; Kailas, S.V. Friction Stir-Welded Dissimilar Aluminum Alloys: Microstructure, Mechanical Properties, and Physical State. *J. Mater. Eng. Perform.* **2013**, *22*, 3890–3901. [CrossRef]

59. Velotti, C.; Squillace, A.; Villano, M.G.; Prisco, U.; Montuori, M.; Giorleo, G.; Astarita, A.; Ciliberto, S.; Giuliani, M.; Bellucci, F. On the critical technological issues of friction stir welding lap joints of dissimilar aluminum alloys. *Surf. Interface Anal.* **2013**, *45*, 1643–1648. [CrossRef]

60. Koilraj, M.; Sundareswaran, V.; Vijayan, S.; Rao, S.K. Friction stir welding of dissimilar aluminum alloys AA2219 to AA5083 – Optimization of process parameters using Taguchi technique. *Mater. Des.* **2012**, *42*, 1–7. [CrossRef]

61. Dinaharan, I.; Kalaiselvan, K.; Vijay, S.; Raja, P.; J, V.S. Effect of material location and tool rotational speed on microstructure and tensile strength of dissimilar friction stir welded aluminum alloys. *Arch. Civ. Mech. Eng.* **2012**, *12*, 446–454. [CrossRef]

62. Palanivel, R.; Mathews, P.K.; Murugan, N.; Dinaharan, I. Effect of tool rotational speed and pin profile on microstructure and tensile strength of dissimilar friction stir welded AA5083-H111 and AA6351-T6 aluminum alloys. *Mater. Des.* **2012**, *40*, 7–16. [CrossRef]

63. Song, S.-W.; Lee, S.-H.; Kim, B.-C.; Yoon, T.-J.; Kim, N.-K.; Kim, I.-B.; Kang, C.-Y. Liquation Cracking of Dissimilar Aluminum Alloys during Friction Stir Welding. *Mater. Trans.* **2011**, *52*, 254–257. [CrossRef]

64. Ghosh, M.; Kumar, K.; Kailas, S.; Ray, A. Optimization of friction stir welding parameters for dissimilar aluminum alloys. *Mater. Des.* **2010**, *31*, 3033–3037. [CrossRef]

65. Kim, N.-K.; Kim, B.-C.; An, Y.-G.; Jung, B.-H.; Song, S.-W.; Kang, C.-Y. The effect of material arrangement on mechanical properties in Friction Stir Welded dissimilar A5052/A5J32 aluminum alloys. *Met. Mater. Int.* **2009**, *15*, 671–675. [CrossRef]

66. Prime, M.; Gnaupel-Herold, T.; Baumann, J.; Lederich, R.; Bowden, D.; Sebring, R. Residual stress measurements in a thick, dissimilar aluminum alloy friction stir weld. *Acta Mater.* **2006**, *54*, 4013–4021. [CrossRef]

67. Miles, M.P.; Nelson, T.W.; Melton, D.W. Formability of friction-stir-welded dissimilar-aluminum-alloy sheets. *Metall. Mater. Trans. A* **2005**, *36*, 3335–3342. [CrossRef]

68. Ouyang, J.H.; Kovacevic, R. Material flow and microstructure in the friction stir butt welds of the same and dissimilar aluminum alloys. *J. Mater. Eng. Perform.* **2002**, *11*, 51–63. [CrossRef]

69. Barbini, A.; Carstensen, J.; Dos Santos, J. Influence of Alloys Position, Rolling and Welding Directions on Properties of AA2024/AA7050 Dissimilar Butt Weld Obtained by Friction Stir Welding. *Metals* **2018**, *8*, 202. [CrossRef]

70. Cavaliere, P.; De Santis, A.; Panella, F.; Squillace, A. Effect of welding parameters on mechanical and microstructural properties of dissimilar AA6082-AA2024 joints produced by friction stir welding. *Mater. Des.* **2009**, *30*, 609–616. [CrossRef]

71. Niu, P.; Li, W.; Li, N.; Xu, Y.; Chen, D. Exfoliation corrosion of friction stir welded dissimilar 2024-to-7075 aluminum alloys. *Mater. Charact.* **2019**, *147*, 93–100. [CrossRef]

72. Rajesh, S.; Badheka, V.J. Process parameters/material location affecting hooking in friction stir lap welding: Dissimilar aluminum alloys. *Mater. Manuf. Process.* **2017**, *33*, 323–332. [CrossRef]

73. Rajesh, S.; Badheka, V.J. Effect of friction stir lap weld and post weld heat treatment on corrosion behavior of dissimilar aluminum alloys. *Proc. Inst. Mech. Eng.* **2017**, *426*. [CrossRef]

74. Rajesh, S.; Badheka, V. Influence of Heat Input/Multiple Passes and Post Weld Heat Treatment on Strength/Electrochemical Characteristics of Friction Stir Weld Joint. *Mater. Manuf. Process.* **2017**, *33*, 156–164. [CrossRef]

75. Mastanaiah, P.; Sharma, A.; Reddy, G.M. Dissimilar Friction Stir Welds in AA2219-AA5083 Aluminium Alloys: Effect of Process Parameters on Material Inter-Mixing, Defect Formation, and Mechanical Properties. *Trans. Indian Inst. Met.* **2015**, *69*, 1397–1415. [CrossRef]

76. Kasman, Ş.; Yenier, Z. Analyzing dissimilar friction stir welding of AA5754/AA7075. *Int. J. Adv. Manuf. Technol.* **2013**, *70*, 145–156. [CrossRef]

77. Forcellese, A.; Simoncini, M.; Casalino, G. Influence of Process Parameters on the Vertical Forces Generated during Friction Stir Welding of AA6082-T6 and on the Mechanical Properties of the Joints. *Metals* **2017**, *7*, 350. [CrossRef]

78. Saeidi, M.; Manafi, B.; Givi, M.B.; Faraji, G. Mathematical modeling and optimization of friction stir welding process parameters in AA5083 and AA7075 aluminum alloy joints. *Proc. Inst. Mech. Eng. Part B: J. Eng. Manuf.* **2015**, *230*, 1284–1294. [CrossRef]

79. Zhu, Z.; Zhang, H.; Yu, T.; Wu, Z.; Wang, M.; Zhang, X. A Finite Element Model to Simulate Defect Formation during Friction Stir Welding. *Metals* **2017**, *7*, 256. [CrossRef]

80. Godhani, P.S.; Patel, V.V.; Vora, J.J.; Chaudhary, N.D.; Banka, R. Effect of Friction Stir Welding of Aluminum Alloys AA6061/AA7075: Temperature Measurement, Microstructure, and Mechanical Properties. In *Innovations in Infrastructure*; Springer: Singapore, Singapore, 2019; pp. 591–598.

81. Zhou, Y.; Chen, S.; Wang, J.; Wang, P.; Xia, J. Influences of Pin Shape on a High Rotation Speed Friction Stir Welding Joint of a 6061-T6 Aluminum Alloy Sheet. *Metals* **2018**, *8*, 987. [CrossRef]

82. Goel, P.; Siddiquee, A.; Khan, N.; Hussain, M.; Khan, Z.; Abidi, M.; Al-Ahmari, A. Investigation on the Effect of Tool Pin Profiles on Mechanical and Microstructural Properties of Friction Stir Butt and Scarf Welded Aluminium Alloy 6063. *Metals* **2018**, *8*, 74. [CrossRef]

83. Patel, V.V.; Badheka, V.; Kumar, A. Friction Stir Processing as a Novel Technique to Achieve Superplasticity in Aluminum Alloys: Process Variables, Variants, and Applications. *Metallogr. Microstruct. Anal.* **2016**, *5*, 278–293. [CrossRef]

84. Patel, V.V.; Badheka, V.; Kumar, A. Effect of polygonal pin profiles on friction stir processed superplasticity of AA7075 alloy. *J. Mater. Process. Technol.* **2017**, *240*, 68–76. [CrossRef]

85. Patel, V.V.; Badheka, V.J.; Kumar, A. Influence of Pin Profile on the Tool Plunge Stage in Friction Stir Processing of Al–Zn–Mg–Cu Alloy. *Trans. Indian Inst. Met.* **2016**, *70*, 1151–1158. [CrossRef]

86. Patel, V.V.; Li, W.Y.; Vairis, A.; Badheka, V.J. Recent Development in Friction Stir Processing as a Solid-State Grain Refinement Technique: Microstructural Evolution and Property Enhancement. *Crit. Rev. Solid State Mater. Sci.* **2019**, accepted. [CrossRef]

87. Dialami, N.; Cervera, M.; Chiumenti, M. Numerical Modelling of Microstructure Evolution in Friction Stir Welding (FSW). *Metals* **2018**, *8*, 183. [CrossRef]

88. Nakamura, T.; Obikawa, T.; Nishizaki, I.; Enomoto, M.; Fang, Z. Friction Stir Welding of Non-Heat-Treatable High-Strength Alloy 5083-O. *Metals* **2018**, *8*, 208. [CrossRef]

89. Aziz, S.A.; Zolkarnain, L.; Rahim, M.A.Z.B.A.; Fadaeifard, F.; Matori, K.A. Effect of the Welding Speed on the Macrostructure, Microstructure and Mechanical Properties of AA6061-T6 Friction Stir Butt Welds. *Metals* **2017**, *7*, 48.

90. Patel, V.V.; Badheka, V.; Kumar, A. Influence of Friction Stir Processed Parameters on Superplasticity of Al-Zn-Mg-Cu Alloy. *Mater. Manuf. Process.* **2015**, *31*, 1–10. [CrossRef]

91. Niu, P.; Li, W.; Vairis, A.; Chen, D. Cyclic deformation behavior of friction-stir-welded dissimilar AA5083-to-AA2024 joints: Effect of microstructure and loading history. *Mater. Sci. Eng. A* **2019**, *744*, 145–153. [CrossRef]

92. Tamadon, A.; Pons, D.J.; Sued, K.; Clucas, D. Thermomechanical Grain Refinement in AA6082-T6 Thin Plates under Bobbin Friction Stir Welding. *Metals* **2018**, *8*, 375. [CrossRef]

93. Wen, Q.; Li, W.; Wang, W.; Wang, F.; Gao, Y.; Patel, V. Experimental and numerical investigations of bonding interface behavior in stationary shoulder friction stir lap welding. *J. Mater. Sci. Technol.* **2019**, *35*, 192–200. [CrossRef]

94. Patel, V.; Li, W.; Xu, Y. Stationary shoulder tool in friction stir processing: a novel low heat input tooling system for magnesium alloy. *Mater. Manuf. Process.* **2018**, *34*, 177–182. [CrossRef]

95. Li, W.; Niu, P.; Yan, S.; Patel, V.; Wen, Q. Improving microstructural and tensile properties of AZ31B magnesium alloy joints by stationary shoulder friction stir welding. *J. Manuf. Process.* **2019**, *37*, 159–167. [CrossRef]

96. Barbini, A.; Carstensen, J.; Dos Santos, J.F. Influence of a non-rotating shoulder on heat generation, microstructure and mechanical properties of dissimilar AA2024/AA7050 FSW joints. *J. Mater. Sci. Technol.* **2018**, *34*, 119–127. [CrossRef]

Article

Electromagnetic Pulse Welding of Tubular Products: Influence of Process Parameters and Workpiece Geometry on the Joint Characteristics and Investigation of Suitable Support Systems for the Target Tube

Koen Faes [1],*, Irene Kwee [1] and Wim De Waele [2]

[1]   Belgian Welding Institute, Department: Research Centre, 9052 Zwijnaarde, Belgium; irene.kwee@bil-ibs.be
[2]   Ghent University, Department of Electrical Energy, Metals, Mechanical Constructions & Systems,
       Faculty of Engineering and Architecture, 9052 Zwijnaarde, Belgium; wim.dewaele@ugent.be
*   Correspondence: koen.faes@bil-ibs.be; Tel.: +32-9-292-14-03

Received: 20 March 2019; Accepted: 29 April 2019; Published: 1 May 2019

Abstract: In this experimental research, copper to steel tubular joints were produced by electromagnetic pulse welding. In a first phase, non-supported target tubes were used in order to investigate the influence of the workpiece geometry on the weld formation and joint characteristics. For this purpose, different joint configurations were used, more specific the tube-to-rod and the tube-to-tube configurations, with target workpieces with different diameters and wall thicknesses. Also, some preliminary investigations were performed to examine a support method for the target tubes. In a second phase, suitable support systems for the target tubes were identified. The resulting welds were evaluated in terms of their leak tightness, weld length and deformation of the target tube. It can be concluded that polyurethane (PU), polymethylmethacrylaat (PMMA), polyamide (PA6.6) and steel rods can be considered as valuable internal supports leading to high-quality welds and a sufficient cross-sectional area after welding. Welds with a steel bar support exhibit the highest cross-sectional area after welding, but at the same time the obtained weld quality is lower compared to welds with a PA6.6 or PMMA support. In contrast, welds with a PA6.6 or PU support show the highest weld quality, but also have a lower cross-sectional area after welding compared to steel internal supports.

Keywords: electromagnetic pulse welding; tubular joints; internal supports

---

1. Introduction

Electromagnetic pulse welding is an innovative solid-state welding technology that belongs to the group of pressure welding processes; it uses electromagnetic forces for deformation and joining of materials. The process can be used to join tubular [1] and sheet metals [2], placed in the overlap configuration. If the workpieces are impacted with high velocity and under a certain angle, a jet is created along the materials' surfaces. This jet removes surface contaminants, such as oxide films, which eliminates the need for pre-process cleaning. In general, no pre-weld cleaning is required.

A wavy or a flat bond interface is formed like in explosion welding. An intermetallic layer can be created as a result of mechanical mixing, intensive plastic deformation and local heating. The temperature increase occurs due to Joule effects and the collision itself. Since the process takes place in a very short lapse of time, heating is not sufficient to generate a temperature increase in a wide area, so there is no significant heat affected zone.

Compared to thermal welding processes, electromagnetic pulse welding offers important advantages since pressure instead of heat is employed to realize the metallic bond. Electromagnetic pulse welding is possible for similar and dissimilar material combinations, including those which are difficult or impossible to join using conventional processes [3–5].

Dissimilar copper (hereafter Cu) to steel (hereafter St) tubular joints are of particular interest for cooling applications in the machine and equipment construction industry. A specific example is a Cu-St tubular joint as part of a refrigeration circuit of a compressed air-dryer, which is currently produced by brazing [6].

Only very few articles discuss electromagnetic pulse welding of copper to steel [7–10]. These publications do not go into much detail however. A more comprehensive description of joining of copper to steel was provided in [11–13].

In Ref. [11], copper flyer tubes (Cu-DHP R290; O.D.: 25 mm) were used in combination with S235JR steel target rods, using different outer diameters to investigate the influence of the standoff distance. It was proven that high-quality welds could be created. The best results were obtained with an overlap distance of 12 mm, a low standoff distance of 1 or 1.5 mm and a high energy level. The field shaper cut resulted in a local decrease of the weld length. The interface was wavy and the wavelength and amplitude increased with increasing energy and standoff distance, as also described in literature about explosion welding.

In Ref. [12], the interface morphology of electromagnetic pulse welding between copper and carbon steel was explored. The interface morphology, diffusion of elements and the hardness distribution were investigated. Wavy and straight bonding areas were found, with weld lengths up to 5 mm. In the wavy bonding area, the wavelength and amplitude are approximately 60 and 20 µm, respectively. The width of mutual diffusion region of Cu and Fe elements was 2 µm in straight weld interfaces and increased up to 6 µm in wavy weld interfaces. The highest hardness appeared in the steel material, near the interface, while the interfacial hardness was in between the values of the 2 base materials.

In Ref. [13], joining of two tubes of pure copper and low carbon steel by electromagnetic pulse welding was described. Satisfactory welds were obtained with an optimal set of parameters. The welded interface revealed a wavy morphology with pockets of intermixed metal vortices. High resolution electron microscopy and microanalysis showed the formation of nano-grains along the interface and evidence of short distance interatomic diffusion across the weld joint respectively. The strain hardening effect due to high energy impact led to significantly higher microhardness on the steel side of the interface.

Joining of tubular parts frequently requires a support of the inner tube in order to avoid undesired deformation or fracture of the joint. Specifically, tubes with a small wall thickness need to be supported, because they can hardly resist radial forces [14]. Joining of tubular parts, for which the inner tube is not supported, has been investigated mainly for aluminum as flyer tube and steel as target (or inner) tube. Applications for aluminum to steel tubular joints are, amongst others, found in the fabrication of tubular space frame structures for automotive vehicles and pipe fittings [15].

In order to avoid deformation of the target tube, several studies have been performed regarding the critical wall thickness of the target tube and the flyer tube [14,16,17]. This critical thickness was defined as the thickness of the tube at which no plastic deformation of the target tube occurred. In addition, it was also shown that the feasibility of joining tubular products was determined by the discharge frequency [14] and the critical discharge voltage (which defines the impact velocity of the flyer tube) [15].

In this experimental research, copper to steel tubular joints were produced by electromagnetic pulse welding. In a first phase, non-supported target tubes were used, in order to investigate the influence of the workpiece geometry on the weld formation and joint characteristics. For this purpose, different joint configurations were used, more specifically the tube-to-rod and the tube-to-tube configurations, with target workpieces with different diameters and wall thicknesses. Also, some preliminary investigations were performed in order to examine a support method for the target tubes.

In a second phase, suitable support systems for the target tubes were investigated. The resulting welds were evaluated in terms of their leak tightness, weld length and deformation of the steel tube.

2. Materials and Methods

2.1. Working Principle of Electromagnetic Pulse Welding

In the electromagnetic welding process, a power supply is used to charge a capacitor bank. When the required amount of energy is stored in the capacitors, it is instantaneously released into a coil, during a very short period of time. The discharge current induces a strong transient magnetic field in the coil, which generates a magnetic pressure, that accelerates a conductive workpiece, named the flyer workpiece, up to a sufficiently high velocity. The flyer workpiece collides with a fixed workpiece (termed target workpiece) and if the conditions of the collision velocity and impact angle are favorable (collision velocity and impact angle), a weld will be created between the two parts. For a sufficiently high velocity and a non-parallel collision, jetting will occur. This phenomenon cleans the surfaces of both materials and removes oxides and other contaminants. After collision, the acting Lorentz force combined with the inertia effect press the atomically clean surfaces of the flyer and target together to form the weld. Bonding between the two materials occurs when the distance between their atoms becomes smaller than the range of their mutual attractive forces [18–21].

The charging of the capacitors typically takes around 5–20 s depending on the installation and required energy level, whereas the actual pulse discharge, acceleration and collision of the flyer only last around 10–20 µs. A schematic illustration of the electromagnetic pulse welding system is shown in Figure 1.

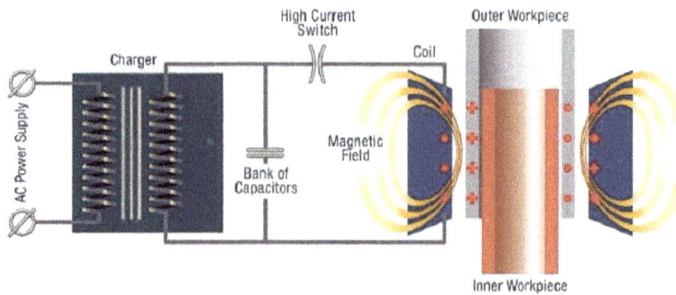

Figure 1. Schematic illustration of the electromagnetic pulse welding process.

2.2. Set-Up of the Electromagnetic Pulse Welding System

Electromagnetic pulse welding of Cu to St tubular joints was performed using a Pulsar model 50/25 system (Bmax, Toulouse, France) with a maximum charging energy of 50 kJ (corresponding with a maximum capacitor charging voltage of 25 kV). The total capacitance of the capacitor banks equals 160 µF. The weldability of copper to steel tubes was investigated using two different coil systems, namely a single turn coil combined with a field shaper and a transformer (ratio 3:1), and a multi-turn coil with 5 turns combined with a field shaper (see Figure 2). The field shaper is a practical tool, which is mainly used for forming and joining of tubular workpieces and serves to concentrate the magnetic flux and to focus the magnetic pressure over the desired area of the workpiece. A radial cut is machined in the field shaper, to lead the induced current to the internal surface of the field shaper. At the location of the field shaper cut, a lower magnetic pressure is acting on the tube, compared to other locations.

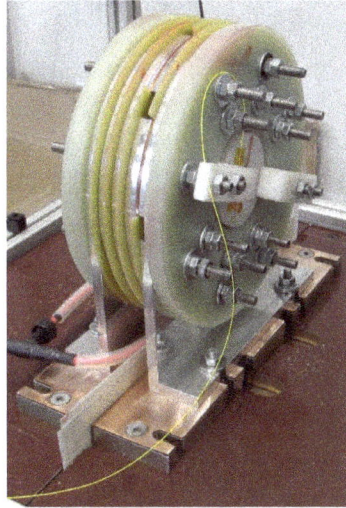

Figure 2. Multi-turn coil used in the experiments.

2.3. Materials and Dimensions

Copper (Cu DHP R220) tubes are welded onto cold-worked carbon steel rods (11SMnPb30 + C) and tubes. The copper tubes have an outer diameter and wall thickness equal to resp. 22.22 and 0.89 mm.

The configuration of the joints is illustrated in Figure 3. The internal parts are machined as shown in this figure, using a shoulder to align the flyer and target tube. The variable welding parameters are the stand-off distance, the overlap length and the free length. The overlap length is a material-dependent parameter that influences the impact angle. The outer diameter of the steel target tube is varied to achieve stand-off distances of 1.0, 1.5 and 2.0 mm. Based on previous experimental research, the length of the tool overlap between the flyer tube and field shaper is fixed at 8 mm and the free length at 15 mm.

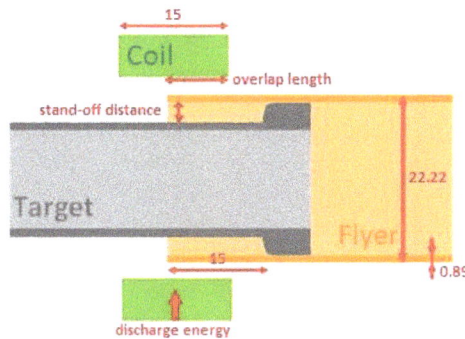

Figure 3. Joint configuration for tube-to-tube joints.

Different joint configurations have been used in the experiments, namely tube-to-rod, tube-to-tube without internal support and tube-to-tube with internal support. An example of a tube-to-tube weld using an internal support is shown in Figure 4.

Figure 4. Example of a tubular connection realised with electromagnetic pulse welding (copper tube outer diameter: 22.22 mm).

2.4. Weld Characterisation Methods

The weld quality was assessed based on a leak test using air and metallographic examination. In order to evaluate the effectiveness of the internal supports, the diameter of the internal part was measured prior and after the joining experiments.

No mechanical properties were measured, such as the tensile strength. It is very difficult to manufacture (standardized) tensile test specimens from the welded tubes, due to the specific shape of the welded samples, and their small size. Bend testing is also not possible, again because of the above-mentioned reasons. Moreover, for the given application, leak tightness and a defect-free weld are the most critical aspects to investigate.

2.4.1. Leak Test with Air

All welds were leak tested using air. The welded specimens were sealed at both ends, submerged into water and pressurized with an air compressor up to 9 bars. Leakage is visually detected by escaping bubbles near the weld interface, which indicate that either some severe imperfections are present, or there is no weld formation at all.

2.4.2. Metallographic Examination

Metallographic examination is performed to determine the actual cause of defective or leaking welds or to assess the quality of leak-tight welds. Hereto, the welded specimens are cross-sectioned in the longitudinal direction at the location of the field shaper cut, as the magnetic pressure is lower at this location. In this way, the weld interface at the field shaper cut as well as at 180° relative to the field shaper cut are investigated. The weld cross-sections are subjected to metallographic preparations, after which the interfacial morphology, the weld length and the reduction of the diameter of the internal part are examined and related to the welding parameters.

For welded tubular specimens, the weld length measured at the field shaper cut and 180° relative to the field shaper cut can be summarized into an arbitrarily defined parameter, called the Weld Quality Indicator (WQI). This parameter was developed by the authors and takes into account both weld lengths and the presence of a non-welded interface, observed during the metallographic examination [22]. The WQI is defined as

$$WQI = \frac{L_1 + L_2 - 0,5 \cdot |L_1 - L_2|}{A + 1}$$

where:

- L_1: the measured weld length near the field shaper cut,
- L_2: the weld length at 180° relative to the field shaper cut,
- A: a parameter that is equal to:
  ○ 0: if both locations contain a welded interface,
  ○ 1: if only one location contains a welded interface (other location is for example cracked),
  ○ 2: if at both locations no weld formation is observed.

The WQI is a measure for the weld length and the continuity of the welded interface along the circumference of the welded tubes. The color scale bar for the WQI in Figure 5 indicates a threshold value of 10, above which a weld is considered to exhibit a sufficiently high quality.

Weld quality indicator

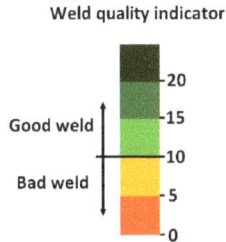

Figure 5. WQI colour scale bar to classify the weld quality, with a threshold value of 10.

2.4.3. Reduction of the Internal Diameter of the Target Tube

Due to the impact of the flyer tube, also the internal part (the target tube) deforms. Figure 6 illustrates the reduction of the inner diameter of the target tube after welding (d_{after}) and the original inner diameter of the target tube (d_{orig}). The cross-sectional area after impact is defined as: $\frac{\pi d_{after}^2}{4}$.

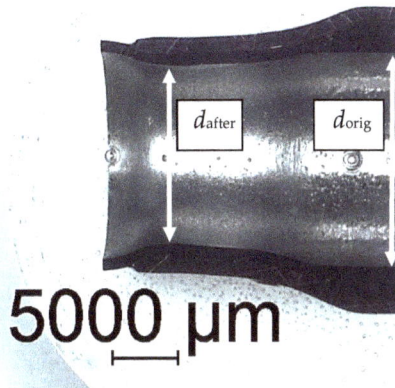

Figure 6. Measurement of the reduction of the inner diameter (d_{after}) and original inner diameter (d_{orig}) of the target tube.

2.5. Internal Supports for the Target Tubes

In order to minimize the radial deformation of the target tube during electromagnetic pulse welding, an internal support is required which is inserted into the target tube. Several types of internal supports have been documented in literature, but the majority were expensive, difficult to remove, or could not resist the severe impact energy [23–26]. If the internal support cannot be removed after the welding process, this can be considered as a process limitation when joining tubular parts for conducting liquids or gases. Therefore, in this experimental research, different other types of internal supports were explored which are preferably inexpensive, allow for easy removal and are possibly re-usable.

Two categories of internal supports have been considered. The first category concerns re-usable internal supports that are able to withstand the impact several times without failure. These internal supports should be extracted after welding by a manual or mechanical operation (e.g., a hydraulic or pneumatic press). The second category are internal supports that are not re-usable, but can be removed without direct access to the support. In this way, the support can also be used for long bended tubular connections within for example a refrigeration circuit.

Possible materials were selected and compared, based on the relevant requirements of the internal support. For the first category, i.e. the re-usable internal supports, a material that has a high fracture toughness and that does not break in a brittle manner is envisioned. Hence, polyurethane (PU) and polyamide (PA6.6) is considered, since both exhibit a high fracture toughness. A steel bar was considered as well.

For the second category, i.e., the non-re-usable internal supports, a first option is that the material can be melted or dissolved in a fluid and hence a low melting point and a high solubility are important. This leads to the selection of ice, which can be melted after welding, and plaster, which could offer the possibility to dissolve into a fluid. The ice was made using normal water. The ice was kept at a temperature of −18 °C prior to the welding experiment and used immediately in order to prevent melting. The plaster had a hardness of 46 N/mm, a porosity level of 46% and a plaster/water mixing ratio of 1.61 kg per liter of water.

Another property relevant for the non-reusable support is the brittleness, so that the material can withstand the first moment of impact, but easily fractures afterwards. In this way, the remains of the material can be removed by pressurized air. Hence, a material with a low fracture toughness is preferred, which leads to the selection of polymethylmethacrylate (PMMA), which is a very brittle composite. For this material, different configurations (rods, series of disks, and tube) were examined. The selected materials for the internal support are summarized in Table 1 and the corresponding configurations are illustrated in Figures 7–10.

Table 1. Materials and dimensions of the internal supports.

Category	Material Requirement	Support Material	Support Type	Izod Impact Strength	Inner Diameter	Outer Diameter	Length	Figure
Re-usable support	High fracture toughness	PU	Tube + M8 bolt	69.9 J/m	8 mm	14/15/ 15.45 mm	30 mm 50 mm	Figure 7
		PA6.6	Rod	160 J/m	not applicable	15.4/ 16.4 mm	35 mm	-
		Steel	Rod	NA	not applicable	15.1 mm 16.4 mm	30 mm	-
Non re-usable support	Low melting point	Ice	Rod	NA	not applicable	-	50 mm	Figure 8
	High solubility	Plaster	Rod	NA	not applicable	-	100 mm	Figure 9
	Low fracture toughness	PMMA	Rod	16 J/m	not applicable	15.4/ 16.4 mm	35 mm 20 mm 15 mm 10 mm	Figure 10
			Disks		not applicable	15.4 mm	4 × 5 mm²	
			Tube		7 mm 9 mm	15.4 mm	35 mm 35 mm	

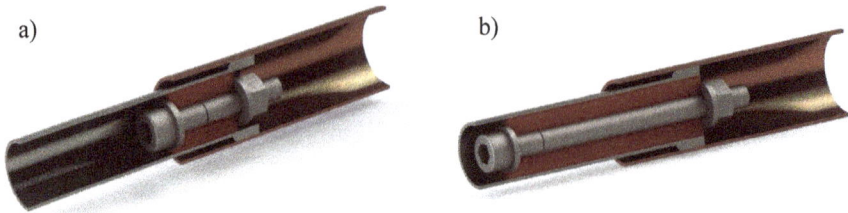

a) b)

Figure 7. Joint configuration with an internal support of PU and M8 bolts (**a**) length of 30 mm; (**b**) length of 50 mm.

Figure 8. Joint configurations with an internal support of ice.

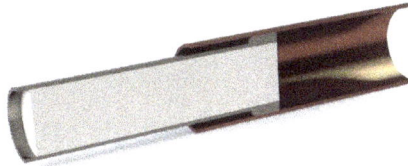

Figure 9. Joint configurations with an internal support of plaster.

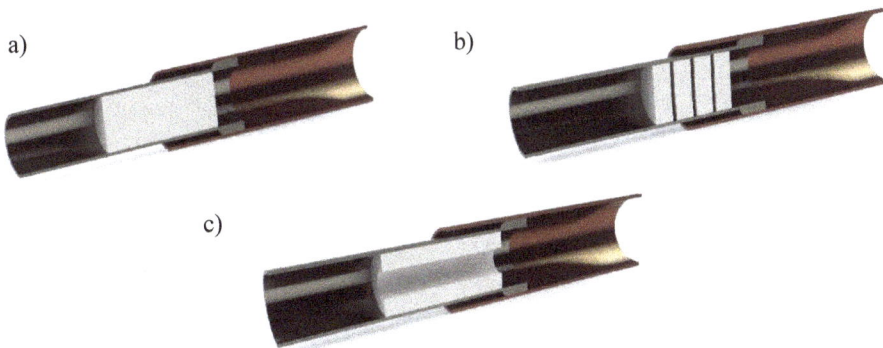

a)

b)

c)

Figure 10. Joint configurations with an internal support of PMMA: (**a**) Rod; (**b**) 4 disks with a length of 5 mm each (gaps between the disks are exaggerated for illustrative purposes); (**c**) tube.

3. Results

3.1. Overview of Experimental Work

During the experimental investigations, different aspects have been examined for manufacturing of tubular Cu-St joints, which were conducted into 2 phases:

- First phase: Influence of the process parameters and the workpiece geometry on the weld formation and joint characteristics. For this purpose, experiments were performed using target rods as a reference and target tubes with wall thicknesses of 1, 2 and 3 mm, without internal support. The purpose was to investigate the effect of the target tube wall thickness on the joint characteristics and the deformation of the target tube during welding. Also, tube-to-tube joints with a target tube with a wall thickness of 1 mm were manufactured using an internal PU-support for comparison.
- Second phase: Investigation of suitable support systems for the target tube with a wall thickness of 1 mm

3.2. Influence of the Process Parameters and the Workpiece Geometry on the Joint Characteristics

For all joint configurations, the internal workpieces were machined in the welding zone to a specific diameter, in order to obtain the desired value for the stand-off distance between the flyer tube and the target workpiece (1.0–1.5 & 2.0 mm). Besides this parameter, also the discharge energy was varied between 18 and 22 kJ. The free length, defined as the overlap distance of the flyer tube and the internal workpiece, was fixed at 15 mm. These parameter values are based on previous experimental work in the frame of the Join'EM project [27]. Also, the overlap between the flyer tube and the field shaper was fixed at 8 mm. The experiments were performed with the single turn coil with field shaper.

The WQI and the reduction of the internal diameter of the target tube were compared for the tube-to-rod configuration and the tube-to-tube configurations with the 3 different wall thicknesses of 1, 2 and 3 mm. In this way, the effect of the wall thickness of the target tube on the WQI and the reduction of the inner diameter of the target tube could be identified.

All of the optimized and semi-optimized welds were leak tight using the air leak test described in Section 2.4.1. Only the welds showing an excessive deformation of the internal part showed small leaks. The investigation described in Section 3.3 was performed using leak-tight welds.

3.2.1. Range of Weld Lengths and Reduction of the Inner Diameter of Leak Tight Welds

Table 2 summarizes the range of the measured weld lengths and the reduction of the internal diameter of the target tube of all leak tight welds created during the experimental investigation. Figure 11 shows a comparison of the metallographic specimens of the different configurations investigated.

Table 2. Range of measured weld lengths and reduction of the internal diameter of the target tube of leak tight welds.

Measurement	Wall Thickness of the Target Tube				
	- (Rod)	3 mm	2 mm	1 mm	1 mm + Support
Range of weld lengths (mm)	2.3–6.7	1.5–5.6	1.4–4.9	1.1–2.9	3.6–4.0
Range of the reduction of the inner diameter of the target tube (mm)	-	0.9–1.3	2.4–3.0	5.7–7.7	2.0–2.4

Figure 11. Metallographic specimens of typical welds: (left to right) Tube-to-rod, tube-to-tube with a target tube with a wall thickness of 3, 2 and 1 mm.

3.2.2. Influence of the Welding Parameters and the Wall Thickness of the Target Tube on the WQI

The effect of the stand-off distance on the weld quality represented by the value of the WQI of the tube-to-rod specimens and the tube-to-tube specimens with the 3 different wall thicknesses of the steel target tube is illustrated in Figure 12, whereas the effect of the discharge energy for a fixed stand-off distance of 2 mm is shown in Figure 13. Similar graphs have been composed for other values of the discharge energy or the stand-off distance.

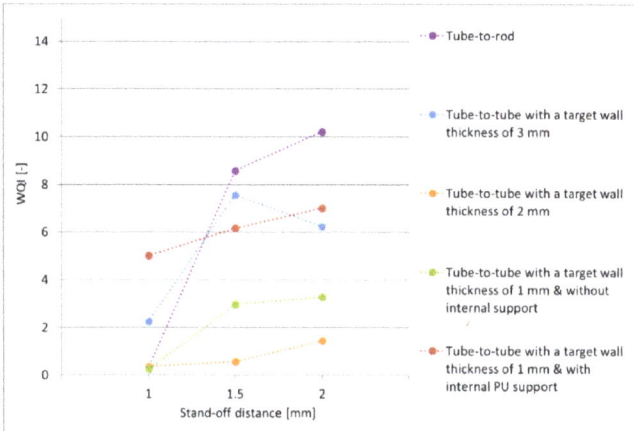

Figure 12. WQI as a function of the stand-off distance. Fixed discharge energy = 20 kJ.

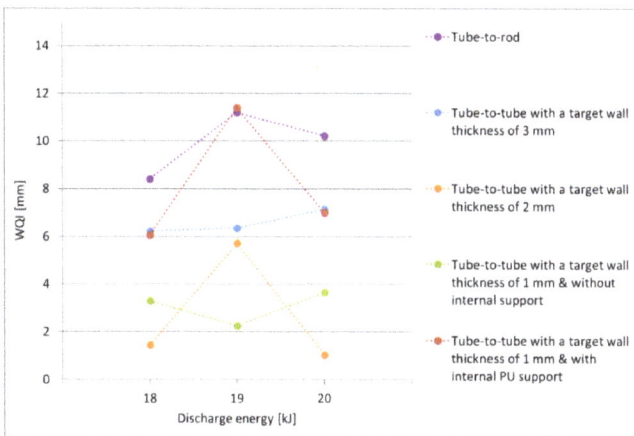

Figure 13. WQI as a function of the discharge energy. Fixed stand-off distance = 2 mm.

In particular, the stand-off distance significantly affects the WQI: a stand-off distance of 2 mm usually results in the largest WQI, as observed in the metallographic examinations. This indicates that 2 mm is a sufficient distance over which the flyer tube can accelerate and reach the required impact velocity for weld formation. In contrast, the shorter weld lengths measured for welds produced with a stand-off distance of 1 and 1.5 mm might indicate that these distances are too small for the flyer tube to reach a sufficient velocity. A stand-off distance of 1 mm usually resulted in a discontinuous weld interface.

In general, no clear correlation between the WQI and the discharge energy is identified (see Figure 13). At a medium discharge energy of 19 kJ, the WQI either reaches a minimum or maximum for the different stand-off distances.

The highest WQI values are obtained for tube-to-rod specimens, for tube-to-tube specimens with a large wall thickness of the target tube or for tube-to-tube specimens with supported target tubes, as observed in the metallographic examination of the weld interface. The use of an internal support also results in an increase of the WQI with a factor up to 4, compared to unsupported tube-to-tube specimens (compare the red and the green curve in Figure 12).

A smaller wall thickness of the target tube leads to a deterioration of the leak tightness of the welds. Moreover, a shift from a continuous to a discontinuous weld interface with a smaller weld length was observed. The center of the welded region appeared to be the weakest part of the weld and when failure occurred, it usually initiated from this location.

Based on these results, it is concluded that a stand-off distance between 1.5 and 2 mm combined with a discharge energy between 18 and 20 kJ leads to high-quality welds. The optimum overlap length equals 8 mm and the free length equals 15 mm.

A general overview of an optimized weld for the tube-to-rod configuration weld is shown in Figure 14. At the weld start (left in Figure 14), an unwelded zone is observed, because at this location, the welding parameters likely are outside of the window of suitable process parameters to obtain a metallic bond. Further to the right, the weld interface is flat or slightly wavy, without any visible intermetallic layers (see Figure 15). Towards the middle region of the welded interface, the waviness and the amount and thickness of intermetallic phases increase (see Figures 16 and 17). Towards the weld end, again an unwelded interface is observed, because of the same reasons at the weld start.

A small area in the middle of the weld interface (Figure 18) was selected for elemental mapping, line scanning and semi-quantitative chemical composition determination using Scanning Electron Microscopy (SEM; JEOL JSM-7600F Analytical Ultrahigh Resolution TFEG-SEM scanning electron microscope, Tokyo, Japan) coupled with Energy Dispersive X-Ray Spectroscopy (EDX).

Figure 14. Cross section of an optimised tube-to-rod weld.

Figure 15. Weld interface at the weld start.

Figure 16. Weld interface at the middle of the weld.

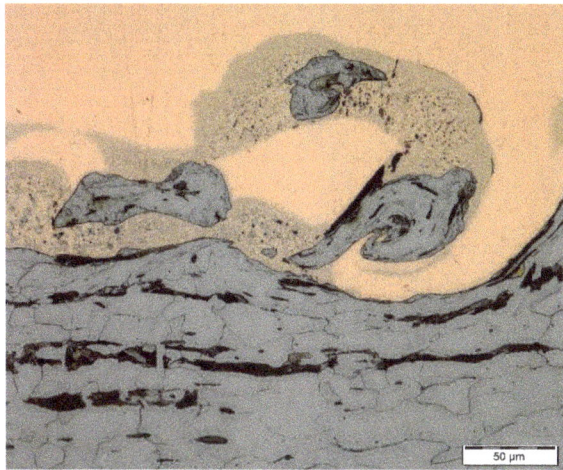

Figure 17. Detail of Figure 16.

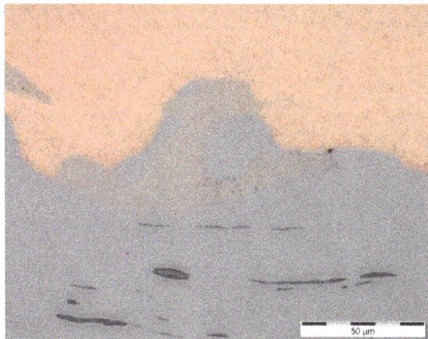

Figure 18. Optical micrograph of the middle of the weld interface.

As illustrated in Figures 18 and 19, an interfacial layer is present in the middle of the weld interface. Elemental mapping of this particular zone illustrates that the intermixed region is randomly dispersed within the steel material and is mainly composed of Cu (see Figures 20 and 21). Moreover, the alloying elements Mn and S are clearly detected in the steel material (see Figures 22 and 23). A line scan of the area shown in Figure 18 confirms that vigorous intermixing has taken place at the weld interface (see Figure 24). The semi-quantitative chemical composition detected by EDX shows that Cu is the dominant element in the interfacial layer (59.7 wt%), compared to Fe (39.8 wt%), as illustrated in the EDX spectrum in Figure 25 and Table 3. In the proximity of the weld interface, Cu (99.1 wt%) and Fe (98.4 wt%) are the main elements at either side of the interface.

Table 3. Semi-quantitative chemical composition by SEM-EDX, performed in the middle of the weld interface in Figure 18.

Position	Cu (wt% ± σ)	Fe (wt% ± σ)	Mn (wt% ± σ)
1	99.1 ± 0.1	0.9 ± 0.1	-
2	59.7 ± 0.2	39.8 ± 0.2	0.5 ± 0.1
3	0.6 ± 0.1	98.4 ± 0.1	0.9 ± 0.1

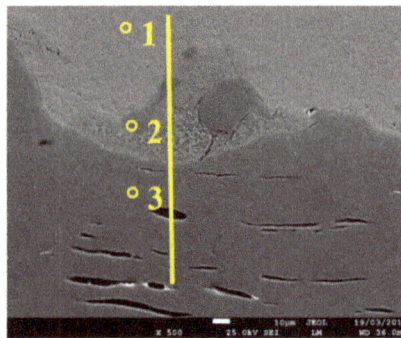

Figure 19. SEM micrograph of the zone shown in Figure 18, micrograph with indication of the line scan and points for semi-quantitative chemical composition.

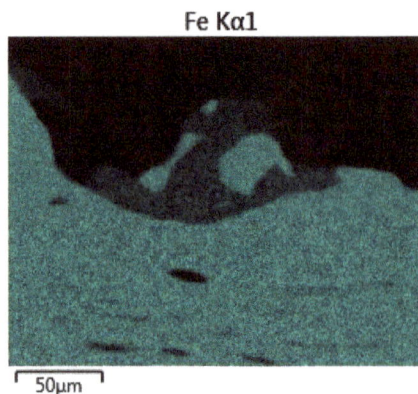

Figure 20. Element mapping of Fe in the zone shown in Figure 18.

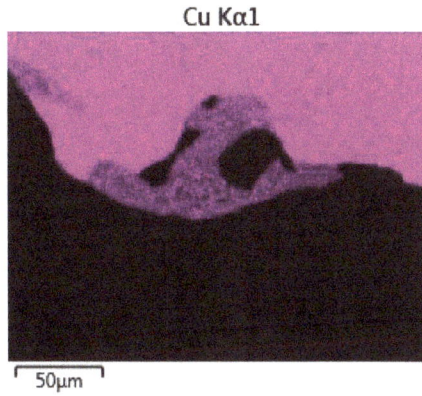

Figure 21. Element mapping of Cu in the zone shown in Figure 18.

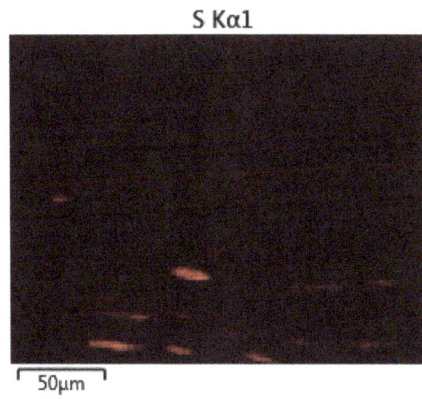

Figure 22. Element mapping of S in the zone shown in Figure 18.

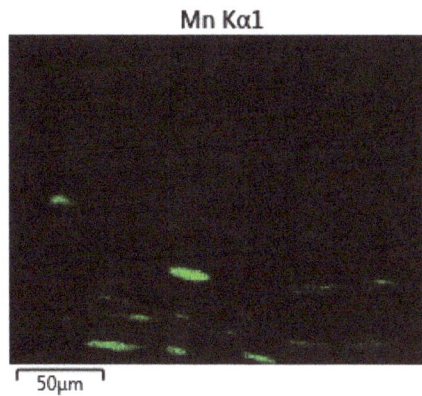

Figure 23. Element mapping of Mn in the zone shown in Figure 18.

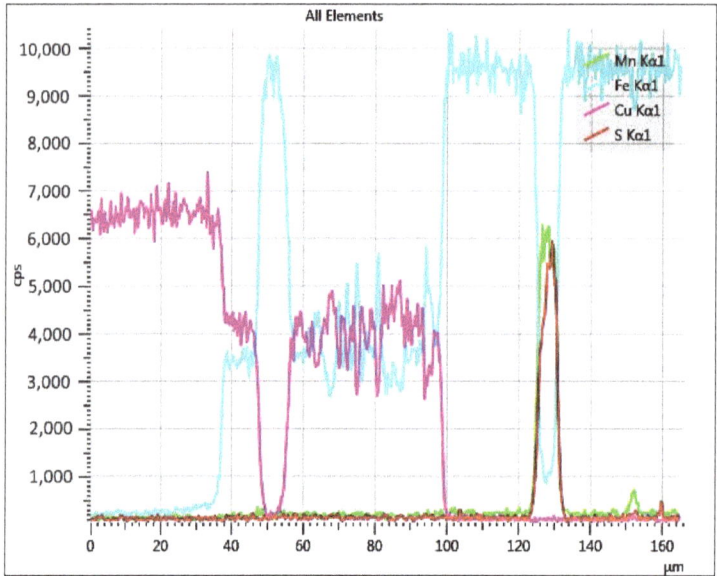

Figure 24. Line scan performed at the middle of the weld interface in Figure 18.

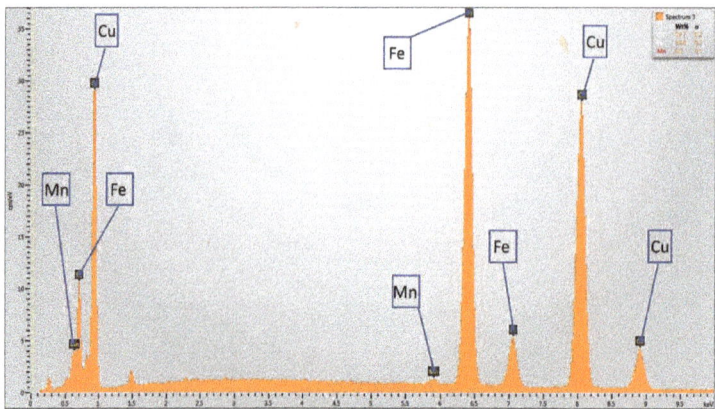

Figure 25. EDX spectrum of the interfacial layer in Figure 18 (point number 2).

The tube-to-tube joints exhibit a large radial deformation, especially when the target tube has a small wall thickness. Due to this, they are more sensitive for failure of the weld. The defective welds resemble as if they fracture after weld formation. Severe shearing of grains, waviness and some intermetallic phases are present at the broken weld interface, similar to successful welds, indicating sufficient energy is present to create a good weld. However, fracture of the weld interface is not caused by intermetallic layers, as they are not present in the majority of the fractured weld interface. The welding speed, also known as the collision point velocity, is much higher than the deformation rate of the target tube in the radial direction. Therefore, it is possible that the weld is created prior to the initiation of the deformation of the target tube in the radial direction [28]. In other words, the target tube only starts to decrease in diameter after the weld has been formed. Based on this, a possible hypothesis is that the weld was not sufficiently strong to withstand the forces caused by the radial inward deformation of the target tube and thus failure at the weld interface occurs. An example of such a failure is shown in Figure 26, compared to a sound weld interface observed in a tube-to-rod specimen, which does not exhibit any radial deformation. In Figure 26a, some porosities might be present, as a consequence of local melting and rapid solidification. Further investigation is however required to identify the cause of these porosities.

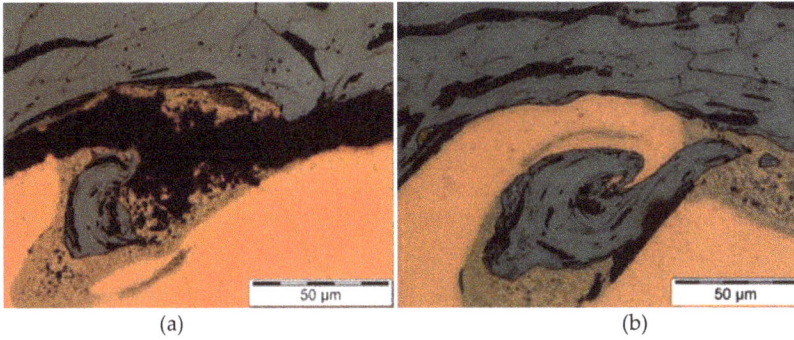

(a)   (b)

Figure 26. Weld produced with a stand-off distance of 2 mm. (**a**) Detail of the weld interface of a fractured tube-to-tube specimen with a target tube with a wall thickness of 2 mm. (**b**) Sound tube-to-rod specimen.

3.2.3. Influence of the Welding Parameters and the Wall Thickness of the Target Tube on the Deformation of the Target Tube

The effect of the stand-off distance on the reduction of the internal diameter of the steel target tubes for the 3 different wall thicknesses is illustrated in Figure 27, whereas the effect of the discharge energy is shown in Figure 28.

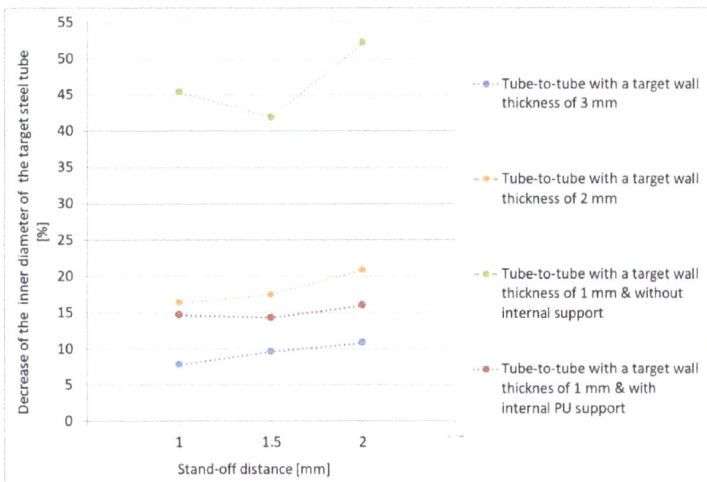

Figure 27. Decrease of the inner diameter of the steel target tube as a function of the stand-off distance. Fixed discharge energy = 20 kJ.

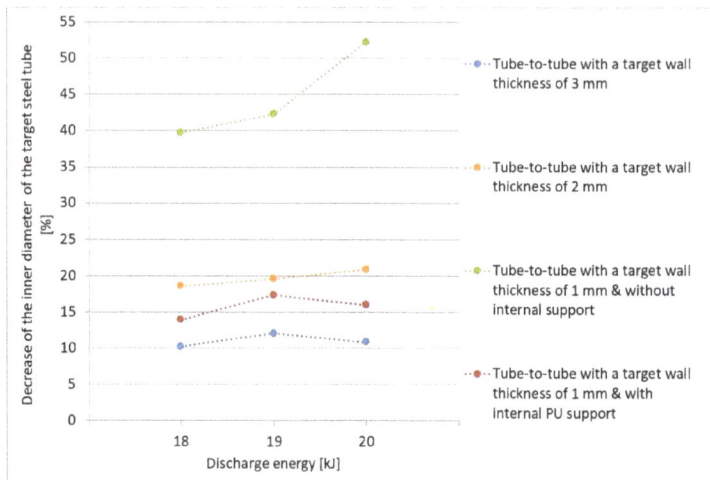

Figure 28. Decrease of the inner diameter of the steel target tube as a function of the discharge energy. Fixed stand-off distance = 2 mm.

A larger stand-off distance results in a larger decrease of the inner diameter, as the flyer can accelerate over a longer distance. This assumes that the flyer has not reached its deceleration point yet over a distance of 2 mm. Also, a higher discharge energy contributes to a larger reduction of the inner diameter, due to more energy being available for deformation.

In general, the inner diameter of the steel target tube after welding decreases for a smaller wall thickness of the target tube, due to the smaller resistance against deformation. For the smallest wall thickness of 1 mm of the target tube, the radial deformation of the target tube becomes irregular for some parameter combinations irregular and the occurrence of buckling is observed.

Moreover, the reduction of the inner diameter of tube-to-tube specimens with a wall thickness of 1 mm is more than twice the amount measured at the target tubes with a wall thickness of 2 and 3 mm. This amount of deformation is most likely unacceptable, as it causes a decrease of the inner diameter of up to 52%. This is equal to a loss of 75% of the inner tube area, which is very critical in for example fluid applications.

Some preliminary experiments were performed with an internal PU-support. This leads to a smaller reduction of the inner diameter of the target tubes with a factor up to 3.4, and an increase of the weld length by a factor up to 4, compared to unsupported tubes. The reduction of the inner diameter for supported tube-to-tube specimens is less dependent on the variation of the stand-off distance and discharge energy.

3.3. Investigation of Suitable Support Systems for the Target Tubes

In the experiments described in the previous section, a single-turn coil combined with a field shaper and a transformer was used. However, most of the welds had a WQI value below the threshold value of 10. Hence, the use of a multi-turn coil was considered, which leads to a higher frequency and hence a lower skin depth, resulting in an improvement of the process efficiency. The use of the multi-turn coil leads to a better weld quality and an improvement of the joint properties, compared to the use of a single turn coil. These investigations fall outside of the scope of this publication.

A high-quality weld produced by electromagnetic pulse welding should meet the following requirements:

- Leak tight
- Interfacial morphology with no or limited porosities and cracks and a sufficiently long welded interface at either side of the specimen. This corresponds with a WQI value of at least 10.
- Smallest possible reduction of the internal diameter of the target tube, such that a large cross-sectional area is maintained after welding. A qualitative support gives rise to a cross-sectional area after impact of at least 100 mm^2. This value was defined based on the application.

For this purpose, different internal supports were investigated, as detailed in Section 2.5. Similar as in the previous test series, for all joint configurations, the stand-off distance between the flyer tube and the target workpiece was varied at 1.0, 1.5 and 2.0 mm. Also, the discharge energy was varied between 18 and 22 kJ. The free length and the overlap between the flyer tube and the field shaper was again fixed at 15 and 8 mm, respectively.

In order to evaluate the quality of the different internal supports, the cross-sectional area after welding is visualized in Figure 29 as a function of the WQI for the different joint configurations that produced leak tight joints. Leaking joints were thus excluded from this graph. The tube-to-tube joint configurations without an internal support are included as a reference. The green area contains leak tight welds that fulfil all the requirements listed above for being classified as a high-quality weld (i.e., WQI > 10 and internal cross section > 100 mm^2). The yellow area contains welds that are leak tight, but only meet the requirement for either the WQI or the cross-sectional area. The red area contains leak tight welds that have a WQI and cross-sectional area below the minimum required values. All results of the different support methods are summarized in Table 4.

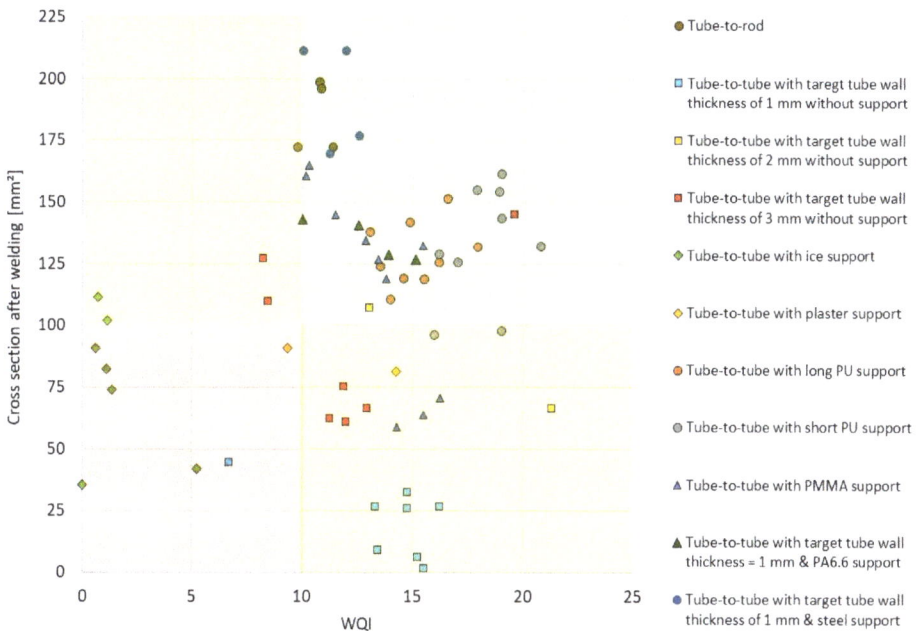

Figure 29. Cross-sectional area after welding as a function of the WQI for the different joint configurations.

Table 4. Evaluation of the different types of internal support.

Internal Support	Weld Quality: (WQI & Weld Length at 0°/180° from Field Shaper Cut	Remaining Internal Cross-Section after Welding (mm²)	Number of Process Steps	Cost	Advantages	Disadvantages
Polyurethane (PU)	++ WQI: 19.1 9.5/9.6 mm	++ 161.5	-	+	• Low material cost • Provides a good support and relatively low reduction of the inner diameter • Large weld lengths	• Time consuming installation compared to other supports, because tightening of the bolts requires time • Risk of the compression-expansion effect at a high discharge energy, leading to radial deformation of the steel target tube next to the impact zone • Difficult to extract after impact. Dimensions of PU-support need to be optimized
Polymethyl-methacrylate (PMMA)	++ WQI: 12.9 6.4/6.6 mm	+ 134.8	+	-	• Easy to insert into the tubes • Provides a good support • Relatively large weld length	• Expensive material cost • Does not break completely after impact and hence relatively difficult to extract • High tolerances needed for PMMA to fit precisely into the target tube • Provides a smell after welding
Ice	- WQI: 0.7 0.0/2.9 mm	- 111.5	-	++	• Very low material cost • Easy to remove after welding	• Very poor weld quality • Large reduction of the internal diameter • Time consuming to apply • Sealings of steel and copper necessary for safety reasons • Risk of expansion of the steel tube next to the impact zone • Ice already starts to melt during the installation of the specimen into the coil • Can cause contamination • Requires an additional operation to clean the tube system after welding
Plaster	++ WQI: 14.3 7.1/7.2 mm	- 81.6	-	+	• Low material cost Large weld lengths	• Large reduction of the internal diameter • Difficult to remove after welding Time consuming to apply prior to the welding process

Table 4. *Cont.*

Internal Support	Weld Quality: (WQI & Weld Length at 0°/180° from Field Shaper Cut	Remaining Internal Cross-Section after Welding (mm^2)	Number of Process Steps	Cost	Advantages	Disadvantages
Solid steel target workpiece + machining afterwards	+ WQI: 10.8 5.4/5.4 mm	++ 251.6	-	+	• No internal support required • Minimal reduction of the internal diameter of the target tube. Relatively large weld length	Accurate machining of target workpiece after impact is required
Steel rod	++ WQI: 12.6 mm 6.2/6.6 mm	++ 176.7	+	+	• Low material cost • Easy to insert into the tube • Provides a good support: minimum reduction of the inner diameter Relatively large weld length	• Difficult to remove manually after welding High tolerances needed for steel rod to fit precisely into the target tube
Polyamide (PA6.6)	++ WQI: 15.5 7.2/8.7 mm	+ 126.7	+	-	• Easy to insert into the tube • Provides a good support • Relatively large weld length Low material cost	• Does not break completely after impact and hence difficult to remove manually High tolerances needed for PA6.6 rod to fit precisely into the target tube

In general, a trade-off exists between the cross-sectional area after welding and the WQI: Welds with a larger remaining internal section after welding exhibit a lower WQI and vice versa. When using a higher discharge energy, a longer weld length is obtained, but also more deformation is observed of the target tube.

For the tube-to-tube joint configuration without internal support (blue, yellow and red square markers in Figure 29), the wall thickness of the target tube has a significant effect on the cross-sectional area after welding. When the target tube has a larger wall thickness, the cross-sectional area prior to impact is smaller, compared to a target tube with a smaller wall thickness. However, at the same time, the reduction of the cross-sectional area is significantly less and hence the cross-sectional area remaining after impact is larger for a target tube with a larger wall thickness. None of the joints performed without internal support fulfil the requirements listed above.

Ice and plaster are excluded as valuable support methods support (green and yellow rhombus markers in Figure 29) because of the low obtained joint quality in the case of ice, or because of the small remaining cross-sectional area after welding in the case of plaster. Other disadvantages of these support methods are mentioned in Table 4.

Tube-to-rod welds (green circle markers in Figure 29) exhibit a continuous weld interface with the highest cross-sectional area after welding (in the assumption that the solid target workpieces are machined to tubular workpieces with a wall thickness of 1 mm) and hence the lowest reduction of the inner diameter. However, the WQI is smaller compared to tube-to-tube configurations with a target tube with an internal support.

It can be concluded that PU, PMMA, PA6.6 and steel rods can be considered as valuable internal supports leading to high-quality welds and a sufficient cross-sectional area after welding. Welds with a steel bar support (blue circle markers in Figure 29) exhibit the highest cross-sectional area after welding, but at the same time their WQI is lower compared to welds with a PA6.6 or PMMA supports (blue and green triangle markers in Figure 29). In contrast, welds with a PA6.6 or PU support show the highest WQI, but also have a lower cross-sectional area after welding compared to steel internal supports.

Furthermore, welds which are situated at the threshold value of the WQI usually contain a discontinuous weld interface. In contrast, no correlation between the cross-sectional area after welding and the quality of the weld interface is identified. A discontinuous weld interface can thus occur for all values of the cross-sectional area after welding ranging from 100 mm^2 up to the maximum measured. In other words, a large reduction of the inner diameter of the target tube does not necessarily have a negative impact on the quality of the weld interface.

Table 4 summarizes the experimental results of the different internal supports, together with their practical advantages and disadvantages. The values for the WQI and the remaining internal cross-section after welding are the highest values achieved for each internal support.

4. Conclusions

The weldability of copper tubes onto steel rods and tubes was investigated. The steel target tubes had different wall thicknesses and the use of different internal supports was examined. The welding process was performed with a single-turn coil with a field shaper and transformer and in a next stage with a multi-turn coil with a field shaper. The welding parameters that were varied were the stand-off distance and discharge energy.

In a first phase, the influence of the workpiece geometry on the weld formation and joint characteristics was investigated. Different joint configurations were used, more specific the tube-to-rod and the tube-to-tube configurations, with target workpieces with different diameters and wall thicknesses. On the one hand, a larger stand-off distance or a higher discharge energy resulted in a larger reduction of the inner diameter of the target tube. On the other hand, a stand-off distance of 1.5 and 2 mm usually lead to a higher weld quality, compared to a stand-off distance of 1 mm. However, the latter usually resulted in a discontinuous weld interface. No clear correlation between the discharge energy and the weld quality (expressed by the WQI) was identified. A smaller wall

Metals **2019**, *9*, 514

thickness of the target tube leads to a deterioration of the leak tightness of the weld. The tube-to-rod specimens exhibited the highest WQI values, compared to the tube-to-tube specimens with target tubes with different wall thicknesses.

In a second phase, suitable internal supports for target tubes with a wall thickness of 1 mm were identified. As a comparison, also experiments with a steel rod as target workpiece were conducted, which were machined afterwards to a steel tube with a wall thickness of 1 mm.

In order to evaluate the usefulness of the different internal supports, the remaining internal cross-section as a function of the WQI for the different joint configurations that produce leak tight joints was plotted. It can be concluded that PU, PMMA, PA6.6 or steel rods can be considered as valuable internal supports leading to high-quality welds with a sufficiently large internal cross- section after welding. In contrast, ice or plaster did not fulfil the requirements because of an inferior weld quality, or because of an excessive deformation of the target workpiece.

Author Contributions: Conceptualization, K.F.; methodology, K.F. and W.D.W.; validation, K.F. and W.D.W.; investigation, K.F. and I.K.; writing—original draft preparation, K.F.; writing—review and editing, W.D.W.; project administration, K.F.; funding acquisition, K.F.

Funding: The results presented in this experimental research were achieved within the project JOINing of copper to steel by ElectroMagnetic fields—"JOIN'EM". This project was funded by the European Union within the framework of the Horizon 2020 research and innovation program under Grant Agreement No. 677660.

Conflicts of Interest: The authors declare no conflict of interest.

References

1. Faes, K.; Zaitov, O.; De Waele, W. Joining of Dissimilar Materials using the Electromagnetic Pulse Technology. In Proceedings of the ASM conference Trends in Welding Research, Chicago, IL, USA, 4–8 June 2012.
2. Kwee, I.; Psyk, V.; Faes, K. Effect of the Welding Parameters on the Structural and Mechanical Properties of Aluminium and Copper Sheet Joints by Electromagnetic Pulse Welding. *World J. Eng. Technol.* **2016**, *4*, 538–561. [CrossRef]
3. Sapanathan, T.; Raoelison, R.N.; Buiron, N.; Rachik, M. Magnetic Pulse Welding: An Innovative Joining Technology for Similar and Dissimilar Metal Pairs. In *Joining Technologies*; Ishak, M., Ed.; IntechOpen: London, UK, 2016.
4. Faes, K.; Kwee, I. Morphological and mechanical characteristics of aluminium-copper sheet joints by electromagnetic pulse welding. In Proceedings of the International Conference of High Speed Forming (ICHSF), Dortmund, Germany, 27–28 April 2016; pp. 299–308.
5. Verstraete, J.; De Waele, W.; Faes, K. Magnetic pulse welding: lessons to be learned from explosive welding. In Proceedings of the conference Sustainable Construction and Design 2011, Department of Mechanical Construction and Production, Ghent University, Ghent, Belgium, 16–17 February 2011.
6. Join'EM project objectives. Available online: http://www.join-em.eu/objectives.html (accessed on 2 April 2019).
7. Kallee, S.; Schäfer, R.; Pasquale, P. Automotive Applications of Electromagnetic Pulse Technology (EMPT). Available online: http://www.english.pstproducts.com/Automotive%20Applications%20of%20EMPT.pdf (accessed on 2 April 2019).
8. Vinitha, G.; Gupta, P.; Kulkarni, M.R.; Saroj, P.C.; Mittal, R.K. Estimation of charging voltage for electromagnetic welding. In Proceedings of the Annual International Conference on Emerging Research Areas: Magnetics, Machines and Drives (AICERA/iCMMD), Kottayam, India, 24–26 July 2014; pp. 1–4. [CrossRef]
9. Kudiyarasan, S.; Vendan, S.A. Magnetic Pulse Welding of Two Dissimilar Materials with Various Combinations Adopted in Nuclear Applications. *Indian J. Sci. Technol.* **2015**, *8*. [CrossRef]
10. Kudiyarasan, S.; Arungalai Vendan, S. Joining of Two Dissimilar Materials in Various Combinations by Using Magnetic Pulse Welding. *Int. J. Interdisciplin. Res. Innovations* **2016**, *4*, 89–100.
11. Verstraete, J. Magnetic pulse welding. Master's Thesis, Department of Mechanical Construction and Production, Faculty of Engineering, Ghent University, Ghent, Belgium, 1 June 2011.

12. Yu, H.; Fan, Z.; Zhao, Y.; Li, C. Research on magnetic pulse welding interface of copper-carbon steel. *Mater. Sci. Technol.* **2015**, *23*, 1–6.

13. Patra, S.; Singh Aroraa, K.; Shome, M.; Bysakh, S. Interface Characteristics and Performance of Magnetic Pulse Welded Copper-Steel Tubes. *J. Mater. Process. Technol.* **2017**, *245*, 278–286. [CrossRef]

14. Lueg-Althoff, J.; Gies, S.; Tekkaya, A.E.; Bellmann, J.; Beyer, E.; Schulze, S. Magnetic pulse welding of dissimilar metals in tube-to-tube configuration. In Proceedings of the 9th International Welding Symposium of Japan Welding Society, Tokyo, Japan, 11–14 October 2016; pp. 87–90.

15. Xu, Z.; Cui, J.; Yu, H.; Li, C. Research on the impact velocity of magnetic impulse welding of pipe fitting. *Mater. Des.* **2013**, *49*, 736–745. [CrossRef]

16. Cui, J.; Sun, G.; Xu, J.; Xu, Z.; Huang, X.; Li, G. A study on the critical wall thickness of the inner tube for magnetic pulse welding of tubular Al-Fe parts. *J. Mater. Process. Technol.* **2016**, *227*, 138–146. [CrossRef]

17. Lueg-Althoff, J.; Schilling, B.; Bellmann, J.; Gies, S.; Schulze, S.; Tekkaya, A.E.; Beyer, E. Influence of the wall thicknesses on the joint quality during magnetic pulse welding in tube-to-tube configuration. In Proceedings of the 7th International Conference on High Speed Forming, Dortmund, Germany, 27–28 April 2016; pp. 259–268.

18. Kwee, I.; Faes, K. Interfacial Morphology and Mechanical Properties of Aluminium to Copper Sheet Joints by Electromagnetic Pulse Welding. *Key Eng. Mater.* **2016**, *710*, 109–114. [CrossRef]

19. Stern, A.; Shribman, V.; Ben-Artzy, A.; Aizenshtein, M. Interface phenomena and bonding mechanism in magnetic pulse welding. *J. Mater. Eng. Perform.* **2014**, *23*, 3449–3458. [CrossRef]

20. Kore, S.D.; Date, P.P.; Kulkarni, S.V. Effect of process parameters on electromagnetic impact welding of aluminum sheets. *Int. J. Impact Eng.* **2007**, *34*, 1327–1341. [CrossRef]

21. Cui, J.; Sun, G.; Li, G.; Xu, Z.; Chu, P.K. Specific wave interface and its formation during magnetic pulse welding. *Appl. Phys. Lett.* **2014**, *105*, 221901. [CrossRef]

22. Simoen, S. Investigation of the weldability of dissimilar metals using the electromagnetic welding process. Master's Thesis, Department of Electrical Energy, Metals, Mechanical Constructions & Systems, Ghent University, Ghent, Belgium, 1 June 2017.

23. Roeygens, L.; De Waele, W. Faes, Experimental investigation of the weldability of tubular dissimilar materials using the electromagnetic welding process. In Proceedings of the conference Sustainable Construction and Design, Department of Mechanical Construction and Production, Ghent University, Ghent, Belgium, 1 June 2017.

24. Lueg-Althoff, J.; Bellmann, J.; Gies, S.; Schulze, S.; Tekkaya, A.E.; Beyer, E. Magnetic Pulse Welding of Tubes: Ensuring the Stability of the Inner Diameter. In Proceedings of the 6th Euro-Asian Pulsed Power Conference, EAPPC 2016: Held with the 21st International Conference on High-Power Particle Beams (BEAMS 2016) and the 15th International Conference on Megagauss Magnetic Field Generation (MG-XV); Estoril, Portugal, 18–22 September 2016; pp. 504–507, ISBN 978-1-5108-4607-4.

25. Lueg-Althoff, J.; Bellmann, J.; Gies, S.; Schulze, S.; Tekkaya, A.E.; Beyer, E. Magnetic Pulse Welding of Dissimilar Metals in Tube-to-Tube Configuration. In Proceedings of the 10th International Conference on Trends in Welding Research and 9th International Welding Symposium of Japan Welding Society, Tokyo, Japan, 11–14 October 2016; Volume 1, pp. 87–90, ISBN 978-1-5108-4403-2.

26. Bertelsbeck, S.; Geyer, M.; Böhm, S. Magnetic impulse welding of flexible tubes. IIW document SC-Auto-56-12. Available online: www.iiwelding.org (accessed on 30 April 2019).

27. JOINing of copper to aluminium by ElectroMagnetic fields (Join'EM). Project with funding from the European Union's Horizon 2020 research and innovation programme under grant agreement No. H2020-FoF-2014-677660 - JOIN-EM. Available online: http://www.join-em.eu/ (accessed on 2 April 2019).

28. Ben-Artzy, A.; Stern, A.; Frage, N.; Shribman, V.; Sadot, O. Wave formation mechanism in magnetic pulse welding. *Int. J. Impact Eng.* **2010**, *37*, 397–404. [CrossRef]

metals

MDPI

Article

Thermal Effects in Dissimilar Magnetic Pulse Welding

Joerg Bellmann [1,2,*], Joern Lueg-Althoff [3], Sebastian Schulze [2], Marlon Hahn [3], Soeren Gies [3], Eckhard Beyer [1,2] and A. Erman Tekkaya [3]

[1] Institute of Manufacturing Science and Engineering, Technische Universitaet Dresden, George-Baehr-Str. 3c, 01062 Dresden, Germany; eckhard.beyer@tu-dresden.de

[2] Business Unit Joining, Fraunhofer IWS Dresden, Winterbergstr. 28, 01277 Dresden, Germany; sebastian.schulze@iws.fraunhofer.de

[3] Institute of Forming Technology and Lightweight Components, TU Dortmund University, Baroper Str. 303, 44227 Dortmund, Germany; joern.lueg-althoff@iul.tu-dortmund.de (J.L.-A.); marlon.hahn@iul.tu-dortmund.de (M.H.); soeren.gies@iul.tu-dortmund.de (S.G.); erman.tekkaya@iul.tu-dortmund.de (A.E.T.)

* Correspondence: joerg.bellmann@tu-dresden.de; Tel.: +49-351-83391-3716

Received: 22 February 2019; Accepted: 14 March 2019; Published: 19 March 2019

Abstract: Magnetic pulse welding (MPW) is often categorized as a cold welding technology, whereas latest studies evidence melted and rapidly cooled regions within the joining interface. These phenomena already occur at very low impact velocities, when the heat input due to plastic deformation is comparatively low and where jetting in the kind of a distinct material flow is not initiated. As another heat source, this study investigates the cloud of particles (CoP), which is ejected as a result of the high speed impact. MPW experiments with different collision conditions are carried out in vacuum to suppress the interaction with the surrounding air for an improved process monitoring. Long time exposures and flash measurements indicate a higher temperature in the joining gap for smaller collision angles. Furthermore, the CoP becomes a finely dispersed metal vapor because of the higher degree of compression and the increased temperature. These conditions are beneficial for the surface activation of both joining partners. A numerical temperature model based on the theory of liquid state bonding is developed and considers the heating due to the CoP as well as the enthalpy of fusion and crystallization, respectively. The time offset between the heat input and the contact is identified as an important factor for a successful weld formation. Low values are beneficial to ensure high surface temperatures at the time of contact, which corresponds to the experimental results at small collision angles.

Keywords: magnetic pulse welding; dissimilar metal welding; solid state welding; welding window; cloud of particles; jet; surface activation

---

1. Introduction

Dissimilar metal welding plays an important role in the fabrication of multi-material parts. Materials with different mechanical, physical, or chemical properties need to be joined in order to fulfill the requirements of lightweight structures and high endurance parts or to save costs. Conventional thermal joining technologies reach their limits because of the formation of brittle intermetallic phases during the welding process. Decreasing the heat input is expedient to generate sound welds and to sustain the properties of the base materials. Impact welding processes like explosive welding (EXW) and magnetic pulse welding (MPW) utilize the oblique high-speed collision between two metallic surfaces to achieve metallurgical connections [1]. MPW is a suitable joining technology for dissimilar metal welding of tubular parts such as hybrid drive shafts [2]. In the initial state, the two joining partners (a movable "flyer" and a stationary "parent") are positioned with some standoff g, which defines the acceleration distance. During EXW, the acceleration of the flyer part is achieved by the detonation of an explosive, which is applied on the flyer. In contrast to that, the flyer acceleration in MPW is driven by a magnetic pressure. This pressure is the result of a sudden discharge of a capacitor bank via a tool coil that is positioned in close vicinity to the flyer. If the flyer material is electrically conductive, opposing eddy currents are induced into the flyer and resulting Lorentz forces drive the flyer away from the tool coil. A surface-related mathematical equivalent of the volume Lorentz forces is the magnetic pressure, which also causes shock loads on the tool coils and limits their lifetime. Comprehensive reviews of EXW and MPW can be found in [3,4], respectively. The weld is formed while both surfaces are pressed together during the collision. Interface pressures, in the order of several GPa, occur for a duration of a few microseconds. The microstructures of explosive welds show grains at the interface that underwent large plastic deformations. Strain rates can reach the order of 10^4 to 10^5 s^{-1} and, consequently, the material behaves like a fluid. The "jetting" effect is a consequence of the hydrodynamic phenomena taking place at the propagating collision point. The jet is often described as a massive material flow and is supposed to clean and activate the surfaces before welding. Within the last decades, the influence of thermal aspects during solid state welding was brought into the focus of many researchers, aiming for the identification of the relevant joining mechanism(s). For example, Ishutkin et al. found for EXW that the temperature of the shock compressed gas in the joining gap reaches several thousand °C and causes surface melting of both joining partners [5]. Together with the prevalent pressure, the overall energy input might be beyond the upper welding boundary and result in bad weld quality due to heat accumulation at the interface. This phenomenon occurs especially at positions that are exposed to the extensive heating for a longer time, i.e. further behind the initial collision point. Additionally, the temperature in the joining gap is strongly influenced by the thermodynamic properties of the medium in the joining gap. Deribas and Zakharenko describe the formation of a "cloud of particles" (CoP) that contributes to the temperature increase [6]. Thus, special interlayers are applied during EXW to avoid excessive intermetallic phase formation, see [7]. Another possibility is to reduce the "levels of temporal and force parameters required for joint formation" if the temperature of the near-contact layer of the welded materials is too high. This suggestion was made by Lysak and Kuzmin according to their pressure-temperature-time model, described in [8]. The latest descriptions of EXW and MPW weld seams have several aspects in common: melted and rapidly cooled metal layers were observed for EXW by Bataev et al. [9] and for MPW by Stern et al. [10]. Sharafiev et al. reported sharp boundaries between the base materials and an intermediate layer of impact welded A-Al joints. New, recrystallized grains in the size of nanometers and high dislocation densities give evidence for melting and rapid solidification [11]. Amorphous structures have also been described by metallurgists [12,13] that can be attributed to solidification with cooling rates in the order of 10^7 K/s [9]. Thus, it can be assumed that "liquid state bonding" is an occurring joining mechanism for EXW as well as MPW [14]. Nevertheless, different theories exist about the most relevant source of the thermal energy. In [9,15], the thermal energy is related to the plastic deformation during the collision, while in [5,16] it is attributed to the shock compressed gas in the joining gap as well. Successful MPW experiments under vacuum conditions revealed that no initial gas is required in the joining gap. Nevertheless, this does not contradict the theory of a compressed cloud of particles in the joining gap. In MPW experiments, it was found that the impact pressure or the normal impact velocity can be reduced as far as the collision angle is small enough [17]. Thus, less energy is needed for the flyer acceleration and the life time of the tool coils is increased. Furthermore, additional energy from an exothermic reaction between the joining partners or interlayers is beneficial for the joint formation, if well-adjusted [18]. This underlines the influence of thermal effects in MPW and the need for a comprehensive model or at least the identification of the most relevant input variables for heating and cooling of the surfaces and the interface. There are already thermal models existing for MPW [9,13,15,19] but they do not take the CoP into consideration as a heat source. The time for solidification has also been calculated in previous publications [9,20,21] but without considering dissimilar materials, the temperature distribution after the heating and not to mention the effect of phase transformations. For the prediction of the weld formation it is important to know the time dependent temperature in the joining zone. The weld formation might be hindered if the surface temperatures are too low at the time of contact or bounce back effects occur [21] before the solidification of the welding interface is completed.

The objectives of this experimental and numerical work can be summarized as follows:

1. Investigate the influence of the flyer kinetics on the material flow.
2. Study the influence of different collision conditions on the formation and properties of the jet or "cloud of particles" (CoP) and the corresponding thermal conditions in the joining gap.
3. Build up a temperature model for the welding interface, based on the heat input by the CoP.

2. Materials and Methods

2.1. Nomenclature

Three different experimental setups and a numerical model with a multitude of parameters will be introduced in the following chapters. Table 1 lists all symbols that are used within this paper in order to shorten the captions of figures and tables.

Table 1. Nomenclature for experimental and numerical setup.

Symbol	Parameter	Symbol	Parameter
A	Area	s	Thickness of the flyer tube
b	Equivalent thickness of the molten layer	S	High voltage switch
c	Heat capacity	t	Time
C	Capacitance; Contact point	T	Temperature
d	Distance to the impact location	t_{con}	Contact time
E	Charging energy	$t_{f,start}$	Flash appearance time
$f_{discharge}$	Discharge frequency	T_{Fly}	Flyer temperature
g	Initial joining gap	T_{fus}	Melting temperature
I	Discharge current	t_{heat}	Heating time
I_f	Intensity of the impact flash	t_{imp}	Impact time
I_{max}	Maximum discharge current	T_{Par}	Parent temperature
k	Thermal conductivity	T_{vap}	Boiling temperature
l	Length of welded zone	t_{wait}	Waiting time
l_c	Collision length	U	Voltage
L_i	Inner inductance of the pulse generator	V	Volume
l_w	Working length	v_i	Impact velocity
m	Mass	$v_{i,r}$	Radial impact velocity
p	Surrounding pressure	w_c	Width of the coil concentration zone
P	Heat input	z	Distance perpendicular to the steel surface
p_m	Magnetic pressure	α	Angle of inclined parent surface
Q	Total heat input	γ	Damping coefficient of $I(t)$
Q_s	Heat input to each surface	ΔH_{fus}	Enthalpy of fusion
Ra	Mean roughness index	Δt	Gap closing time
R_i	Inner resistance of the pulse generator	ρ	Density

2.2. Experiments

The material combination aluminum-steel was chosen for this study due to its relevance for current and future lightweight concepts in the transportation sector. For a good comparability with previous studies, the outer tube consists of the aluminum alloy EN AW-6060, while the inner rod is made of C45 [17,22]. The chemical compositions of the alloys are given in Table 2. Both parts were cleaned in ethanol before the joining experiments to remove debris from their surfaces.

Table 2. Aluminum EN AW-6060 alloy composition adapted from [23] and steel C45 (1.0503) alloy composition adapted from [24].

Flyer Part EN AW-6060 [1], Quasi-Static Yield Strength Approximately 60 MPa [2]		Parent Part C45 (1.0503), Normalized, Quasi-Static Yield Strength Approximately 490 MPa [3], Surface Polished (Ra = 1)	
Element	Weight %	Element	Weight %
Mg	0.35–0.6	C	0.42–0.5
Mn	≤0.1	Mn	0.5–0.8
Fe	0.1–0.3	P	<0.045
Si	0.3–0.6	S	<0.045
Cu	≤0.1	Si	<0.4
Zn	≤0.15	Ni	<0.4
Cr	≤0.05	Cr	<0.4
Ti	≤0.1	Mo	<0.1

[1] T66 heat treated: one hour at 500 °C and naturally aged, [2] determined by tube tensile test, [3] adapted from [25].

In the basic setup, MPW experiments with different charging energies were performed in order to identify the minimum energy required for a continuous weld seam along the circumference. Therefore, the setup depicted in Figure 1 was connected to two different pulse generators, resulting in two different resonant circuits with their characteristic values listed in Table 3. The current $I(t)$ was measured for each trial using a Rogowski current probe CWT 3000 B from Power Electronic Measurements Ltd. (Nottingham, UK) and the maximum current amplitude I_{max} was evaluated. The pulse generator MPW 50/25 (Bmax, Toulouse, France) was connected to a single turn working coil, while the magnetic pressure at the EmGen setup was generated by a five-turn coil with a field shaper. During the high speed collision between the flyer and the parent part, a characteristic flash occurs which is called impact flash [26]. The time-dependent course of the light emission was measured with the flash measurement system described previously [22]. It was triggered by the current signal of the generator, where the rise of the current was defined as $t = 0$ µs, see schematic oscilloscope in Figure 1. Thus, the starting time of the flash $t_{f,start}$, its duration and maximum intensity were analyzed. The welding result was checked with a manual peel test at four positions at the circumference: $0°, 90°, 180°$ and $270°$. The position at the slot of the coil (0°) is of special interest during process adjustment due to the reduced magnetic field intensity. However, the influence of the slot is limited to approximately 10% of the circumference. Hence, the metallographic analysis of the welding interface was performed at the opposed position (180°), since it is representative for almost the complete circumference [27]. Selected welding trials were performed with anodized flyer tubes (layer thickness 5 µm), which enable the reconstruction of the material flow [28].

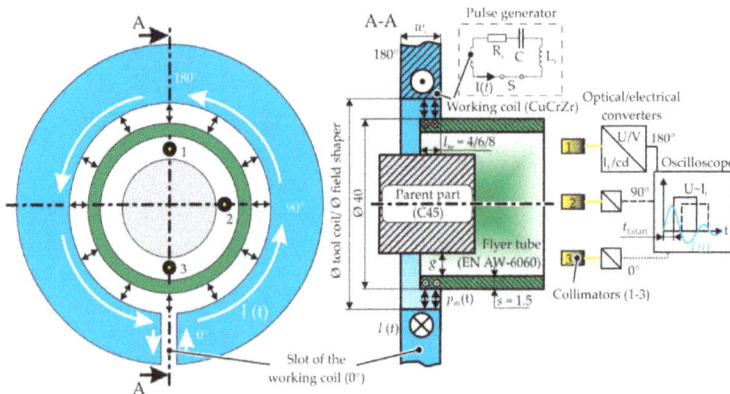

Figure 1. Basic setup for magnetic pulse welding (MPW) of tubes to cylinders (all values in millimeters, not true to scale, position of all parts is fixed during MPW).

Table 3. Characteristics of the resonant circuits and the deployed pulse generators.

Setup	Unit	Bmax MPW 50/25	EmGen
Capacitance	µF	160	140
Inductance [1]	nH	372	2700
Maximum charging energy	kJ	32	40
Maximum charging voltage	kV	20	24
Applied charging energy—E	kJ	4.5–9.6	7.0–22.7
Discharge frequency [1]—$f_{discharge}$	kHz	~21	~9
Damping coefficient γ [1]	1/s	16,500	2700

[1] for the complete resonant circuit with working coil, field shaper and workpieces.

The flyer kinetics at the lower process boundary were studied for both pulse generators with the modified setup shown in Figure 2 while the parameter d and, thus, the initial collision point was increased stepwise from zero to five millimeters from the flyer edge. After the flyer was sheared from the parent part, the length and location of the weld seam were measured at the 90°, 180° and 270° position for each test.

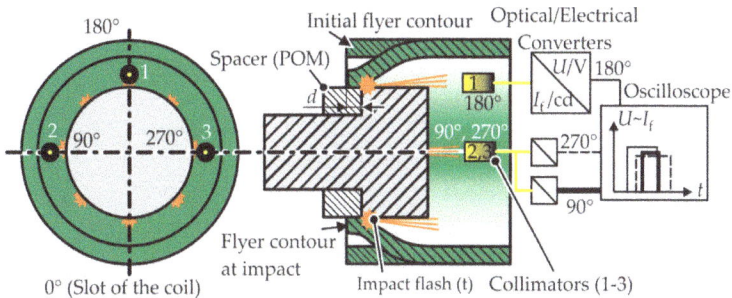

Figure 2. Modified MPW setup for investigation of the flyer kinetics with collimators (1–3) for flash detection (not true to scale, position of all parts is fixed during MPW).

For the third experimental part, the MPW process was carried out in a vacuum chamber as depicted in Figure 3, using the Bmax pulse generator. The surrounding pressure p, the collision length l_c, the working length l_w as well as the contour of the parent part were varied. The camera Canon EOS 700D with an exposure time of six seconds, a fixed aperture of F13 and the light intensity ISO 100 was placed behind a translucent Plexiglas disc to take longtime exposures and to identify the color of the process glare. The average R/G/B value of 5x5 pixels was converted to the 2-D xy-color chart. This enables the estimation of the temperature in the joining gap, if an ideal black body emission is assumed. Again, the flash measurement system was used to detect the temporal course of the light intensity. Furthermore, a translucent plastic disc was placed inside of the flyer tube to study the interaction with the jet or cloud of particles, respectively. Qualitative conclusions can be drawn from the location where the debris sticks on the disc or if the plastic sheet is fractured. Additionally, the locations of the jet residues at the inner flyer surfaces were analyzed.

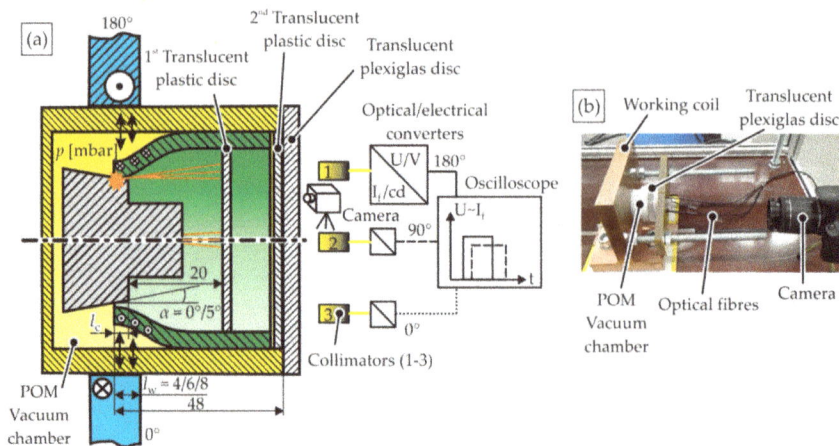

Figure 3. Modified setup for MPW at different surrounding pressures p with collimators (1–3) for flash detection and camera (**a**) schematic (not true to scale) and (**b**) photograph (position of all parts is fixed during MPW).

2.3. Numerical Simulations

The aim of the numerical investigations is to estimate the heating and cooling of the surfaces of the joining partners and predict the weldability based on the liquid state bonding theory. Experiments will be described in the following chapters, which reveal temperatures in the closing joining gap that are far beyond the fusion temperatures of both joining partners. The CoP is assumed to be a main heat source before the contact and, thus, the model should answer the question, whether liquid state bonding can occur under these conditions. Although the temperature of the CoP is experimentally estimated in this paper, its heat transfer to the surfaces is very hard to determine. The density or mass, respectively, and the surface coefficient for heat transfer are difficult to access and, thus, the strategy performed in [15,29,30] is applied: The amount of melted flyer and parent material is quantified in polished cross sections and serves as an upper boundary of the heat input. It is cross-checked with the kinetic flyer energy, which determines the overall limit of the introduced energy. Furthermore, the following assumptions were made for the model:

1. The thermal energy of the CoP is responsible for the surface activation before both joining partners get into contact. In order to simplify the numerical model, this is assumed to be the only heat source. The heat input by plastic deformation after the collision is not considered in the model.
2. The thermal energy of the CoP is equally distributed to both joining partner surfaces, which seems admissible for small collision angles.
3. At the welding interface, just solid and liquid phases are present at the time of contact. If the surface temperature would lead to vaporization before contact, the material in the gaseous phase would be pressed out of the joining gap together with the CoP during MPW, provided a sufficient collision angle.
4. The influence of the temperature on the materials' densities, heat capacities and thermal conductivities is not considered in the simulations.

The one-dimensional numerical model for the calculation of the time-dependent surface temperatures was set up within the commercial software COMSOL Multiphysics® (Version 5.2, COMSOL Multiphysics GmbH, Goettingen, Germany). The model was simplified compared to the real MPW process that consists of three stages: The initial collision depicted in Figure 4a, the CoP formation, shown in Figure 4b and the movement of the collision point C, see Figure 4c. During the simulation, the flyer and parent parts were fixed with a constant gap of 2 µm, as shown in Figure 4d.

The CoP formation was not implemented in the model, but its heat input to both joining partners as well as the heat losses by conduction in the parts were taken into account. In order to recreate the moving contact point C during the real MPW process with the fixed joining partners, the following strategy was applied in the numerical model: During the surface activation by the CoP in Figure 4b, any heat transfer through the gap was suppressed. But then, at the contact time t_{con} in the real MPW process, the heat conductivity of the gap was set to an extremely high value of 10^{10} W/(mK) and, thus, the intimate contact between both joining partners as shown in Figure 4c was imitated without the need for moving parts or meshes within the simulation. The implemented temporal course of the heat input P and the contact time t_{con} are depicted in Figure 4e. The following time steps were chosen for the numerical simulations: 0.02 µs for 0 µs < t < 10 µs and 0.5 µs for 10.5 µs < t < 100 µs. In order to study the temperature distribution in the close vicinity to the welding interface, the element size was set to 0.1 µm. Due to the one dimensional character of the model, the heat sources were defined with a certain area density at two points on both surfaces, see Figure 4d. During the simulations, different heat quantities Q_S, heating times t_{heat} and material combinations were investigated. The relevant material parameters are listed in Table 4. The heating time at a certain point C strongly depends on the velocity of the CoP and the distance to the initial point of impact. Due to the accumulation of the CoP during the weld front propagation, the heat input and heating duration vary, too. Furthermore, a waiting time t_{wait} was defined between the end of the heat input and the contact time t_{con}. By setting t_{wait} to 0 µs, the heat input generated by the CoP is transferred to the plates in the vicinity of the contact line. This corresponds to the case when the CoP and the collision point C have the same velocity. If the CoP travels faster than the collision point, the waiting time is increased in the simulation. This procedure abstracts the effects in the moving interaction zone of a real MPW process. Nevertheless, it simplifies the modeling and reveals the most relevant influencing factors. Of course, these parameters have to be transferred to the kinematic process variables like collision front velocity or collision angle in a further step.

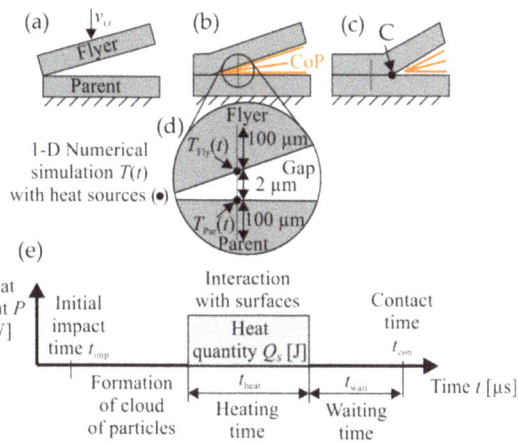

Figure 4. Process steps during MPW showing: (**a**) the initial collision, (**b**) the cloud of particles (CoP) formation and surface activation and (**c**) the surface contact with the moving contact point C, (**d**) dimensions of the 1-D model, (**e**) modeling scheme with the temporal course of the heat input and contact time.

Table 4. Material specific values (temperature independent during the simulations).

Physical Quantity	Symbol	Unit	EN AW-6060	C45	Cu [31]
Density	ρ	kg/m^3	2700 [32]	7700 [24]	8960
Heat capacity	c	J/kgK	898 [32]	470 [24]	390
Thermal conductivity	k	W/mK	210 [32]	42.6 [24]	384
Melting temperature	T_{fus}	°C	659 [31], pure aluminum	1536 [31], pure iron	1083
Boiling temperature	T_{vap}	°C	2467 [31], pure aluminum	3070 [31], pure iron	2595
Enthalpy of fusion	ΔH_{fus}	kJ/kg	356 [31], pure aluminum	276 [31], pure iron	213

3. Results and Discussion

3.1. Effect of the Flyer Kinetics on the Material Flow

During the experimental study, the charging energy E was increased stepwise until a circumferential weld seam was proved in the manual peel test. The minimum radial impact velocity for the Bmax setup was measured via photon Doppler velocimetry (PDV) and found to be approximately 340 m/s [17]. The corresponding cross sections of the 180° location are shown in Figure 5 and reveal small waves at the joining interface.

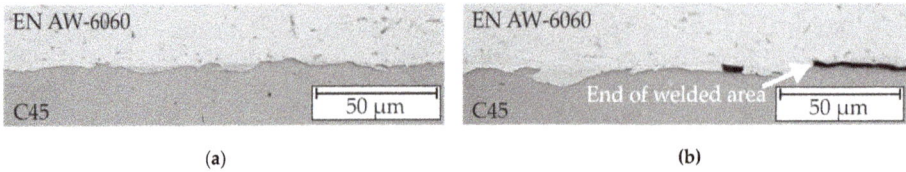

(a)   (b)

Figure 5. Polished cross section of the MPW sample joined with the Bmax setup (g = 1.5 mm, E = 5.8 kJ, I_{max} = 451 kA, Ø coil = 42 mm, w_c = 10 mm, l_w = 6 mm, welding direction from left to right, $v_{i,r} \approx$ 340 m/s measured with photon Doppler velocimetry (PDV) [17]): (a) in the middle and (b) at the end of the joining zone.

Although the required energy at the EmGen machine was higher compared to the Bmax setup, the minimum radial impact velocity was lower at approximately 250 m/s. One of the reasons is the difference in the flyer forming and collision behavior, which heavily depends on the discharging frequency [17]. The welding interface is almost smooth, see Figure 6.

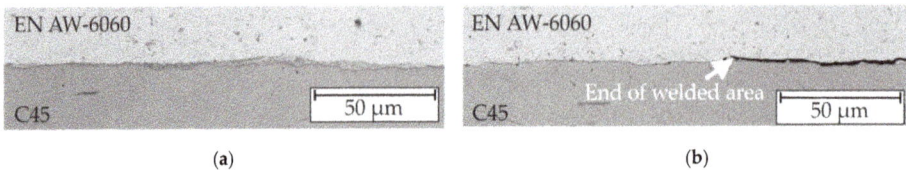

(a)   (b)

Figure 6. Polished cross section of the MPW sample joined with the EmGen setup (g = 1.5 mm, E = 22.7 kJ, I_{max} = 113 kA, Ø coil = 41 mm, w_c = 15 mm, l_w = 6 mm, welding direction from left to right, $v_{i,r} \approx$ 250 m/s calculated according to [22] with measured flash appearance time equal to 12.6 µs): (a) in the middle and (b) at the end of the joining zone.

The application of a 5 µm thick anodized layer on the flyer tube prevented the welding effect at the EmGen setup on the same energy level, but enabled the reconstruction of the material flow. Figure 7a shows the fragmented anodized layer that mainly stayed on its original position. The flyer material has been extruded through the interspaces. At the end of the joining zone, this effect was significantly reduced.

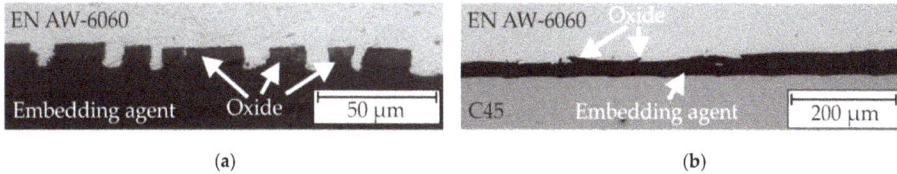

(a)   (b)

Figure 7. Polished cross section of the unwelded MPW sample joined with the EmGen setup (aluminum flyer anodized with 5 μm thickness, g = 1.5 mm, E = 22.7 kJ, I_{max} = 117 kA, Ø coil = 41 mm, w_c = 15 mm, l_w = 6 mm, welding direction from left to right, $v_{i,r} \approx$ 250 m/s calculated according to [22] with measured flash appearance time equal to 12.6 μs): (**a**) in the middle (according to [17]) and (**b**) at the end of the joining zone.

If the radial impact velocity was increased to 460 m/s, the anodized layer in the middle of the joining zone was completely removed and a sound weld was generated, see Figure 8a. At the end of the joining zone, the anodized layer was embedded into the flyer material, which illustrates the massive material flow in Figure 8b. This phenomenon can occur at high speed impact situations and is called "jet". Jetting is associated with a high degree of plastic deformation and surface enlargement. It is also often described as a prerequisite for the surface cleaning and removing of oxides before the surfaces get in contact. The experiments described so far give clear evidence that this large amount of plastic deformation is not required for successful MPW, unless certain surface layers have to be removed or roughness peaks to be overcome. Furthermore, welding with velocities below the "jetting"-regime requires less impact velocity, which is beneficial for the lifetime of the tool coils due to the reduced mechanical and thermal loads. Thus, it is worthwhile to analyze the physical conditions that enable MPW at the lower process boundary.

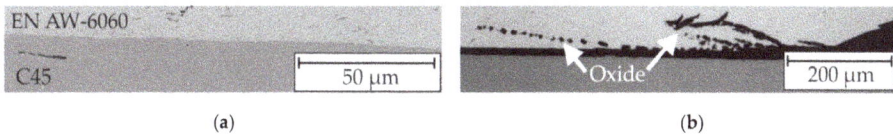

(a)   (b)

Figure 8. Polished cross section of the MPW sample joined with the Bmax setup (aluminum flyer anodized with 5 μm thickness, g = 2 mm, E = 9.6 kJ, I_{max} = 533 kA, Ø coil = 42 mm, w_c = 10 mm, l_w = 6 mm, welding direction from left to right, $v_{i,r} \approx$ 460 m/s measured with PDV [33]): (**a**) in the middle (according to [33]) and (**b**) at the end of the joining zone.

During all successful MPW experiments a flash was visible. In a former study, it was shown that the high velocity impact flash occurred approximately 0.5 μs after the initial impact [22] and it was correlated with the weld seam formation [34]. Thus, the flash detection was utilized as a measurement system for the parameter adjustment and quality assurance during MPW [28,35]. The setup shown in Figure 2 was used to analyze the flyer kinetics and weld formation by increasing the parameter d stepwise from zero to five millimeters. Thus, the initial collision point was shifted and the effect on the flash appearance time and weld formation was studied. In Figure 9, the average values of the flash appearance times are plotted for both MPW setups. The location of the slot of the coil or field shaper was not considered, since it differs significantly from the remaining circumference due to the reduced magnetic field intensity. The flash at the EmGen setup occurred more than 8 μs later. The measured flash appearance time of 19.5 μs corresponds to a radial impact velocity $v_{i,r}$ of approximately 160 m/s. Nevertheless, the gap closing time Δt is much smaller for the EmGen setup, which seems to be an important factor for the weld establishment at these comparatively low radial impact velocities. At the EmGen setup, no flash was detected for d = 5 mm and consequently, no weld seam was generated, see Figure 10. An experiment with d = 10 mm, where the collision occurred between the flyer and the plastic spacer, did also not lead to an impact flash. This shows that the impact of the two metals itself must be responsible for the flash initiation and not the compressed air within the closing gap between the flyer and the plastic spacer due to the high speed compression.

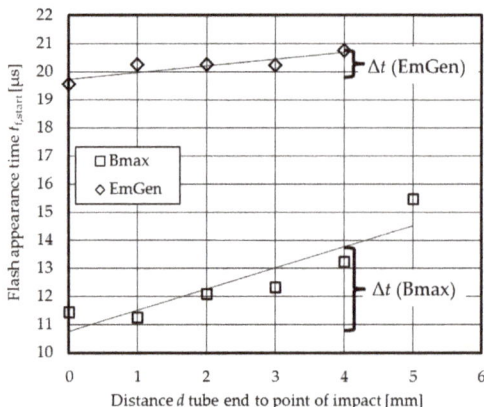

Figure 9. Mean values of flash appearance time $t_{f,start}$, referring to the rising tool coil current at $t = 0$ µs as depicted in the schematic oscilloscope in Figure 1, at 90°, 180°, and 270° for varied values d for two MPW setups at the specific lower process boundary (Bmax: $g = 1.5$ mm, $E = 4.5$ kJ, $I_{max} = 380$ kA, Ø tool coil = 41 mm, $w_c = 10$ mm, $l_w = 6$ mm, EmGen: $g = 1.5$ mm, $E = 7$ kJ, $I_{max} = 70$ kA, Ø field shaper = 41 mm, $w_c = 10$ mm, $l_w = 6$ mm).

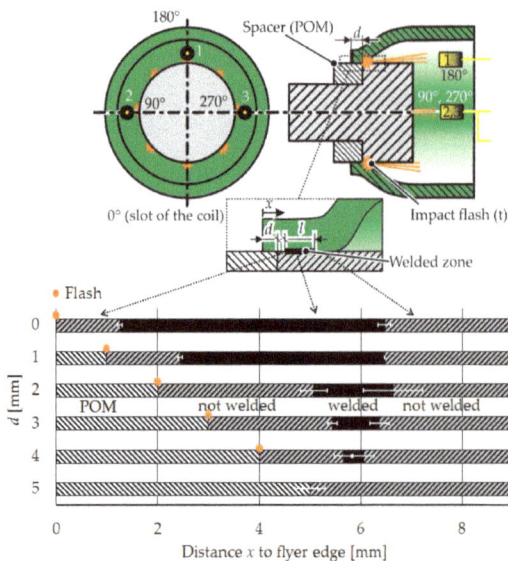

Figure 10. Mean values of welding start and end at 90°, 180°, and 270° for varied values d.

After the MPW experiment, the flyer was sheared from the parent part and the average weld seam length l and its location were measured for each test, see Figure 10. The weld starts 1.2 mm after the first contact between flyer and parent, at the earliest. If the point of the initial metal impact was postponed by the parameter d, the weld formation was shifted, too. Thus, it seems reasonable that the high speed impact between the metallic partners and the corresponding impact flash are closely related to another necessary welding criterion, most likely the surface activation of the adjacent areas to be welded.

Basically, the necessity of the surface activation before welding is in good correlation with the "traditional" view of the role of the jetting effect and the corresponding surface enlargement. However, as shown in the first experiments, MPW is also possible at lower impact velocities where jetting and material flow are not initiated. To explain these findings, the mechanism presented by Deribas [6] and Ishutkin [5] seems to be a reasonable approach. They identified the appearance of a "cloud of particles" (CoP) at lower impact velocities. Obviously, the CoP plays an important role at the MPW process and, thus, it seems worthwhile to study this phenomenon in detail. The following section describes the investigation of the CoP with modified MPW experiments.

3.2. Characteristics of the "Cloud of Particles" (CoP)

The setup depicted in Figure 3 was used for a comprehensive investigation and characterization of the CoP. Therefore, the influence of the collision conditions on the appearance of the CoP and effect on the weld formation was studied. Furthermore, attendant effects like residues on the flyer and tempering colors on the parent surfaces outside the joining zone were recorded, see Figure 11.

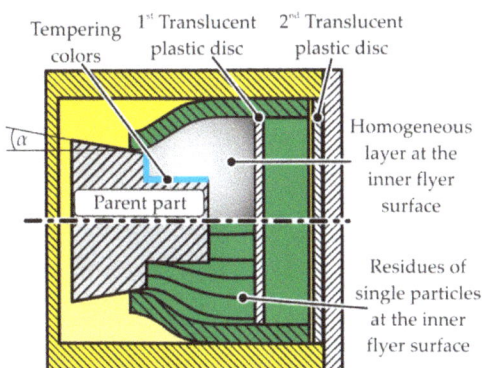

Figure 11. Effects on the flyer and parent surface (position of all parts is fixed during MPW).

A chamber was designed that enables MPW experiments on the Bmax-setup at different surrounding pressures. The maximum intensity of the process glare in vacuum is significantly reduced compared to the process in ambient atmosphere, see also [36]. Nevertheless, it is still visible, as shown in Figure 12.

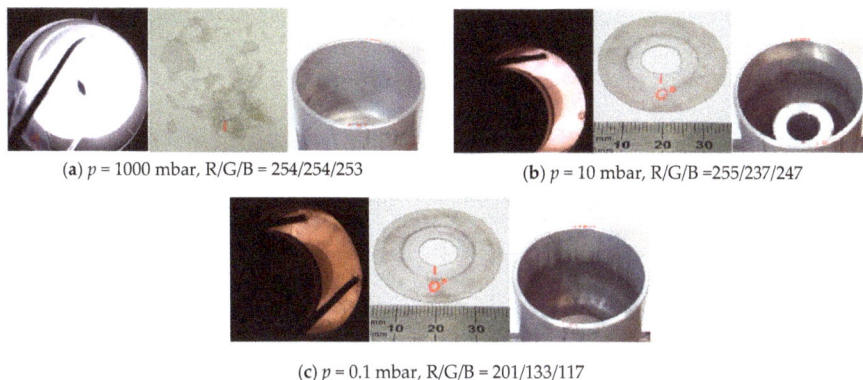

(**a**) p = 1000 mbar, R/G/B = 254/254/253

(**b**) p = 10 mbar, R/G/B =255/237/247

(**c**) p = 0.1 mbar, R/G/B = 201/133/117

Figure 12. Long time exposures of the process glare, first plastic disc and flyer tube after MPW experiments at the Bmax setup (l_c = 3 mm, α = 0°, g = 1.5 mm, E = 4.5 kJ, I_{max} = 365 kA, Ø coil = 41 mm, w_c = 10 mm, l_w = 6 mm): (**a**) p = 1000 mbar, (**b**) p = 10 mbar, (**c**) p = 0.1 mbar.

The rise time until the intensity reached its maximum at ambient atmosphere was 0.4 µs and it increased up to 3 µs in vacuum, see Figure 15. Since welding was possible in both cases, it can be concluded that a surrounding gas is not mandatory for welding. Probably, the CoP is formed at the initial impact, independent of the surrounding medium. The density of the medium in the joining gap determines the velocity of the jet [37] and it seems likely that shock compression of the surrounding gas occurs, which leads to the immediate light emission [16]. Furthermore, the shock compressed gas is the reason for the complete fracture of the first plastic disc during MPW in ambient atmosphere. In vacuum, there is no interaction with other gases and the CoP expands freely in the welding direction. At the same time, the residues at the inside of the flyer tube are noticeably increased, as well as the intensity of the tempering colors at the parent part. The starting time of the flash was almost identical for p = 1000 mbar ($t_{f,start}$ = 11.07 µs) and p = 0.1 mbar ($t_{f,start}$ = 10.96 µs). Thus, the flyer forming behavior is assumed to be independent of the gas pressure p.

The CoP formation was analyzed at p = 0.1 mbar by varying the collision length l_c between 1 and 3 mm. Similar to [22], at ambient atmosphere, no flash was detectible at a collision length of 1 mm. No debris was deposited on the first plastic disc, as seen in Figure 13a. An increase of the collision length to 3 mm lead to a process glare, contaminants on the first plastic disc as depicted in Figure 13b and tempering colors on the parent part. Thus, the CoP was formed and heated up between 1 and 3 mm after the initial impact. The collision length l_c was set to 3 mm for the following experiments.

(a) l_c = 1 mm, R/G/B = 0/0/0   (b) l_c = 3 mm, R/G/B = 236/186/182

Figure 13. Long time exposures of the process glare, first plastic disc and flyer tube after MPW experiments at the Bmax setup (p = 0.1 mbar, α = 0°, g = 1.5 mm, E = 4.5 kJ, I_{max} = 365 kA, Ø coil = 41 mm, w_c = 10 mm, l_w = 6 mm): (a) l_c = 1 mm (b) l_c = 3 mm.

After investigating the CoP initiation at constant collision conditions, the appearance at different impact velocities and angles was studied systematically. Again, the process glare and attendant phenomena were recorded and compared, see Figure 14.

The collision angle was decreased by increasing the working length l_w. The intensity of the process glare increased significantly and finally its color changed from red to light blue for l_w = 8 mm. The first plastic disc fractured at the outer circumference at the largest collision angle at l_w = 4 mm and no weld was achieved. Furthermore, a circular crater was left at the second translucent plastic disc resulting from the harsh impact of single ejected particles that were guided by the inner flyer contour (see Figure 11). At a working length l_w of 6 mm, a weld was generated. The marks at the first plastic disc and the inner flyer tube indicate a more vapor-like appearance of the CoP, since the residues are homogeneously distributed. This trend is continued for l_w = 8 mm. Here, the first plastic disc is almost clean, but the inside of the flyer tube is uniformly covered with a dark grey layer. Assuming an ideal black body emission, the R/G/B values enable the estimation of the CoP temperature. It is increased from 5500 K (l_w = 6 mm) to 8000 K (l_w = 8 mm), which is also reflected by the tempering colors of the parent parts. Due to the increased density of the CoP within the smaller volume of the joining gap, the heating, the fine dispersion of the CoP and, finally, the weld formation are supported. Similar effects were detected in the MPW experiment with increased charging energy E = 6.5 kJ, see Figure 14d. Here, the CoP was finely dispersed and on a high temperature level, too. The rise of the light intensity is shown in Figure 15 by the dotted line. The time interval Δt between the formation of the CoP and the reaction with the first plastic disc is much shorter than in the reference experiment with E = 4.5 kJ. Thus, the average velocity of the CoP is higher and in the order of 10 km/s.

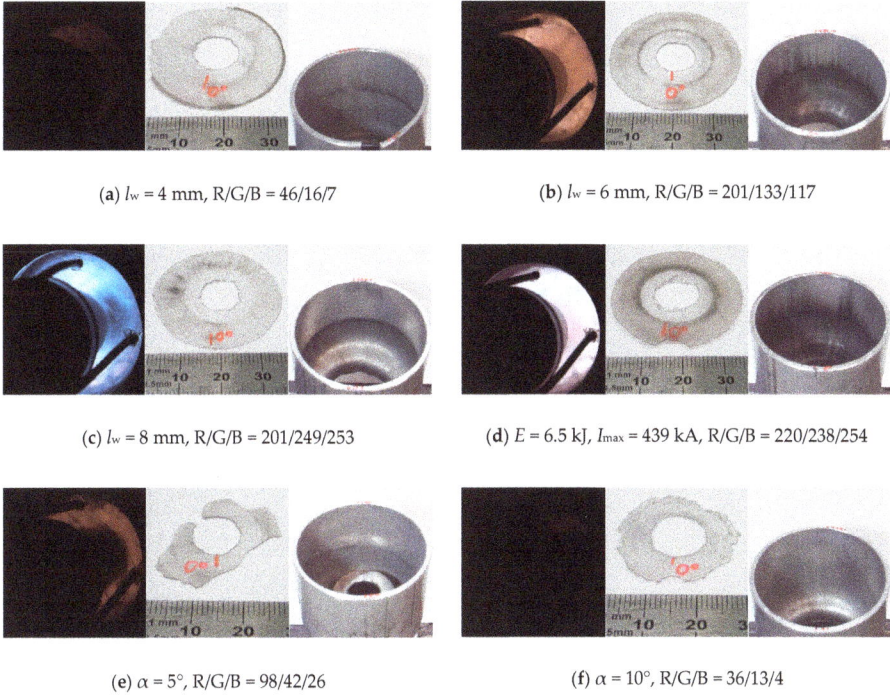

(**a**) l_w = 4 mm, R/G/B = 46/16/7

(**b**) l_w = 6 mm, R/G/B = 201/133/117

(**c**) l_w = 8 mm, R/G/B = 201/249/253

(**d**) E = 6.5 kJ, I_{max} = 439 kA, R/G/B = 220/238/254

(**e**) α = 5°, R/G/B = 98/42/26

(**f**) α = 10°, R/G/B = 36/13/4

Figure 14. Long time exposures of the process glare, first plastic disc and flyer tube after MPW experiments at the Bmax setup (basic parameters: l_c = 3 mm, α = 0°, g = 1.5 mm, E = 4.5 kJ, I_{max} = 365 kA, Ø coil = 41 mm, w_c = 10 mm, l_w = 6 mm): (**a**–**c**) for different working length l_w, (**d**) charging energy, and (**e**,**f**) parent angles α.

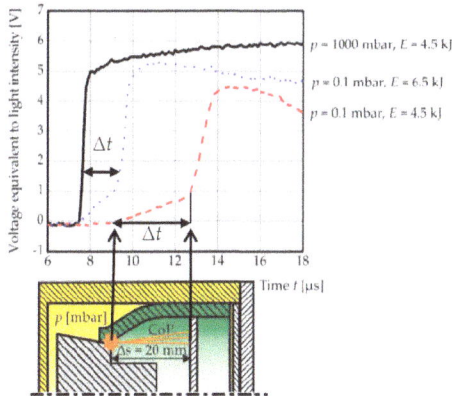

Figure 15. Light intensities for different surrounding pressures p and charging energies at the Bmax setup (t refers to the rising tool coil current as depicted in the schematic oscilloscope in Figure 1, l_c = 3 mm, α = 0°, g = 1.5 mm, Ø coil = 41 mm, w_c = 10 mm, l_w = 6 mm).

Chamfers at the parent part as depicted in Figure 3 lead to the destruction of the first plastic disc, a weaker process glare and a non-weld for α = 10° (see Figure 14e,f). In these cases, the degree of CoP compression and, thus, the temperature were lower.

3.3. Temperature Model

Although the temperature of the CoP was estimated in the previous chapter, the quantification of the heat transfer to the surfaces is still difficult since the density or mass of the CoP, as well as the surface coefficient for heat transfer, are hard to access. The knowledge of the heat input is essential for the following temperature model and, thus, a strategy based on the metallographic analysis of the welding result is applied. In a first step, the macroscopic distortion of the flyer tube and the parent part after MPW was analyzed as shown in Figure 16a. The initial wall thickness of 2 mm was reduced to 1.4 mm next to the initial impact zone of the free flyer edge and increased to 2.3 mm at the end of the close-fit zone. The indentation depth of the flyer into the parent was measured in analogy with [27] at different positions in the welding direction. The maximum value of 44 μm was found to be 0.5 mm behind the beginning of the welded zone and clearly below 10 μm at the end of the welded zone. At other MPW samples, the weld seams were even well established at positions without any indentation of the flyer into the parent material. Thus, plastic deformation might not be a necessary welding criterion, but is just a side effect of the high speed collision. In a second step and in order to study the thermal effects as another important welding criterion, the amount of melted flyer and parent material is quantified in polished cross sections. This value can vary along the welding direction, depending on the prevalent collision conditions as shown in Figure 16. The weld seam exhibits an almost smooth interface next to the start of the weld seam in Figure 16b and a characteristic wavy shape at the end of the welded zone in Figure 16c. The cross section of the pocket highlighted in Figure 16d is equal to a continuous melted layer with a thickness of approximately 6 μm. There are a few single iron particles visible in this pocket. As a consequence of the extreme high cooling rates that occur during welding, the volume in the pocket is "frozen" in a non-equilibrium state. Thus, a mixture of different phases can occur, including non-stoichiometric or metastable phases as reported by Bataev et al. [9] for different material combinations joined by EXW. Although the metallurgy was not studied in detail here, it is very likely that these types of phases occurred, too. The analysis of the chemical composition revealed an average ratio of 80 weight percent aluminum and 20 weight percent iron at the position of the line scans indicated in Figure 16d. Based on this ratio, the mass of molten aluminum m_{Al} per area A can be derived from Equations (1) to (6), where b is the equivalent thickness of the molten layer, m the mass, V the volume, and ρ the density of aluminum, iron, or both elements, respectively. For this calculation, the material specific values in Table 4 are applied, while the index Al corresponds to EN AW-6060 and Fe to C45.

$$m_{Al}/m = m_{Al}/(m_{Al} + m_{Fe}) = 0.8 \tag{1}$$

$$m_{Fe} = (1/0.8 - 1) \times m_{Al} \tag{2}$$

$$V = V_{Al} + V_{Fe} = m_{Al}/\rho_{Al} + m_{Fe}/\rho_{Fe} = m_{Al}/\rho_{Al} + (1/0.8 - 1) \times m_{Al}/\rho_{Fe} \tag{3}$$

$$m_{Al} = V/(1/\rho_{Al} + (1/0.8 - 1)/\rho_{Fe}) \tag{4}$$

$$V = b \times A \tag{5}$$

$$m_{Al}/A = b/(1/\rho_{Al} + (1/0.8 - 1)/\rho_{Fe}) = 0.0149 \text{ kg/m}^2 \ (m_{Fe}/A = 0.00372 \text{ kg/m}^2) \tag{6}$$

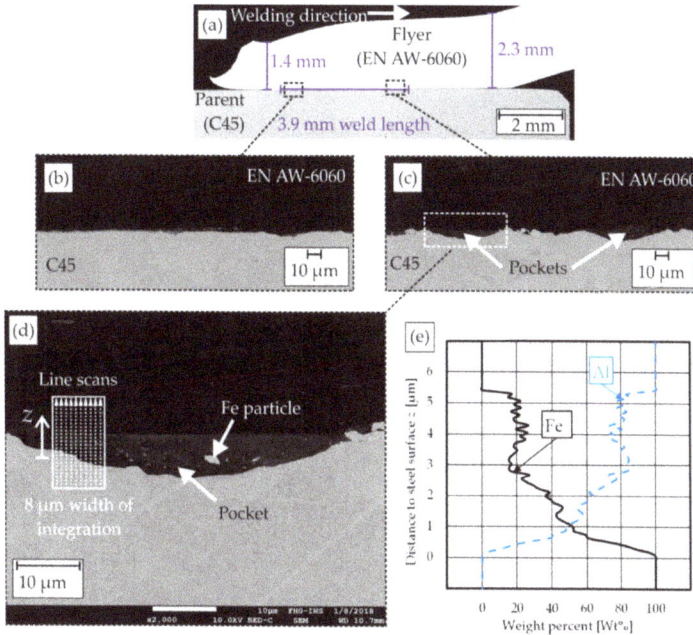

Figure 16. (**a**) Polished cross section of the joining zone after MPW at the Bmax setup (l_w = 6 mm, α = 0°, g = 1.5 mm, E = 8.0 kJ, I_{max} = 503 kA, Ø coil = 41 mm, w_c = 10 mm, flyer tube thickness s = 2 mm, flyer tube material in T66 condition), secondary electron image of the interface, (**b**) approximately 0.5 mm behind weld seam beginning with 500× magnification and approximately 3.5 mm behind weld seam beginning with (**c**) 500× magnification, (**d**) 2000× magnification and location of ten parallel line scans for EDS-analysis and (**e**) average element distribution perpendicular to the steel surface.

Now, the maximum heat input Q_{Al} per area needed for the melting of aluminum can be calculated by applying Equations (7) [31] and (8) , taking the enthalpy of fusion ΔH_{fus} into account, and assuming a temperature-independent heat capacity c. Furthermore, the highest reachable temperature is the boiling temperature of the material, since possible metal vapor is assumed to be spewed out of the joining gap.

$$Q_{Al}/A = m_{Al}/A \times (c_{Al} \times \Delta T + \Delta H_{fus}) = 38.7 \text{ kJ/m}^2 \ (Q_{Fe}/A = 6.2 \text{ kJ/m}^2) \tag{7}$$

$$\Delta T \text{ [K]} = T_{vap} - 20\,°C \tag{8}$$

The total heat input Q per area is now calculated according to (9) and is in good accordance with previous publications [15].

$$Q/A = Q_{Al}/A + Q_{Fe}/A = 44.9 \text{ kJ/m}^2 \tag{9}$$

To add plausibility to this calculated value, the kinetic energy of a flyer with a thickness s of 1.5 mm and a radial impact velocity $v_{i,r}$ of 340 m/s according to Figure 5 by Equation (10) [31]. In this case, the total heat input is approximately one fifth of the kinetic flyer energy.

$$E_{kin}/A = s/2 \times \rho_{Al} \times (v_{i,r})^2 = 234.1 \text{ kJ/m}^2 \tag{10}$$

The total heat input calculated in (9) serves as an upper boundary in the one-dimensional thermodynamic COMSOL model and is assumed to be distributed equally to both surfaces. Thus, half of the total heat input Q is introduced into the aluminum and steel surfaces, respectively, and named as Q_S.

In the following section, the influence of certain input parameters for the numerical simulations is described. In each diagram, the results of a reference setup with the parameters given in Table 5, are plotted for a better comparison. Here, the maximum heat quantity is introduced into the surfaces within 0.5 µs, followed by an immediate contact of both joining partners.

Table 5. Parameters of the reference setup.

Physical Quantity	Symbol	Unit	Value
Heat quantity at each surface	Q_S/A	J/m^2	22,450
Heating time	t_{heat}	µs	0.5
Waiting time	t_{wait}	µs	0
Flyer material	-	-	EN AW-6060
Parent material	-	-	C45
Consideration enthalpy of fusion?	-	-	true

In the first simulation, the influence of the phase transitions on the surface temperatures was investigated, see Figure 17. The surface temperatures during the heating stage are almost identical, while they differ significantly during the cooling stage, exactly at the time when the melting point of aluminum is reached, approximately 1 µs after the contact t_{con}. The interface stays in the liquid phase for 0.6 µs, before the temperature decreases. This period might be critical for thin flyers, since the bounce-back effect occurs quite early.

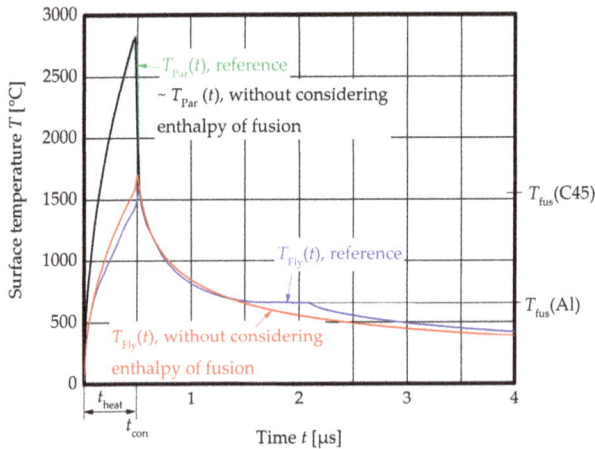

Figure 17. Comparison of $T_{Fly}(t)$ and $T_{Par}(t)$ for the reference setup (heat quantity at each surface 22,450 J/m^2, heating time 0.5 µs, waiting time 0 µs, flyer material EN AW-6060, parent material C45) with and without considering the enthalpy of fusion.

In Figure 18, the heat quantity Q_S is reduced to 9100 J/m^2 at each surface, which was identified as a lower boundary in the analytical investigation of the melted volume in the interface. This value is insufficient to reach the melting temperature of steel and, thus, welding in the liquid phase is probably hindered.

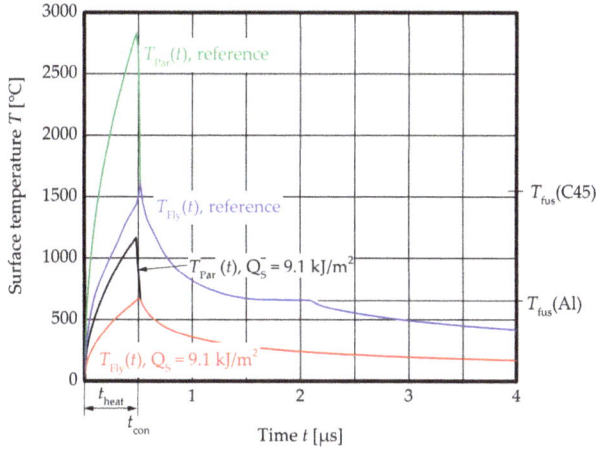

Figure 18. Comparison of $T_{Fly}(t)$ and $T_{Par}(t)$ for the reference setup and the setup with 9100 J/m^2 heat quantity at each surface.

Since the heating time t_{heat} of the CoP is hard to assess experimentally, the influence was studied in the numerical simulation, see Figure 19. The increase of the heating time to 1 μs at a constant heat quantity leads to lower maximum temperatures after the heating period. Nevertheless, both materials are in the liquid phase at the time of contact t_{con} and show a similar cooling behavior like the reference setup.

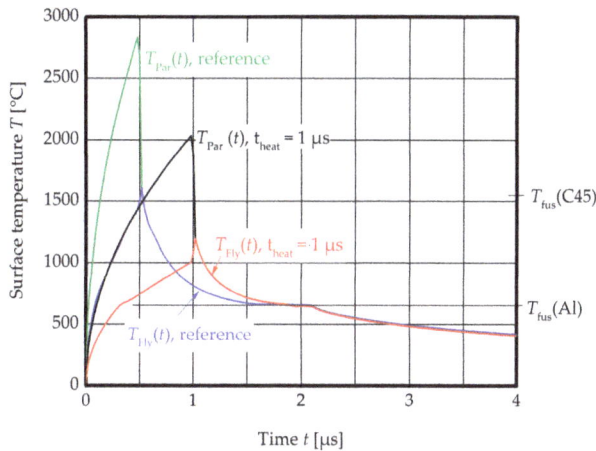

Figure 19. Comparison of $T_{Fly}(t)$ and $T_{Par}(t)$ for the reference setup and the setup with 1 μs heating time.

The waiting time t_{wait} between the end of the heating and the contact t_{con} was identified as an important factor. The cooling rate of the steel surface in the numerical simulation is in the order of 10^9 K/s and, thus, it is solidified approximately 0.3 μs after the heating time. Figure 20 shows that a waiting time of 2 μs and even 1 μs until contact is too long for a liquid phase bonding, since the steel surface is already solidified at t_{con1} and t_{con2}, respectively. The immediate contact after heating is necessary for this material combination, which corresponds to short gap closing times, high collision front velocities, or small collision angles, respectively. As explained in the previous chapter, this is linked to an increased compression of the CoP and, thus, higher temperatures in the joining gap and heat input to both surfaces.

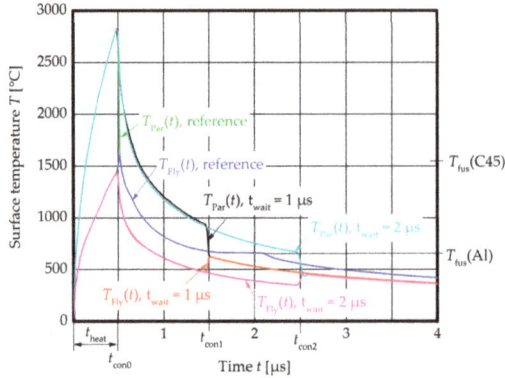

Figure 20. Comparison of $T_{\text{Fly}}(t)$ and $T_{\text{Par}}(t)$ for the reference setup and the setups with 1 µs and 2 µs waiting time.

The last set of simulations investigated the influence of the flyer material. The results for copper and C45 as flyer materials are plotted in Figure 21. From the theoretical point of view, liquid state bonding can be achieved in both cases. The surface of the copper reaches the melting temperature at the time of contact t_{con} and enables a very fast cooling due to the high heat conductivity. The similar material combination C45 to C45 fulfills the requirements for liquid state bonding as well. Nevertheless, to establish a CoP with the same heat input like in the reference setup aluminum to steel, a higher impact velocity is needed due to the increased hardness of both joining partners. In such configurations, a soft interlayer material as described in [38] might be beneficial to reach a strong CoP at lower impact velocities.

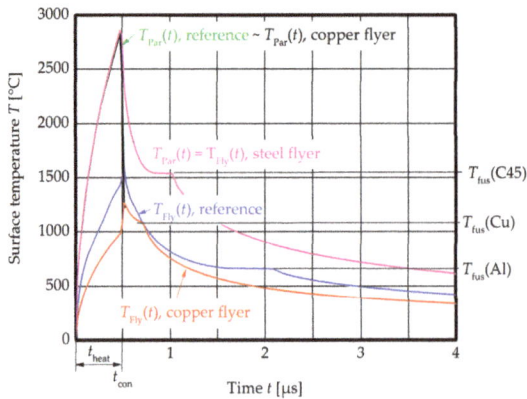

Figure 21. Comparison of $T_{\text{Fly}}(t)$ and $T_{\text{Par}}(t)$ for the reference setup and the setups with copper and C45 as flyer materials.

The results of the numerical simulations could be stated more precisely, if some of the input data would have been accessible during the experiments. Nevertheless, this numerical parameter study revealed the most important factors for MPW. An extension of the model will also include the reaction enthalpy for dissimilar metal welding, which was previously found to be beneficial for the weld formation at the lower process boundary [18]. Furthermore, the combination with mechanical models that predict the time of bounce back effects will allow for a comprehensive prediction of the welding result.

4. Research Highlights

1. The experiments showed that jetting in the type of a strong material flow is not mandatory for a successful MPW process. A cloud of particles (CoP), which is ejected during the impact with lower velocities, enables welding, too. Compared to the "real" jet in the style of a massive material flow at higher impact velocities, the CoP cannot remove thick surface layers or facilitate welding with rough surfaces. In this case, an adapted surface preparation and cleaning process is essential to ensure a sufficient surface activation.

2. The appearance of the CoP and its effect on the weld formation is determined by the prevalent collision conditions, especially the collision angle. This factor can be adjusted by various machine related factors and the part geometries. Vacuum experiments show that the CoP is established during the first metal to metal impact with a certain minimum impact velocity. Afterwards, it is compressed in the closing joining gap, successively heated up and finally ejected in welding direction.

3. For small collision angles, the level of compression and the internal friction of the CoP are higher and, thus, the temperature in the joining gap increases. In this configuration, the CoP is finely distributed like a metal vapor, which activates the surfaces of the joining partners homogeneously and can be seen inside the flyer tube after the experiment. If the collision angle is increased, the temperature decreases and single macroscopic particles are ejected. These particles seem to have a reduced surface activation effect, compared to the finely dispersed metal vapor described previously and thus, inhibit welding.

4. These findings allow for an optimization of the energy input during MPW. If a small collision angle is ensured, the initial impact velocity can be reduced. Thus, less mechanical energy is required for the forming process and the loading on the tool coils is reduced with positive effects on their lifetime.

5. Normally, the MPW process is performed in ambient atmosphere, where the free CoP ejection is hindered by the surrounding air. This leads to a shock compression and sudden heat up of the gas and results in a very strong process glare. This strong lightning can be utilized for the quality assurance during industrial production [39].

6. The numerical simulations of the surface temperatures of both joining partners revealed a strong influence of the waiting time between the end of the heat input by the CoP and the contact of both joining partners. Especially for dissimilar metal welding, this time needs to be very low to avoid solidification before the contact. This finding is important for the theory of liquid state bonding and in good correlation with the experimental results. Small collision angles, or gap closing times, respectively, are beneficial for the weld formation during MPW.

5. Patents

The flash measurement system enables the parameter identification and quality assurance during production. It was patented for different impact welding processes [39,40].

Author Contributions: Conceptualization, J.B.; Funding acquisition, E.B. and A.E.T.; Investigation, J.B.; Methodology, J.B.; Project administration, E.B. and A.E.T.; Validation, J.L.-A., S.S., M.H., and S.G.; Writing—original draft, J.B.; Writing—review & editing, J.L.-A., S.S., M.H., and S.G.

Funding: This research was funded by the Deutsche Forschungsgemeinschaft (DFG, German Research Foundation), grant number BE 1875/30-3 and TE 508/39-3. This work is based on the results of subproject A1 of the priority program 1640 ("joining by plastic deformation"). We acknowledge support by the Open Access Publication Funds of the SLUB/TU Dresden.

Acknowledgments: We would like to acknowledge the effort for the sample preparation and SEM and EDS analysis at Fraunhofer IWS Dresden.

Conflicts of Interest: The authors declare no conflict of interest.

References

1. Wang, H.; Wang, Y. High-Velocity Impact Welding Process: A Review. *Metals* **2019**, *9*, 144. [CrossRef]
2. Bmax. Magnetic Pulse Welding—The Ultimate Solution for Driveshaft Manufacturers. Available online: https://www.bmax.com/wp-content/uploads/2018/03/wp2015_mpw_driveshaft3.pdf (accessed on 2 March 2016).
3. Crossland, B. *Explosive Welding of Metals and Its Application*; Clarendon Press: Oxford, UK, 1982.
4. Kapil, A.; Sharma, A. Magnetic Pulse Welding: An efficient and environmentally friendly multi-material joining technique. *J. Clean. Prod.* **2015**, 35–58. [CrossRef]
5. Ishutkin, S.N.; Kirko, V.I.; Simonov, V.A. Thermal action of shock-compressed gas on the surface of colliding plates. *Fizika Goreniya i Vzryva* **1979**, *16*, 69–73. [CrossRef]
6. Deribas, A.; Zakharenko, I.D. Surface effects with oblique collisions between metallic plates. *Translated from Fizika Goreniya i Vzryva* **1974**, *10*, 409–421. [CrossRef]
7. Verbraak, A.C.; Boes, J.M.; Chirer, E.G.; Visser, L.; Verkaik, A. Method for Explosive Welding and Explosive Welded Products. U.S. Patent No. 3,305,922, 28 February 1967.
8. Lysak, V.; Kuzmin, S. Lower Boundary in Metal Explosive Welding: Evolution of Ideas. *J. Mater. Process. Technol.* **2012**, 150–156. [CrossRef]
9. Bataev, I.A.; Lazurenko, D.V.; Tanaka, S.; Hokamoto, K.; Bataev, A.A.; Guo, Y.; Jorge, A.M. High cooling rates and metastable phases at the interfaces of explosively welded materials. *Acta Mater.* **2017**, *135*, 277–289. [CrossRef]
10. Stern, A.; Shribman, V.; Ben-Artzy, A.; Aizenshtein, M. Interface Phenomena and Bonding Mechanism in Magnetic Pulse Welding. *J. Mater. Eng. Perform.* **2014**, *23*, 3449–3458. [CrossRef]
11. Sharafiev, S.; Wagner, M.F.-X.; Pabst, C.; Groche, P. Microstructural characterisation of interfaces in magnetic pulse welded aluminum/aluminum joints. *IOP Conf. Ser. Mater. Sci. Eng.* **2016**, *118*. [CrossRef]
12. Watanabe, M.; Kumai, S. Welding Interface in Magnetic Pulse Welded Joints. *Mater. Sci. Forum* **2010**, *654–656*, 755–758. [CrossRef]
13. Li, J.; Yu, Q.; Zhang, Z.; Xu, W.; Sun, X. Formation mechanism for the nanoscale amorphous interface in pulse-welded Al/Fe bimetallic systems. *Appl. Phys. Lett.* **2016**, *108*, 201606. [CrossRef]
14. Ghosh, P.; Patra, S.; Chatterjee, S.; Shome, M. Microstructural evaluation of magnetic pulse welded plain carbon steel sheets. *J. Mater. Process. Technol.* **2018**, *254*, 25–37. [CrossRef]
15. Zakharenko, I.D.; Sobolenko, T.M. Thermal effects in the weld zone in explosive welding. *Combust. Explos. Shock Waves* **1971**, *7*, 373–375. [CrossRef]
16. Koschlig, M.; Veehmayer, M.; Raabe, D. Production of Steel-Light Metal Compounds with Explosive Metal Cladding. In Proceedings of the 3rd International Conference on High Speed Forming, Dortmund, Detuschland, 11–12 March 2008; pp. 23–32.
17. Lueg-Althoff, J.; Bellmann, J.; Gies, S.; Schulze, S.; Tekkaya, A.E.; Beyer, E. Influence of the Flyer Kinetics on Magnetic Pulse Welding of Tubes. *J. Mater. Process. Technol.* **2018**, 189–203. [CrossRef]
18. Bellmann, J.; Lueg-Althoff, J.; Schulze, S.; Gies, S.; Beyer, E.; Tekkaya, A.E. Effects of Reactive Interlayers in Magnetic Pulse Welding. In Proceedings of the 8th International Conference on High Speed Forming, Columbus, OH, USA, 13–16 May 2018.
19. Marya, M.; Marya, S. Interfacial microstructures and temperatures in aluminium–copper electromagnetic pulse welds. *Sci. Technol. Weld. Join.* **2004**, *9*, 541–547. [CrossRef]
20. Hassani-Gangaraj, M.; Veysset, D.; Nelson, K.A.; Schuh, C.A. Melting Can Hinder Impact-Induced Adhesion. *Phys. Rev. Lett.* **2017**, *119*, 175701. [CrossRef]
21. Kuzmin, V.I.; Laysak, V.I.; Kriventsov, A.N.; Yakovlev, M.A. Critical conditions of the formation and failure of welded joints in explosive welding. *Weld. Int.* **2004**, 223–227. [CrossRef]
22. Bellmann, J.; Lueg-Althoff, J.; Schulze, S.; Gies, S.; Beyer, E.; Tekkaya, A.E. Measurement of Collision Conditions in Magnetic Pulse Welding Processes. *J. Phys. Sci. Appl.* **2017**, *7*, 1–10. [CrossRef]
23. Seeberger. Datasheet AlMgSi (EN AW-6060). Available online: http://www.seeberger.net/_assets/pdf/werkstoffe/aluminium/de/AlMgSi.pdf (accessed on 8 September 2016).

24. Günther. Schramm. Datasheet C45 (1.0503). Available online: http://www.guenther-schramm-stahl.de/files/datasheets/C45(1.0503).pdf (accessed on 8 September 2016).

25. Deutsche Edelstahlwerke. Unlegierter Vergütungsstahl 1.1191/1.1201: C45E/C45R. Available online: https://www.dew-stahl.com/fileadmin/files/dew-stahl.com/documents/Publikationen/Werkstoffdatenblaetter/Baustahl/1.1191_1.1201_de.pdf (accessed on 21 February 2019).

26. Eichhorn, G. Analysis of the hypervelocity impact process from impact flash measurements. *Planetary Space Sci.* **1976**, *24*, 771–781. [CrossRef]

27. Bellmann, J.; Beyer, E.; Lueg-Althoff, J.; Gies, S.; Tekkaya, A.E.; Schettler, S.; Schulze, S. Targeted Weld Seam Formation and Energy Reduction at Magnetic Pulse Welding (MPW). *eBIS* **2017**, *2017*, 91–102. [CrossRef]

28. Bellmann, J.; Lueg-Althoff, J.; Schulze, S.; Gies, S.; Beyer, E.; Tekkaya, A.E. Measurement and analysis technologies for magnetic pulse welding: Established methods and new strategies. *Adv. Manuf.* **2016**, 322–339. [CrossRef]

29. Geng, H.; Xia, Z.; Zhang, X.; Li, G.; Cui, J. Microstructures and mechanical properties of the welded AA5182/HC340LA joint by magnetic pulse welding. *Mater. Charact.* **2018**, *138*, 229–237. [CrossRef]

30. Bergmann, O.R. The Scientific Basis of Metal Bonding with Explosives. In Proceedings of the 8th International ASME Conference on High Energy Rate Fabrication, San Antonio, TX, USA, 17–21 June 1984; pp. 197–202.

31. Fischer, U. *Tabellenbuch Metall (in German)*; 44., neu bearbeitete Aufl., 6. Druck; Europa Lehrmittel: Haan-Gruiten, Germany, 2008; ISBN 9783808517246.

32. Bikar-Metalle GmbH. EN AW-6060. Available online: https://www.bikar.com/fileadmin/download/6060-komplett.pdf (accessed on 21 February 2019).

33. Bellmann, J.; Lueg-Althoff, J.; Göbel, G.; Gies, S.; Beyer, E.; Tekkaya, A.E. Effects of Surface Coatings on the Joint Formation During Magnetic Pulse Welding in Tube-to-Cylinder Configuration. In Proceedings of the 7th International Conference on High Speed Forming, Dortmund, Germany, 27–28 April 2016; pp. 279–288.

34. Bellmann, J.; Lueg-Althoff, J.; Schulze, S.; Gies, S.; Beyer, E.; Tekkaya, A.E. Magnetic Pulse Welding: Solutions for Process Monitoring within Pulsed Magnetic Fields. In Proceedings of the Euro-Asian Pulsed Power Conference & Conference on High-Power Particle Beams, Cascais, Portugal, 18–22 September 2016.

35. Bellmann, J.; Lueg-Althoff, J.; Schulze, S.; Gies, S.; Beyer, E.; Tekkaya, A.E. Parameter Identification for Magnetic Pulse Welding Applications. *Key Eng. Mater.* **2018**, *767*, 431–438. [CrossRef]

36. Pabst, C.; Groche, P. The Influence of Thermal and Mechanical Effects on the Bond Formation During Impact Welding. In Proceedings of the 7th International Conference on High Speed Forming, Dortmund, Germany, 27–28 April 2016; Tekkaya, A.E., Kleiner, M., Eds.; 2016; pp. 309–320.

37. Pabst, C.; Groche, P. Identification of process parameters in electromagnetic pulse welding and their utilisation to expand the process window. *Int. J. Mater. Mech. Manuf.* **2018**, *6*, 69–73.

38. Yablochnikov, B.A. Method of Magnetic Pulse Welding an End Fitting to a Driveshaft Tube of a Vehicular Driveshaft. U.S. Patent No. 7,015,435 B2, 21 March 2006.

39. Bellmann, J. Method and Device for Monitoring the Process for a Welding Seam Formed by Means of Collision Welding. W.O. 2018/050569 A1, 8 September 2017.

40. Bellmann, J. Verfahren und Vorrichtung zur Prozessüberwachung bei einer mittels Kollisionsschweißen gebildeten Schweißnaht. D.E. 10 2016 217 758 B3, 16 September 2016.

metals

MDPI

Article

Arc-Assisted Laser Welding Brazing of Aluminum to Steel

Xiaoquan Yu [1], Ding Fan [2,*], Jiankang Huang [1,*], Chunling Li [1] and Yutao Kang [1]

[1] School of Materials Science and Engineering, Lanzhou University of Technology, Lanzhou 730050, China; yuxiaoquangood@163.com (X.Y.); pyw910108@163.com (C.L.); m18709408129@163.com (Y.K.)

[2] State Key Laboratory of Advanced Processing and Recycling of Non-ferrous Metal, Lanzhou University of Technology, Lanzhou 730050, China

* Correspondence: fand@lut.cn (D.F.); sr2810@163.com (J.H.); Tel.: +86-133-2122-4851 (D.F.); +86-138-9338-4906 (J.H.)

Received: 2 March 2019; Accepted: 27 March 2019; Published: 31 March 2019

Abstract: Using laser beam as main heat source, and trailing arc as an assisted role, aluminum alloy was joined to galvanized steel in a butt configuration. Under suitable welding parameters, a sound welding seam was obtained. The interface intermetallic compounds layer and wetting behavior of weld joint were studied. The assisted arc can improve the wetting and spreading ability of weld pool duo to large temperature field. There are two different types of IMCs: near to the steel side one is Fe_2Al_5 with tooth-like shape and near to the weld seam side is the other one Fe_4Al_{13} with flocculent-like shape. The highest tensile strength can reach 163 MPa when the fracture occurred at the weld seam.

Keywords: welding-brazing; arc assisted laser method; aluminum-steel butt joint; mechanical properties

1. Introduction

In recent years, low carbon emission and lightweight design have been concerned by the automobile industry [1]. Using the aluminum and steel hybrid structure to replace the traditional single steel structure can effectively reduce the weight of the car body [2]. However, there are some differences of physical properties between aluminum and steel, such as the melting point and the linear expansion coefficient, etc. It is difficult to realize the metallurgical bonding using the conventional fusion welding.

To realize a high quality welding of aluminum and steel welding, the welding brazing method was proposed. [3]. In the welding brazing process, the heat source is mainly used for melting the base metal with lower melting point, and then the brazing joint was formed by wetting of molten metal [4]. There are various investigations on the aluminum and steel welding brazing process, including laser beam welding [4–6], CMT (cold metal transfer) welding [7–9], and TIG (tungsten inert gas) arc welding [10,11]. However, there are two major issues that have limited the welding brazing of steel and aluminum; one is the poor wetting ability of weld pool, and the other is the generation of brittle and hard intermetallic compounds (IMC), which can introduce deleterious effect on the welded joints [12]. At present, the main welding heat sources used in the aluminum/steel welding brazing are laser and arc, the welding method which uses the laser as the heat source can achieve high efficiency and accurate welding [13,14]. But the laser source is concentrated at one point, and the capability of wetting and spreading of the molten pool becomes poor [15]. The wetting problem becomes more serious during the welding-brazing process of the butt configuration. To improve the wetting and spreading ability of weld pool, Laukant et al. [16] used dual-spot laser beam to joined steel plate to aluminum alloy plate, and they suggested that a larger heating area can provide a better back formation of weld seam, and the better wetting and spreading ability resulted in a higher strength of the joint. Using laser welding-brazing to join a 6016 aluminum plate and a low carbon steel plate, Alexandre et al. [17] studied the relationship between the tensile strength, and wetting length (L), and wetting angle (θ). They found that a larger L/θ ratio could improve the tensile strength of the joint.

In addition to wetting and spreading factor, the interfacial IMC also had a significant influence on the mechanical properties of weld joint. Sun et al. [18] found that the IMC layer was composed by Fe_2Al_5 phase and $FeAl_3$ phase during Al-steel laser welding brazing. They discovered that the total thickness of IMC would become larger with the increase of laser power, and the highest tensile strength of weld joint could be obtained when the laser power of 3.05 kW. Using hot dip aluminizing on the steel surface, Shahverdi et al. [19] analyzed the microstructure of the IMC layer, and they found that the growth speed of Fe_2Al_5 phase was higher than $FeAl_3$ phase. Zhang et al. [20] suggested that the IMC layer was inevitably generated at the interface, and when the thickness of the IMC layer exceeded a certain range, the joint mechanical properties would be greatly deteriorated. The composition distribution and thickness of the IMC were controlled, the welding process parameters could be optimized to improve the mechanical properties of the joint.

In the present study, to improve the wetting ability of weld pool, we proposed an arc assisted laser welding brazing method. The aluminum alloy was joined to the galvanized steel by Tungsten inert gas (TIG) arc-assisted laser welding brazing. In the process, the laser beam was put in front of the assisted arc for melting the aluminum alloy, the assisted arc changed the temperature field, and increased the wetting ability of the molten metal. This study focused on investigating the interfacial microstructure and weld seam formation in the arc assisted laser welding-brazing process. The mechanical properties and fracture behavior of joints at different welding parameters were also discussed.

2. Experimental Details

2.1. Materials

The materials used for joining are ST04Z galvanized steel and 5A06 aluminum alloy. The dimensions of steel and aluminum plates are $150 \times 50 \times 1$ mm^3 and $150 \times 50 \times 2$ mm^3, respectively. Table 1; Table 2 list the chemical compositions of galvanized steel and 5A06 Al alloy, respectively. Pre-placed powder and flux were properly dissolved in acetone and applied on the surface of work-pieces. The main compositions of powder are Mg 5–8, Si 1–3, Mn 1–3, B 0.5–1, Zn 5–10, Al 75–88.5 in weight percent (wt. %).

Table 1. Chemical compositions of ST04Z galvanized steel (wt. %).

Mn	Si	P	S	Cu	Zn	C	Ni	Fe
0.4	≤0.40	0.02	≤0.30	≤0.15	≤0.15	0.08	≤0.15	Bal.

Table 2. Chemical composition of 5A06 aluminum alloy (wt. %).

Mg	Si	Mn	Ti	Cu	Zn	Fe	Al
5.8–6.8	0.4	0.5–0.8	0.02–0.10	0.1	0.20	0.4	Bal.

2.2. Experimental Procedure

A high-power transverse flow CO_2 laser equipment (GS-TFL-10K) (Wuhan Hans Goldensky laser system Co., Ltd., Wuhan, China) and an AC-TIG welder (TSP300) (Shenzhen Huayilong Electric Co., Ltd., Shenzhen, China) were used to join the Al alloy plate and the steel plate, in which the TIG welder provided the arc to assist the laser beam. The main parameters of CO_2 laser equipment are as follows: the maximum output power of 10.0 KW, laser beam diameter is 0.4 mm. Before the welding, a mechanical grinding was used to remove the oxides layer on the surface of Aluminum plate, and the pre-placed powder and flux were mixed with acetone and then applied on the surface of the work-pieces. The aluminum plate and the galvanized steel plate were placed in the same horizontal plane and then fixed on the self-made welding fixture; the butt joint was adopted in this experiment. The welding schematic is shown in Figure 1.

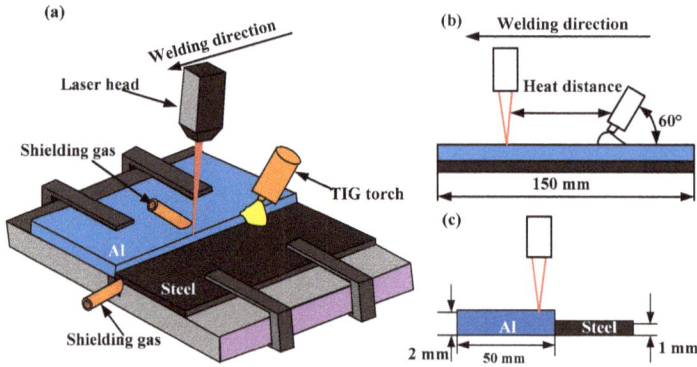

Figure 1. Schematic diagram of the arc-assisted laser welding; (a) overview, (b) front view, and (c) side view.

During the welding process, argon was used to shield the molten pool from air; shielding gas flow rate was 10 L/min, welding speed kept unchanged at 10 mm/s, and the heat input can be expressed as: $H = P/v + UI/v$, where the p is laser power, U is the arc voltage, I is the arc current, and the v is the welding speed. The main welding parameters are listed in Table 3.

Table 3. The main parameters of arc assisted laser welding-brazing.

Heat Input (KJ/cm)	Arc Current (A)	Heat Distance (mm)
1.0	15	15
1.2	15	15
1.5	15	15
1.7	15	15
1.2	10	15
1.2	15	15
1.2	18	15
1.2	20	15
1.2	15	5
1.2	15	10
1.2	15	15
1.2	15	20

After the welding, the surface of the work-piece was lightly ground by an abrasive paper. Specimens for the microstructure analysis were cut from the Al-Steel weld joint; the surface of the specimens was mechanically ground and polished to have mirror-like quality. Microstructure and composition of the interfacial layer were identified using JSM-6701F scanning electron microscopy (SEM) (JEOL (BEIJING) CO., Ltd., Beijing, China) equipped with energy dispersive spectrometer (EDS) (JEOL (BEIJING) CO., Ltd., Beijing, China). The phase composition in the interfacial layer was identified by TN-570X X-ray diffraction (XRD) (Shimadzu (China) CO., Ltd., Beijing, China) using Cu–Kα radiation with step size of 0.02° and step time of 60 s. The tensile test was carried out at room temperature by the WDW-300J tensile testing machine (Nanjing OuChengjing Testing Equipment

Co., Ltd., Nanjing, China). According to the weld joints strength test from the reference [21,22], we designed the tensile specimens as shown in Figure 2. The tensile strength can be calculated as:

$$\delta = \frac{F}{A} = \frac{F}{l \times \alpha} \tag{1}$$

here F is the maximum load, A is cross section area of fracture position, l is the width of tensile strength test sample (10 mm), α is the height of fracture surface. The height (α) and width (l) are measured using a Vernier caliper. The engineering strain is defined as $\Delta L/L_0$, where ΔL and L_0 are the elongation length and initial lengths respectively.

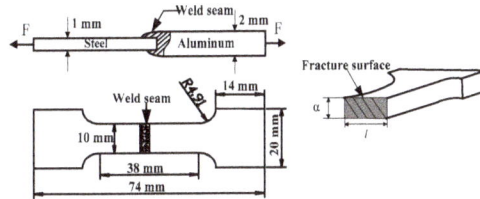

Figure 2. The geometry of tensile specimens and fracture surface.

2.3. Numerical Analysis of Temperature Field

A FEM analysis was used to calculate the temperature field. The flow of the molten pool was ignored in the calculation of the temperature distribution during welding.

The governing equation can be expressed by the following equation [23]

$$\frac{\partial(\rho\varphi)}{\partial t} + \frac{\partial(\rho\mu\varphi)}{\partial x} + \frac{\partial(\rho\omega\varphi)}{\partial y} + \frac{\partial(\rho w\varphi)}{\partial z} = \frac{\partial}{\partial x}(\Gamma\frac{\partial\varphi}{\partial x}) + \frac{\partial}{\partial y}(\Gamma\frac{\partial\varphi}{\partial y}) + \frac{\partial}{\partial z}(\Gamma\frac{\partial\varphi}{\partial z}) + S \tag{2}$$

where ρ is the density of base material, μ, v, and ω is fluid velocity in x, y, and z directions respectively, and φ is solving variables such as temperature, speed, etc. Γ is generalized diffusion coefficient, and s is source term such as mass source term, energy source term, and momentum source term.

The thermal boundary condition of the upper surface of work piece can satisfy the following formulations [24]:

$$-\lambda\frac{\partial T}{\partial Z} = q_{arc} + q_{laser} - h_c(T - T_0) - \beta\varepsilon(T^4 - T_0^4) - \omega H_V \tag{3}$$

here λ is thermal conductivity, h_c is the heat transfer coefficient, β is the Boltzmann constant, ε is the radiation coefficient of surface, H_V is latent heat of vaporization, q_{arc} and q_{laser} represent the arc heat flux and laser heat flux which can be expressed by the following equations [25]:

For the front heat source:

$$q_{arc} = \frac{a_{f1}}{a_{f1} + a_{r1}} \frac{6\eta U_1 I_1}{\pi a_{f1} b_{h1}} \times \exp(-\frac{3(x - v_0 t)^2}{a_{f1}^3}) \exp(-\frac{3y^2}{b_{h1}^2}) \tag{4}$$

$$q_{laser} = \frac{a_{f2}}{a_{f2} + a_{r2}} \frac{6\eta Q_2}{\pi a_{f2} b_{h2}} \times \exp\left(-\frac{3(x + 0.015 - v_0 t)^2}{a_{f2}^3}\right) \exp(-\frac{3y^2}{b_{h2}^2}) \tag{5}$$

For the rear heat source:

$$q_{arc} = \frac{a_{r1}}{a_{r1} + a_{r1}} \frac{6\eta U_1 I_1}{\pi a_{r1} b_{r1}} \times \exp\left(-\frac{3(x - v_0 t)^2}{a_{r1}^3}\right) \exp\left(-\frac{3y^2}{b_{h1}^2}\right) \tag{6}$$

$$q_{laser} = \frac{a_{r2}}{a_{r2} + a_{r2}} \frac{6\eta Q_2}{\pi a_{r2} b_{h2}} \times \exp\left(-\frac{3(x + 0.015 - v_0 t)^2}{a_{r2}^3}\right) \exp(-\frac{3y^2}{b_{h2}^2}) \tag{7}$$

here η is the heat efficiency, U_1 is welding voltage, I_2 is the welding current, Q_2 is laser power; a_{f1}, a_{r1}, b_{h1}, are arc heat source model parameters; a_{f2}, a_{r2}, b_{h2} are laser heat source model parameters; v_0 is heat source velocity.

3. Results and Discussion

3.1. Arc Effect on the Formation of Weld Seam

In the welding brazing process, the wetting and spreading behavior of Al molten pool on the steel surface determined the formation of weld seam [26]. In the present study, the weld seam morphology was mainly evaluated based on its continuity and back formation. Figure 3 shows the morphologies of top and back surfaces of the welding seams formed by single laser and arc assisted laser welding processes, respectively. Both the two methods can get continuous front formation of the weld seam, however, the back formation of arc assisted laser method exhibited a better wetting and spreading morphology than single laser welding brazing. This result suggests that assisted arc improves the wettability and fluidity of Al molten pool and helps the Al molten pool spread to the back surface of the steel plate.

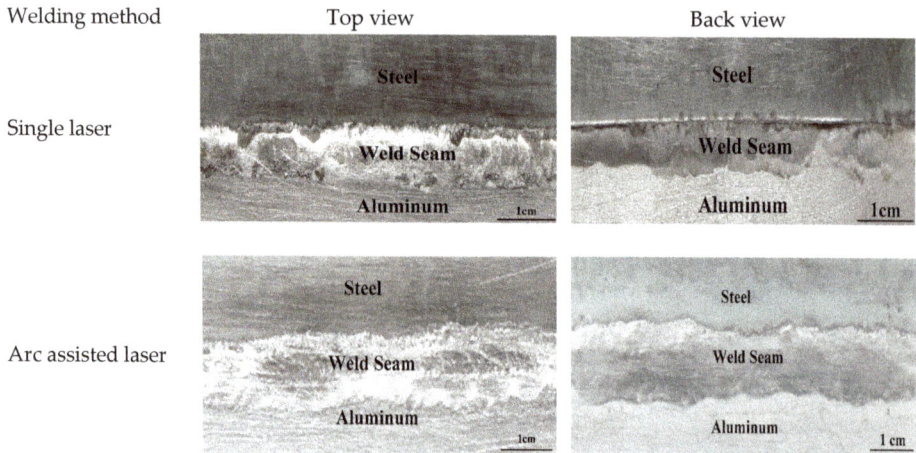

Welding method	Top view	Back view
Single laser		
Arc assisted laser		

Figure 3. Surface morphologies of the weld seams formed by single laser and arc assisted laser welding processes.

Figure 4 shows SEM images of the cross-sections of Al-steel butt joints formed by single laser and arc-assisted laser welding processes. The length of bottom wetting zone in arc assisted laser was about 2 mm, while it was 0.5 mm in single laser welding. In general, the welding seam produced by the arc assisted laser welding is better than that by the single laser welding. Especially, the wetting and spreading of back formation. Such trend is likely due to the arc-induced change of the heat conduction during the cooling (solidification) stage.

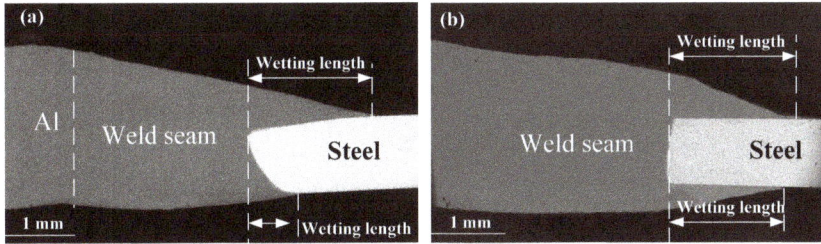

Figure 4. SEM images of the cross-sections of Al-steel butt joints formed by two different welding processes; (**a**) single laser welding brazing, and (**b**) arc assisted laser welding brazing

To understand the wetting and spreading behavior of motel pool, a FEM model of temperature field was established. Figure 5 displays the thermal cycles of point A and point B on the top surface of the steel, the thermal profiles calculated by the FEM at different positions were in good agreement with the experimentally measured values.

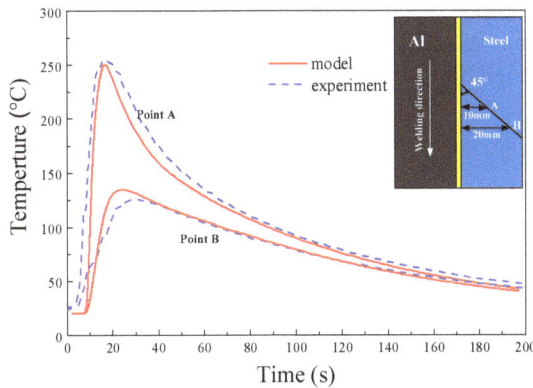

Figure 5. Comparison between experimental and calculated thermal cycles.

Figure 6 shows the temperature field of molten pool (X-Y plane) in the single laser and arc assisted laser welding process. As can be seen from Figure 6, the temperature field distribution of the molten pool under the action of a single laser and arc assisted laser is similar to the double ellipse distribution. The temperature distributions on both sides of aluminum/steel were asymmetric. As shown in Figure 6b, due to the addition of the assisted arc, the temperature field distribution of the base metal surface becomes larger than that in the single laser process, and the maximum temperature of the molten pool reaches 1002.54 K. The heat input from the arc can increase the spreading time of the molten metal in the welding pool, thus forming a better joint. Adding arc in the laser welding helps enhances the heat conduction, which allows the molten metal to have enough time to wet on the surfaces of the workpiece and to solidify in a little longer time.

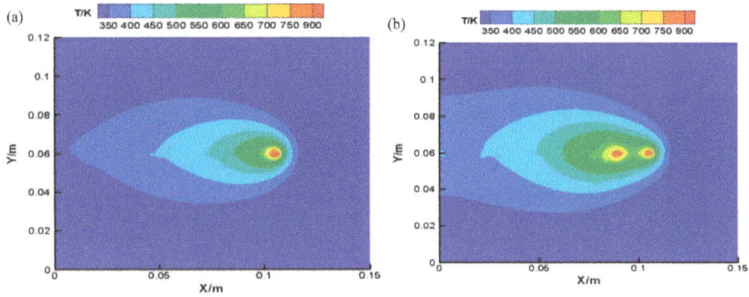

Figure 6. The temperature field of molten pool: (**a**) single laser welding brazing process, and (**b**) arc assisted laser welding brazing.

3.2. Microstructures and Phase Identification of IMCs Layer

It is known that the structure and thickness of IMC layers play important roles in determining the mechanical behavior of welding joints [27]. A typical cross-section of weld joint was selected to observe the IMC layer. The specimen was obtained using the following process parameters: laser power of 1200 W, welding speed of 10 mm/s, heat distance of 15 mm, and arc current of 15 A.

Figure 7 shows the IMC layers formed at three spots of A, B, and C around a welding joint. One can easily observe the non-smooth interfaces of the IMC layer formed between the steel and the Al alloy weld seam. The thickness of the IMC layer changes with the location along the interfaces. For example, the average thickness of the IMC layer is 13 μm over the spot of B and 9 μm over the spot of C. In general, the IMC layer around the edge containing the spot B is thicker than those along the edges containing spots of A and C, since it takes more time for the completeness of solidification along the edge containing the spot B.

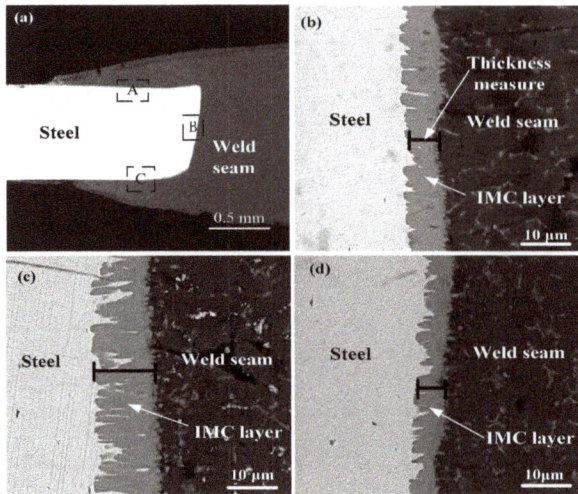

Figure 7. SEM images of the intermetallic compounds IMC layers at different locations of the Al-steel joint; (**a**) overview of the joint, (**b**) the IMC layer around the spot A, (**c**) the IMC layer around the spot B, and (**d**) the IMC layer around the spot C.

To better understand the thickness changes of IMC layer at the interface, the temperature profiles of three positions at the interface were obtained by FEM calculated model. Figure 8 shows temperature curves of interface obtained from calculated results at heat input of 1.2 KJ/cm. The liquid/solid reaction time of the three curves are essentially equal, but the peak temperatures from three positions are difference and the thickness of IMC layer becomes thick with the temperature increase. It indicated that the IMC thickness is mainly depended on the peak temperature.

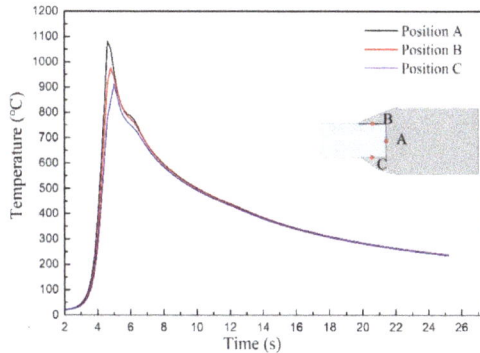

Figure 8. Temperature profiles of different positions at Al/Steel interface.

In order to identify the phase compositions of the IMC layer, an EDS line scanning analyses of the IMC layer was performed, as shown in Figure 9. It is evident that both Fe and Al uniformly distributed in the corresponding regions, i.e., the steel and the Al alloy, as expected, since no IMC can be observed in the steel matrix, and there are only a small amounts of IMC randomly distributed in the welding seam. According to the EDS pattern of the line scan, the IMC layer can be divided into two layers of *I* and *II*. In the layer *I*, both the fractions of Al and Fe remain relatively unchanged with the change of the distance to the steel/IMC interface. In the layer *II*, the fraction of Fe decreases with the decrease of the distance to the interface between the IMC layer and the welding seam, while the fraction of Al increases with the decrease of the distance to the IMC/weld seam interface. Such behavior suggests that the IMC in the layer *I* has different structure with that in the layer *II*. It is worth pointing out that there are little amounts of other elements presented in the IMC layers. The EDS point analysis of the IMCs was performed for the positions A and B as labeled in Figure 9, the atomic ratio of Al to Fe is in the range of 65/27 (~5:2) to 77/22 (~3:1), which suggests the formation of Fe_2Al_5 and Fe_4Al_{13} IMC.

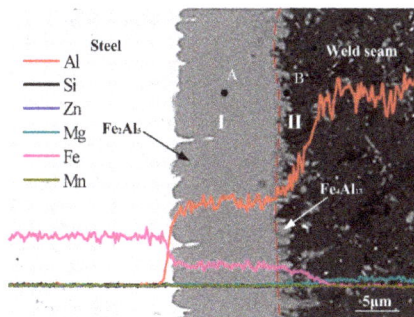

Figure 9. SEM images of the Al/Fe interface and the corresponding EDS pattern of the line scan.

The XRD analysis results of fracture plane in steel side, as shown in Figure 10. The XRD results show that the reaction layer included two different kinds IMC: Fe_4Al_{13} and Fe_2Al_5. According to Fe-Al equilibrium phase diagram [28], six non-stoichiometric IMC of Fe_3Al, $FeAl$, $FeAl_2$, Fe_2Al_3, Fe_2Al_5, and Fe_4Al_{13} possibly form during reaction between iron and aluminum. Previous studies about the welding of aluminum alloy to steel indicated that the formed compound near welded seam was intermetallic compounds FeAl₃ phase [29,30]. Van et al. [31] state that the η phase has two term name of "Fe₄Al₁₃" and "FeAl₃", while Fe₄Al₁₃ is more accurate expression. The IMC layer can be divided into two sublayers, as above analysis; one consists of Fe₂Al₅ IMC, and the other consists of Fe₄Al₁₃ IMC. The interface between the Fe₂Al₅ phase and the steel is presented in the needle-like shape, and the interface between the Fe₄Al₁₃ phase and the Al alloy is presented in the flocculent-like shape.

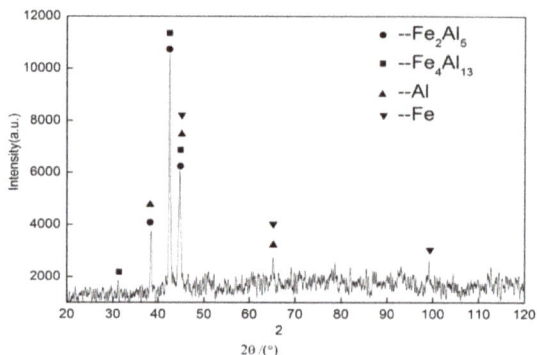

Figure 10. XRD patterns of the fractured surface on the steel substrate.

The hardness of the microstructure near the interface was measured by a Vickers indenter. Figure 11 shows the distribution of the Vickers hardness across the interface. The Al has the lowest Vickers hardness, as expected. It worth to point that the IMC layer has the highest Vickers hardness and the hardness of IMC is 5–6 times higher than the Al base metal. This result indicates that comparing to the base metals, the intermetallic compounds are more brittle and have large resistances to the penetration of an indenter onto the surface. The crack was easily generated in the IMC layer.

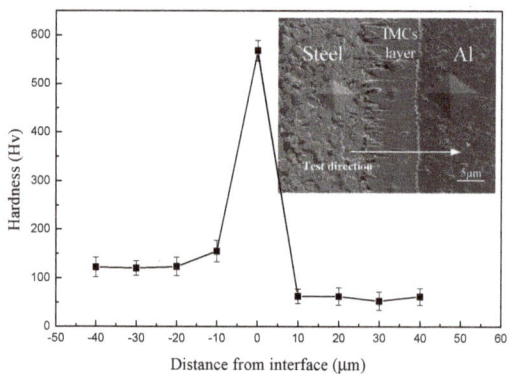

Figure 11. Distribution of Vickers hardness across the interface.

3.3. Mechanical Properties of Joints

To compare the mechanical property between single laser welding and arc assisted laser welding, tensile tests of the tensile specimens were performed. Figure 12 shows the engineering stress-strain curves for the welded joint prepared by using single laser and arc assisted laser in the same welding parameters. The tensile strength value of the weld joint obtained by arc-assisted laser welding brazing was about 1.3 times than that in single laser welding brazing process. The highest value was near to 163 MPa, which almost 5A06 aluminum alloy strength for 74% (Under the same test conditions, the tensile strength of 5A06 aluminum alloy is 219 MPa), it indicates that the addition of arc can improve the tensile strength of aluminum/steel butt joint.

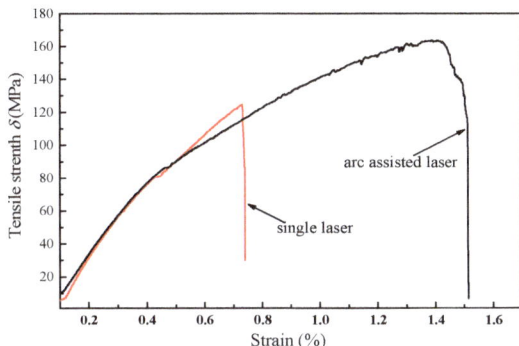

Figure 12. Engineering stress engineering strain curves of single laser welding and arc assisted laser welding.

To understand the relationship between the wetting width and tensile strength, the fracture mode and wetting width under different welding heat input were analyzed. The total wetting width can be expressed:

$$W_t = W_b + W_f \tag{8}$$

here W_t is the total wetting width, W_b is the wetting width of back formation, and the W_f is the wetting width of front formation. Figure 13 shows the typical failure modes of joint that was obtained in different welding heat input for arc assisted laser welding. The total wetting width increase with the heat input increase. When the heat input was lower, the specimen was failure at the wetting and spreading zone due to the poor wetting ability of the weld pool. The highest tensile strength was 163 MPa when the weld joint failure at the weld seam. The worst case of tensile properties was fractured along the aluminum/steel interface with average tensile strength of 98 MPa when the total heat input exceeds 1.5 KJ/cm. In this case, the total width was about 5.7 mm, however, duo to the heat input was higher, the thickness of IMC exceeded permissible value (about 10 μm) [32] and the mechanical properties become worse. Note that a smooth edge morphology of fracture joint was appeared when the heat input exceeds 1.5 KJ/cm, it can be attributed to a brittle fracture caused by IMC layer. To improve the tensile strength of welded joints, one needs to limit the thickness of IMC.

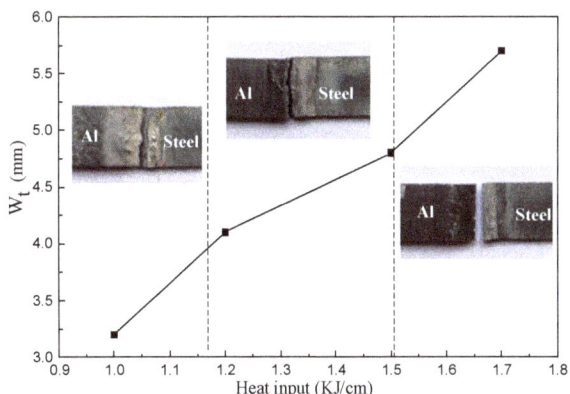

Figure 13. Fracture modes of the weld joints in different welding heat input.

As discussed in introduction part, the IMC layer plays an important role in determining the structural durability of the structures consisting of welded joints. To examine the effect of the IMC layer (welded joints) on the Al/steel weld joint, tensile tests of the tensile specimens were performed.

As the tensile strength and IMC were mainly affected by the heat source, we selected three main parameters (laser heat input, assisted arc current, and heat source distance) as variables. The tensile strength of the joint and the thickness of the IMC were measured under the same welding parameters. As shown in Figure 14, the thickness of the IMC increases linearly with the increase of the process parameters, while the tensile strength reaches certain value and then drops rapidly. This implied that due to the high heat input, the excessive growth of IMC layer led to the poor tensile strength of the joint. Therefore, in order to improve the strength of joint, it should control the heat input. The maximum tensile strength can reach 163 MPa When the IMC thickness between 8 μm and 12 μm. Tensile strength decreased rapidly when the thickness of the compound was more than 13 μm. Comparing Figure 14a,c to Figure 14b, Figure 14b had a more slowly decreasing trend in the process of tensile strength reduction. It indicates that the appropriate heat distance between laser and arc can improve tensile strength due to good weld appearance.

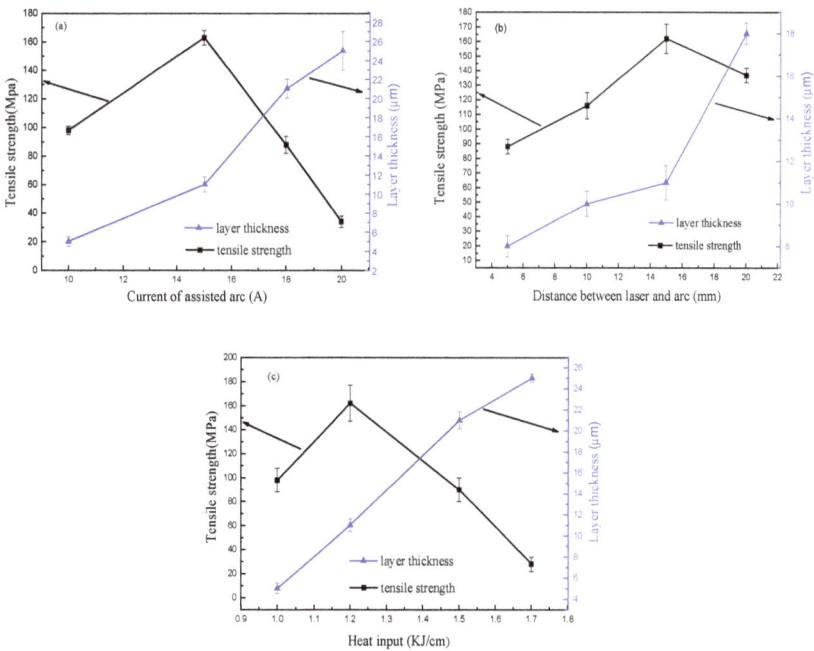

Figure 14. Effects of welding parameters on the IMC layer thickness and the tensile strength of the butt joint: (**a**) effects of assisted arc current, (**b**) effects of distance between heat source, (**c**) effects of heat input.

In order to further understand the fracture behavior of the arc assisted laser welding brazing tensile specimens, the fractured surfaces were observed by SEM, the typical fracture surface images were shown in Figure 15. As shown in Figure 15a, there are a large number of dimples in the fracture surface, and exhibited the typical ductile fracture when the fracture occurred at the weld seam of the aluminum alloy. Figure 15b shows the morphology of the tensile specimen detached along the IMC layer which exhibited brittle fracture pattern. As shown in Figure 15b, the fracture surface presented typical cleavage fracture with river pattern strips, it can be attributed to the lattice mismatch inside the IMC layer [33]. At a suitable heat input, the weld joint will fracture at the weld seam and shows a high tensile strength. On the other hand, a higher heat input will cause the excessive growth of the IMC layer, and the crack will generate at the IMC layer and rapidly propagate along the interface, which results in a poor tensile mechanical properties of the joint.

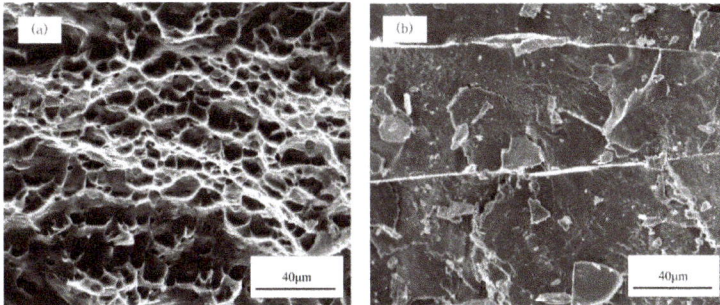

Figure 15. SEM fractography of the ruptured specimens: (**a**) fractured at weld seam, (**b**) detached between galvanized steel and brazed seam interface.

4. Conclusions

Dissimilar metals of 5A06 aluminum and galvanized steel were butt joined by arc assisted laser welding brazing technique. Major conclusions of this study could be summarized as followings:

1. Using arc assisted laser welding brazing method, the galvanized steel was joined to the aluminum alloy with butt joint. In this welding process, a sound weld seam formation could be obtained on the back and front side, and the addition of arc could improve the wetting and spreading of weld pool and enhanced the tensile strength of weld joint.
2. Compared to the single laser welding brazing method, the arc assisted laser welding brazing method had lager temperature distribution due to the addition of arc, the wetting width increased with the increasing of total heat input.
3. Two different IMC phases were formed at the Al/steel interface, which was composed of Fe_2Al_5 near to the steel base metal and Fe_4Al_{13} near to the aluminum welding brazing seam.
4. There were there failure modes in tensile strength test: the wetting zone fracture, weld seam fracture, and Al/steel interface fracture. The maximum tensile strength of weld joint was 163 MPa, which was nearly 74% of 5A06 aluminum alloy when the fracture occurred at the weld seam.

Author Contributions: Conceptualization, D.F. and J.H.; methodology, X.Y.; formal analysis, Y.K.; data curation, X.Y. and C.L.; writing—original draft preparation, X.Y.; writing—review and editing, D.F.; supervision, D.F. and J.H.

Funding: This study was financially supported by the National Natural Science Foundation of China (No. 51465031).

Conflicts of Interest: The authors declare no conflict of interest.

References

1. Liedl, G.; Bielak, R.; Ivanova, J.; Enzinger, N.; Figner, G.; Bruckner, J.; Pasice, H.; Pudar, M.; Hampel, S. Joining of Aluminum and Steel in Car Body Manufacturing. *Phys. Procedia* **2011**, *12*, 150–156. [CrossRef]
2. Kouadri-David, A.; PSM Team. Study of metallurgic and mechanical properties of laser welded heterogeneous joints between DP600 galvanized steel and aluminum. *Mater. Des.* **2014**, *54*, 184–195. [CrossRef]
3. Wang, P.; Chen, X.; Pan, Q.; Madigan, B.; Long, J. Laser welding dissimilar materials of aluminum to steel: An overview. *Int. J. Adv. Manuf. Technol.* **2016**, *87*, 3081–3090. [CrossRef]
4. Li, L.; Xia, H.; Tan, C.; Ma, N. Influence of laser power on interfacial microstructure and mechanical properties of laser welded-brazed Al/steel dissimilar butted joint. *J. Manuf. Process.* **2018**, *32*, 160–174. [CrossRef]
5. Meco, S.; Pardal, G.; Ganguly, S.; Williams, S.; Mcpherson, N. Application of laser in seam welding of dissimilar steel to aluminum joints for thick structural components. *Opt. Lasers Eng.* **2015**, *67*, 22–30. [CrossRef]

6. Liu, J.; Jiang, S.; Shi, Y.; Kuang, Y.; Huang, G.; Zhang, H. Laser fusion–brazing of aluminum alloy to galvanized steel with pure Al filler powder. *Opt. Laser Technol.* **2015**, *66*, 1–8. [CrossRef]
7. Zhang, H.T.; Feng, J.C.; He, P. Interfacial phenomena of cold metal transfer (CMT) welding of zinc coated steel and wrought aluminum. *Mater. Sci. Technol.* **2008**, *24*, 1346–1349. [CrossRef]
8. Kang, M.; Kim, C. Joining Al 5052 alloy to aluminized steel sheet using cold metal transfer process. *Mater. Des.* **2015**, *81*, 95–103. [CrossRef]
9. Yang, S.; Zhang, J.; Lian, J.; Lei, Y. Welding of aluminum alloy to zinc coated steel by cold metal transfer. *Mater. Des.* **2013**, *49*, 602–612. [CrossRef]
10. Dong, H.; Yang, L.; Dong, C.; Kou, S. Improving arc joining of Al to steel and Al to stainless steel. *Mater. Sci. Eng. A* **2012**, *534*, 424–435. [CrossRef]
11. Yagati, K.P.; Bathe, R.N.; Rajulapati, K.V.; Rao, K.B.S.; Padmanabham, G. Fluxless arc weld-brazing of aluminium alloy to steel. *J. Mater. Process. Technol.* **2014**, *214*, 2949–2959. [CrossRef]
12. Martinsen, K.; Hu, S.J.; Carlson, B.E. Joining of dissimilar materials. *CIRP Ann.-Manuf. Technol.* **2015**, *64*, 679–699. [CrossRef]
13. Qiu, R.; Satonaka, S.; Iwamoto, C. Effect of interfacial reaction layer continuity on the tensile strength of resistance spot welded joints between aluminum alloy and steels. *Mater. Des.* **2009**, *30*, 3686–3689. [CrossRef]
14. Kashani, H.T.; Kah, P.; Martikainen, J. Laser Overlap Welding of Zinc-coated Steel on Aluminum Alloy. *Phys. Procedia* **2015**, *78*, 265–271. [CrossRef]
15. Dharmendra, C.; Rao, K.P.; Wilden, J.; Reich, S. Study on laser welding–brazing of zinc coated steel to aluminum alloy with a zinc based filler. *Mater. Sci. Eng. A* **2011**, *528*, 1497–1503. [CrossRef]
16. Laukant, H.; Wallmann, C.; Korte, M.; Glatzel, U. Flux-less joining technique of aluminum with zinc-coated steel sheets by a dual-spot-laser beam. *Adv. Mater. Res.* **2005**, *6*, 163–170. [CrossRef]
17. Alexandre, M. Dissimilar material joining using laser (aluminum to steel using zinc-based filler wire). *Opt. Laser Technol.* **2007**, *39*, 652–661. [CrossRef]
18. Sun, J.; Yan, Q.; Gao, W.; Huang, J. Investigation of laser welding on butt joints of Al/steel dissimilar materials. *Mater. Des.* **2015**, *83*, 120–128. [CrossRef]
19. Shahverdi, H.R.; Ghomashchi, M.R.; Shabestari, S.; Hejazi, J. Microstructure alanalysis of interfacial reaction between molten aluminium and solid iron. *J. Mater. Process. Technol.* **2002**, *124*, 345–352. [CrossRef]
20. Zhang, M.; Chen, G.; Zhang, Y.; Wu, K. Research on microstructure and mechanical properties of laser keyhole welding–brazing of automotive galvanized steel to aluminum alloy. *Mater. Des.* **2013**, *45*, 24–30. [CrossRef]
21. Zhang, Y.; Li, F.; Guo, G.; Wang, G.; Wei, H. Effects of different powders on the micro-gap laser welding-brazing of an aluminium-steel butt joint using a coaxial feeding method. *Mater. Des.* **2016**, *109*, 10–18. [CrossRef]
22. Cao, R.; Yu, G.; Chen, J.H.; Wang, P.C. Cold metal transfer joining aluminum alloy-to-galvanized mild steel. *J. Mater. Process. Technol.* **2013**, *213*, 1753–1763. [CrossRef]
23. Wang, R.; Lei, Y.; Shi, Y. Numerical simulation of transient temperature field during laser keyhole welding of 304 stainless steel sheet. *Opt. Laser Technol.* **2011**, *43*, 870–873. [CrossRef]
24. Cho, D.W.; Cho, W.I.; Na, S.J. Modeling and simulation of arc: Laser and hybrid welding process. *J. Manuf. Process.* **2014**, *16*, 26–55. [CrossRef]
25. Meng, X.; Qin, G.; Su, Y.; Fu, B.; Ji, Y. Numerical simulation of large spot laser + MIG arc brazing–fusion welding of Al alloy to galvanized steel. *J. Mater. Process. Technol.* **2015**, *222*, 307–314. [CrossRef]
26. Gatzen, M.; Radel, T.; Thomy, C.; Vollertsen, F. Wetting and solidification characteristics of aluminium on zinc coated steel in laser welding and brazing. *J. Mater. Process. Technol.* **2016**, *238*, 352–360. [CrossRef]
27. Sun, J.; Huang, J.; Yan, Q.; Li, Z. Fiber laser butt joining of aluminum to steel using welding-brazing method. *Int. J. Adv. Manuf. Technol.* **2016**, *85*, 2639–2650. [CrossRef]
28. Kattner, U.R. *Binary Alloy Phase Diagrams*; ASM International: Materials Park, OH, USA, 1990.
29. Meco, S.; Ganguly, S.; Williams, S.; McPherson, N. Effect of laser processing parameters on the formation of intermetallic compounds in fe-al dissimilar welding. *J. Mater. Eng. Perform.* **2014**, *23*, 3361–3370. [CrossRef]
30. Reisgen, U.; Otten, C.; Schönberger, J. Investigations about the influence of the time–temperature curve on the formation of intermetallic phases during electron beam welding of steel–aluminium material combinations. *Weld. World* **2014**, *58*, 443–454. [CrossRef]

31. Van Alboom, A.; Lemmens, B.; Breitbach, B.; De Grave, E.; Cottenier, S.; Verbeken, K. Multi-method identification and characterization of the intermetallic surface layers of hot-dip Al-coated steel: $FeAl_3$, or Fe_4Al_{13}, and Fe_2Al_5, or Fe_2Al_{5+x}. *Surf. Coat. Technol.* **2017**, *324*, 419–428. [CrossRef]
32. Song, J.; Lin, S.; Yang, C.; Fan, C. Effects of Si additions on intermetallic compound layer of aluminum–steel TIG welding–brazing joint. *J. Alloys Compd.* **2009**, *488*, 217–222. [CrossRef]
33. Xia, H.; Tan, C.; Li, L.; Ma, N. In Situ SEM Observations of Fracture Behavior of Laser Welded–Brazed Al/Steel Dissimilar Joint. *J. Mater. Eng. Perform.* **2018**, *27*, 1047–1057. [CrossRef]

metals

MDPI

Article

Dissimilar Metals Laser Welding between DP1000 Steel and Aluminum Alloy 1050

António B. Pereira [1,*], Ana Cabrinha [1], Fábio Rocha [1], Pedro Marques [1,2], Fábio A. O. Fernandes [1] and Ricardo J. Alves de Sousa [1]

[1]   TEMA–Centre for Mechanical Technology and Automation, Department of Mechanical Engineering, University of Aveiro, Campus de Santiago, 3810-193 Aveiro, Portugal; anacabrinha@ua.pt (A.C.); fabio.jr@ua.pt (F.R.); pdbcm@ua.pt (P.M.); fabiofernandes@ua.pt (F.A.O.F.); rsousa@ua.pt (R.J.A.d.S.)
[2]   Industrial Engineering and Management, Universidade Lusófona, Campo Grande 376, 1749 Lisbon, Portugal
*   Correspondence: abastos@ua.pt; Tel.: +351-234370827

Received: 16 December 2018; Accepted: 14 January 2019; Published: 18 January 2019

Abstract: The welding of dissimilar metals was carried out using a pulsed Nd: YAG laser to join DP1000 steel and an aluminum alloy 1050 H111. Two sheets of each metal, with $30 \times 14 \times 1 \text{ mm}^3$, were lap welded, since butt welding proved to be nearly impossible due to the huge thermal conductivity differences and melting temperature differences of these materials. The aim of this research was to find the optimal laser welding parameters based on the mechanical and microstructure investigations. Thus, the welded samples were then subjected to tensile testing to evaluate the quality of the joining operation. The best set of welding parameters was replicated, and the welding joint obtained using these proper parameters was carefully analyzed using optical and scanning electron microscopes. Despite the predicted difficulties of welding two distinct metals, good quality welded joints were achieved. Additionally, some samples performed satisfactorily well in the mechanical tests, reaching tensile strengths close to the original 1050 aluminum alloy.

Keywords: laser welding; pulsed Nd:YAG laser; DP1000 steel; 1050 aluminum alloy; dissimilar materials welding; steel/aluminum joint

---

1. Introduction

Currently, there is a growing interest across various industries to join different metals or alloys [1,2]. This interest is justified by the flexibility that such options would provide in terms of mechanical project and design, for instance, by utilizing high-quality alloys for critical structural points and low-quality ones for less significant areas. Additionally, considering for instance the automotive industry, this solution makes it possible to manufacture lighter components [3–6]. This would also significantly reduce fuel consumption, thereby reducing CO_2 emissions, which is of the utmost importance in complying with environmental demands and policies, and also to improve manufacturers/brands in their green marketing campaigns.

Steel and aluminum are two materials that are extremely difficult to join by welding. They have different chemical compositions, as well as physical and mechanical properties, which translates into a homogeneity problem. These materials have completely different melting points. Aluminum can melt during the welding process, whereas steel can remain solid [2]. In other words, the laser power necessary to melt steel is excessive for aluminum. On the other hand, aluminum's reflexivity and the high melting temperature of its surface oxides make its melting process quite difficult. Thus, specific welding strategies must be adopted in order to properly weld these materials.

Moreover, the thermal conductivities and thermal expansion coefficients of steel and aluminum alloys are also completely different, which makes welding of these materials a true challenge [4–6]. Therefore, during the welding process, thermal stresses and residual stresses develop on the welding joint [2]. Additionally, the occurrence of intermetallic compounds (IMCs) of Fe–Al on the welding joint during solidification is possible [7]. IMCs are very hard and fragile, decreasing the joint tenacity and plasticity of the joint, i.e., basically turning it brittle [2]. The IMC layers formation is highly dependent on the temperature and time [8], which must be controlled to successfully join steel and aluminum [9]. Consequently, there is a series of problems in joining these two materials, and some solutions have been proposed. Torkamany et al. [10] reported a successful welding of low-carbon steel ST14 (0.8 mm thick) and 5754 aluminum alloy (2.0 mm thick), using a pulsed Nd:YAG laser. It was found that parameters such as power and pulse duration had a significant influence on the overall welding process. Torkamany et al. [10] reported the expected formation of IMCs. An increase in both parameters meant an equal increase in the terms of the IMCs formation. Moreover, with the increase in laser power, the extent of spatter, cracks, and pores also increased. On the other hand, for low laser power values, penetration was insufficient. Sun et al. [11] studied weld butt joints between Q235 low-carbon steel and an AA6013 aluminum alloy, also using an Nd:YAG laser and adding the aluminum alloy 4043 as filler material. As in Reference [10], the formation of the IMCs: Fe_2Al_5 and $FeAl_3$ occured. Failure analysis of the joints showed a typical cleavage-type fracture with crack initiation starting within the $FeAl_3$ phase. The maximum registered tensile strength of the joint was 120 MPa.

These were the most recent and relevant studies found in the literature. Other authors have used fusion-brazing welding [12] or dual laser beam fusion welding [9] or even explosion welding [13–15], which are relatively if not widely different from the investigation at hand, especially the latter. Nevertheless, despite considering these techniques, we did not find any study relating to DP1000 steel and aluminum alloy 1050 welding. In fact, even in well performed literature reviews by Sun and Ion [1], Wang et al. [2], Shah and Ishak [4], and Katayama [3], no study was found regarding the laser welding of these metals. In this work, the authors aimed to determine the optimal set of laser welding parameters, based on mechanical testing and microstructure investigations.

Nowadays, laser technology is used in many applications. Concerning mechanical technology, this can be used to perform precise welding or even precise cutting of high strength steels [16,17]. Laser welding is a joining method that is still relatively recent and makes it possible to obtain better results compared to other welding technologies. This is the main reason behind the replacement of resistance spot welding in several industries. For instance, considering the automotive industry, laser welding has been extensively used, since it makes it possible to perform smaller welding joints (small heat-affected zone) with higher precisions, penetrations, and flexibility. Additionally, it only requires one side of the base metal to accomplish the joining process and it is a technique characterized by high welding speeds, making it an excellent alternative to the conventional welding processes [2,18]. Currently, laser welding is widely used in metals such as titanium alloys [19,20] and dual-phase steels [21].

2. Materials and Methods

2.1. Laser Welding Machine

The machine used was a SWA300 (SISMA, Vicenza, Italy), Figure 1a. The Nd:YAG laser uses, as an active medium, a crystalline solid made of Nd:YAG (neodymium-doped yttrium aluminum Garnet - Y3Al5O12). A parametric study was performed to find a set of welding parameters that made it possible to obtain quality welding joints between DP1000 steel and 1050 aluminum alloy. To evaluate the welded samples, the same methodology adopted in a previous study was also adopted [21], as well as performing tensile and microhardness tests.

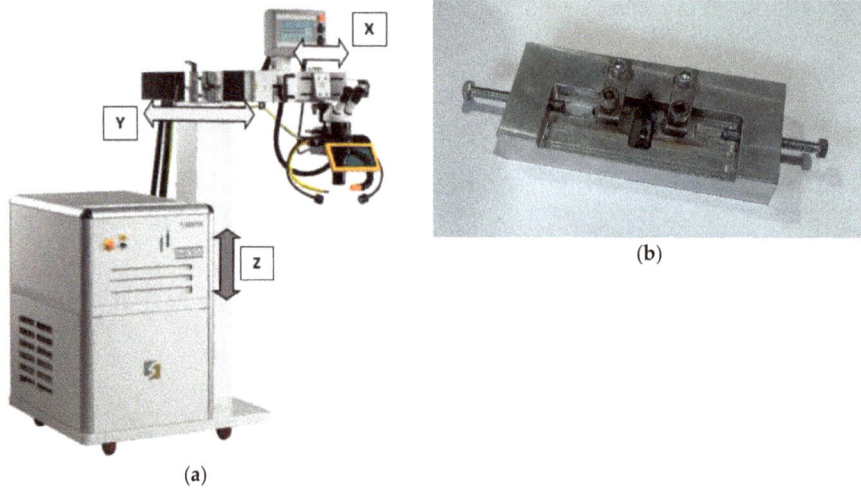

Figure 1. (**a**) SISMA SWA300 Nd:YAG laser welding machine; (**b**) Fixation/support system.

The SISMA SWA300 was primarily designed to perform mold reparation and maintenance, with or without the use of a filler material. Since this machine was designed for mold reparation, it does not have an original fixation system. Thus, a simple and effective solution was designed as depicted in Figure 1b.

The support system has a hole for inserting the protection gas tube. From left to right, Figure 2 illustrates the procedure necessary to properly fix the samples. The first step was to insert the protection gas tube, ensuring that the protective gas was properly covering the samples. The second step consisted of putting the samples in their positions. In order to perform lap welds, a third metal sheet with the same thickness was placed under the top one, which meant it was under the steel sample. The third step was to ensure that both parts were properly placed, using screws at the sides. The last step was to guarantee the complete fixation by using the upper screws, thereby closing the gap between the metal sheets as much as possible.

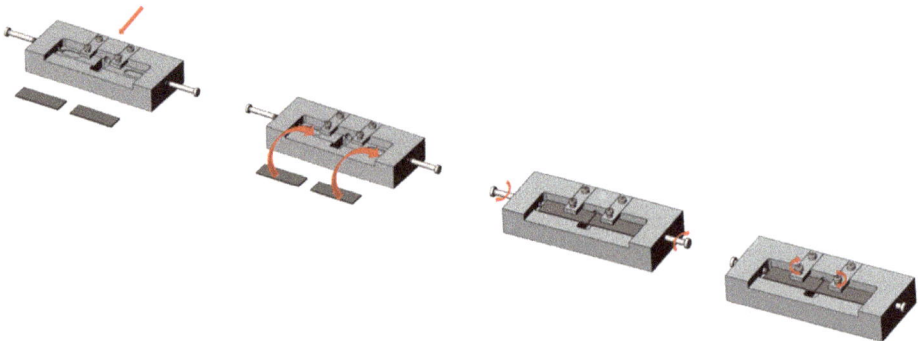

Figure 2. Illustration of the support/fixation system operation.

In order to prevent oxidation, an isolating box was designed and placed on this device. This box was made of glass, due to its transparency and its better thermal resistance (compared to acrylic). Figure 3 shows the glass parts that were glued to create the referred to protective box, for use during the joining process. The protection gas used was Argon ARCAL TIG MIG, supplied by Air Liquide (Paris, France).

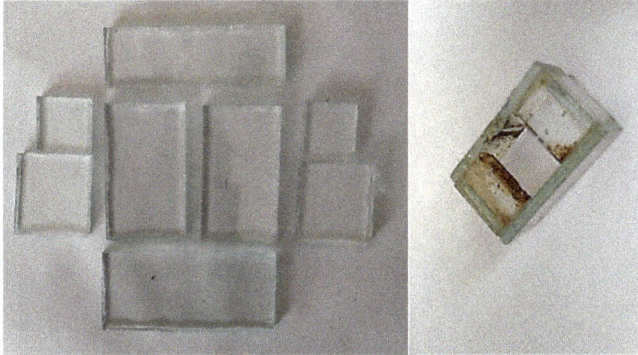

Figure 3. Isolating glass box.

2.2. Materials and Samples

As described, the dissimilar metals chosen were the DP1000 steel and the AA1050 aluminum alloy H111. Both samples had a thickness of 1mm and dimensions of 30mm × 14 mm. The 1mm thick samples were cut using a guillotine (GUIFIL, Portugal).

These materials had completely different melting points. Table 1 presents a simple comparison between iron and aluminum. Aluminum can melt during the welding process, whereas steel can remain solid [2]. In other words, the laser power necessary to melt steel was overly excessive compared to the aluminum. Therefore, the welding was performed only on the steel part, whilst slightly penetrating the aluminum one, without destabilizing it. Tables 2 and 3 present the chemical compositions of both materials.

Table 1. Properties of the Fe and Al at room temperature [22].

Metal	Melting Temperature [K]	Density [kg·m^{-3}]	Thermal Conductivity [W·m^{-1}·K^{-1}]	Specific Heat Capacity [J·kg^{-1}·K^{-1}]	Thermal Expansion Coefficient [K^{-1}]
Fe	1809	7870	78	456	12.1×10^6
Al	933	2700	238	917	23.5×10^6

Table 2. Chemical composition of the AA1050 aluminum alloy [wt.%] [23].

Ti	Zn	Mn	Fe	Si	Mg	Cu
0.05	0.07	0.05	0.4	0.25	0.05	0.05

Table 3. Chemical composition of DP1000 steel [wt.%] [24].

C	Si	Mn	Cu	Al	Cr	Ni	Nb	V
0.141	0.49	1.47	0.02	0.041	0.03	0.04	0.016	0.01

The samples were lap welded with the steel on top (Figure 4). It was verified that as the welding progressed, naturally the samples would heat, and the penetration would increase longitudinally. Thus, for the second welding joint, the sample was rotated 180° and the welding process proceeded in the same direction. This way, less penetration achieved on the initial regions of the first welding joint would be compensated by the second, which would accomplish just that.

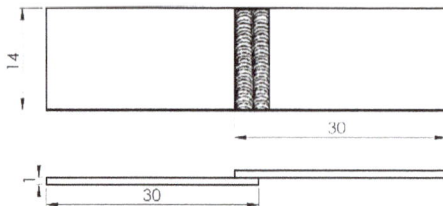

Figure 4. Schematic of the welding performed on the samples.

2.3. Laser Welding Parameters

Regarding the laser parameters that could be analyzed using the SISMA SWA300, the following parameters were considered: laser power (percentage of peak pulse); pulse duration; frequency/superposition; and laser beam diameter. Since there was no data in the literature regarding DP1000–AA1050 aluminum welding, the ideal welding parameters were found by trial and error, as well as taking into consideration the conclusions of other studies previously carried out using the same machine. From these conclusions, the following were considered:

- Laser beam power: the penetration was proportional to the laser beam power. The ideal values were situated between 6 kW and 7.20 kW of the maximum peak power of 12 kW.
- Pulse duration: the penetration was also proportional to higher pulse durations. Previous values were between 12–16 ms.
- Superposition: It did not influence penetration until values around 80%.
- Laser beam diameter: the penetration seemed to be inversely proportional to the welding spots diameter. A constant value of 1 mm was used.
- Welding speed: a high welding speed meant high penetration. The speed was then limited by the amount of power/energy and pulse duration used. For higher values of pulse duration and/or power (and thus energy), smaller values of speed were allowed for a proper welding process.

Similar to Torkamany et al. [10], which evaluated the influence of power and pulse duration on the welding process, in the first stage, only power was varied to find its ideal value. Then, other parameters were changed based on visual analysis and tensile testing. This evolution regarding the welding parameter values is described thoroughly in the results section.

Table 4 shows the parameter values used for each welded sample. A square wave was defined as the pulse type for all the samples. The welding speed was the lowest possible concerning machine restrictions, because for higher values, the penetration on the aluminum sample would be excessively high, destroying any chance of a properly welded joint. Additionally, a constant laser beam diameter and superposition of 1 mm and 60% were used, respectively.

Table 4. Parametric study for steel–aluminum welding.

Sample	Power [kW]	Pulse Duration [ms]
1	8.40	14
2	7.20	14
3	7.08	14
4	6.96	14
5	6.84	14
6	6.72	14
7	6.60	14
8	6.48	14
9	6.36	14
10	6.24	14
11	6.12	14
12	6.00	14

Table 4. *Cont.*

Sample	Power [kW]	Pulse Duration [ms]
13	6.48	16
14	6.48	15
15	6.48	13
16	6.48	12
17	6.48	11
18	6.60	13
19	6.60	15
20	6.60	16

2.4. Tensile Testing

To determine the mechanical properties of the welded samples, mainly the tensile strength and respective deformation of each set of parameters, tensile testing was performed using an universal testing, 10 kN machine (Shimadzu, Kyoto, Japan), Figure 5.

Figure 5. Shimadzu 10 kN tensile testing machine.

Considering lap welding, deflection phenomena are expected to occur during tensile testing. We then used the following equation to compute stress:

$$S = \frac{F}{A_0} + \frac{Mx \times y}{Ix},$$

(1)

where F is the magnitude of the forces applied, A_0 is the initial cross section area of the joint, M_x and I_x are the torque and moment of inertia about the X-axis, respectively, and y is the maximum distance about the Y-axis considering a xOy plane as Figure 6:

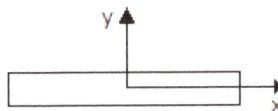

Figure 6. Position of the Cartesian coordinate system plane on the sample.

Since the superposition of the welding spots had a value of 60%, the area of the joint was approximated to one of a 14 × 1 mm² rectangle. However, the deflection stresses were negligible

(<1.5 MPa on every test) due to the small dimensions of the samples. Thus, in order to determine the nominal stress, we used the following approximation:

$$S \approx \frac{F}{A_0},$$ (2)

where S is the nominal stress, F is the magnitude of the applied force, and A_0 is the cross-sectional area of the sample (since fracture should occur on the aluminum part). The normal strain was determined through:

$$e = \frac{\Delta L}{L_0},$$ (3)

where e is the nominal deformation, ΔL is the change in length, and L_0 is the original length considered for the sample at hand, in this case it was 30 mm for each test. The true stress was calculated by:

$$\sigma = S(1+e),$$ (4)

where σ is the true stress, whilst the true strain ε was determined as:

$$\varepsilon = \ln(1+e),$$ (5)

2.5. Microstructure

For a deeper analysis of the welding, a new sample based on the optimum welding parameters found in the tensile testing was re-created to analyze its microstructure. Firstly, the sample was transversely cut for the observation of a substantial part of its welding joint, as well as longitudinally, to have a 25 mm length to fit the mold (Figure 7). Cutting was precisely performed using a Struers cutting machine.

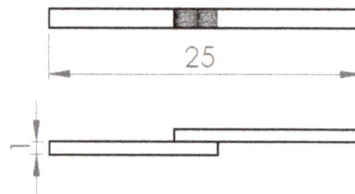

Figure 7. Illustration of the sample for microstructural observation.

After this, the sample was inserted on a cylindrical mold of 25 mm diameter and then epoxy resin was injected into it, as well as a hardener. Once solidified (Figure 8), the injected sample was removed from the mold and then polished for observation. Polishing diamond papers (3 μm) of 180, 320, 800, 1200, 2500, and 4000 grit were used. The machine used was a rotary polishing machine, the RotoPol-21 (Struers, Cleveland, OH, USA).

Then, the injected sample was subjected to chemical etching using nital (4%) (Ethanol + Nitric Acid), to help the microstructure observation. The nital etching was executed, lasting 20 seconds. At first, it was also subjected to a Keller's reagent (HNO_3 2.5 mL; HCl 1.5 mL; H_2F_2 1 mL, H_2O 95 mL). However, it would cause oxidation due to the iron (Fe) present in the sample, complicating the observation. Thus, at the end, only the nital etching was considered. The microstructural observations were performed using an optical microscope and a scanning electron microscope TM4000Plus (SEM, Hitachi, Japan) with an integrated back-scattered electron detector (BSE). Elemental composition mapping was performed using an energy-dispersive X-ray spectroscopy (EDS, Hitachi, Japan).

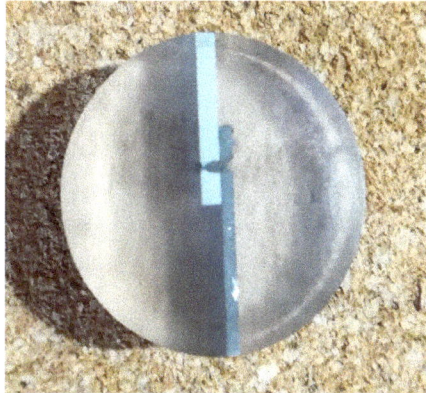

Figure 8. Solidified sample after injection of epoxy resin + hardener.

3. Results

3.1. Laser Welding

Considering the values presented in Table 4 for each welded sample, after tensile testing, it was verified that considering the speed used, the ideal interval was from 6 kW to 7.20 kW. More specifically, the best performance on tensile testing was achieved at a laser power of 6.48 kW (sample eight from 12 samples, Table 4). It became clear that for welding powers higher than 7.20 kW, the laser would fully penetrate the aluminum part, resulting in brittle joints, which were easily breakable by hand. Then, the duration of the pulse was altered, and the laser power fixed (samples 13 to 16). On the final samples (17 to 20), these two parameters were slightly changed from the ideal values. Figure 9 presents the 20 welded samples.

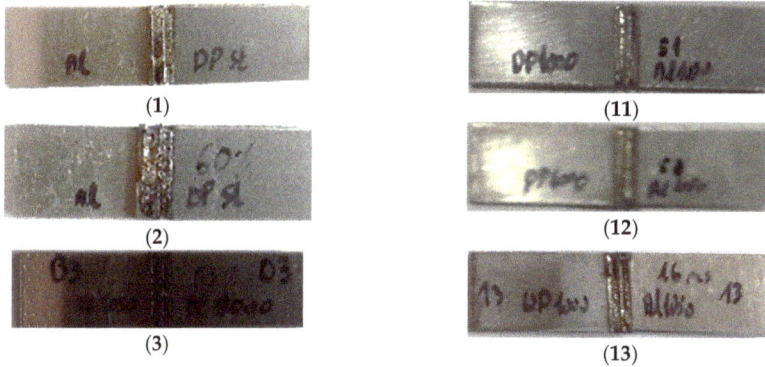

(1)

(11)

(2)

(12)

(3)

(13)

Figure 9. *Cont.*

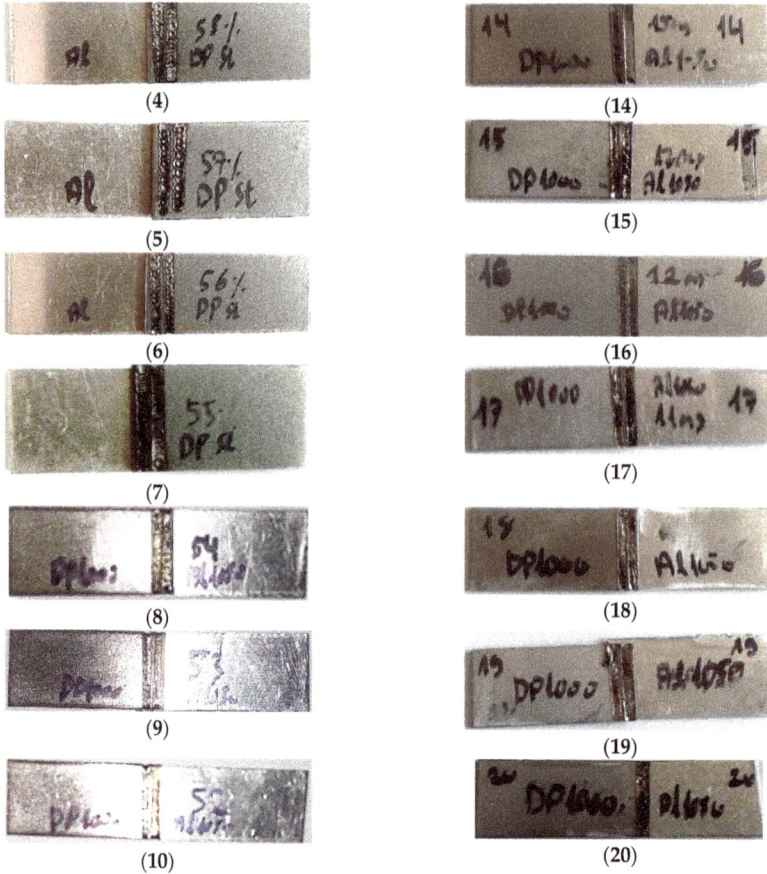

Figure 9. Steel-aluminum welded samples.

3.2. Tensile Testing

Figure 10 presents the samples after tensile testing, as well as a reference sample made of aluminum (60 × 14 mm), since ideally, the fracture occurs on the aluminum part, due to its lower tensile strength. We considered the normalization of this reference sample. Nevertheless, it was not done for a better similarity with the welded samples. Additionally, it is worth mentioning that sample 20 was not subjected to tensile testing, since we noticed that the laser significantly penetrated the aluminum, with some wide penetrations in some regions. Therefore, the sample would fracture easily while testing, producing negligible data. Thus, this specific sample was removed from the experiment.

Figure 10. *Cont.*

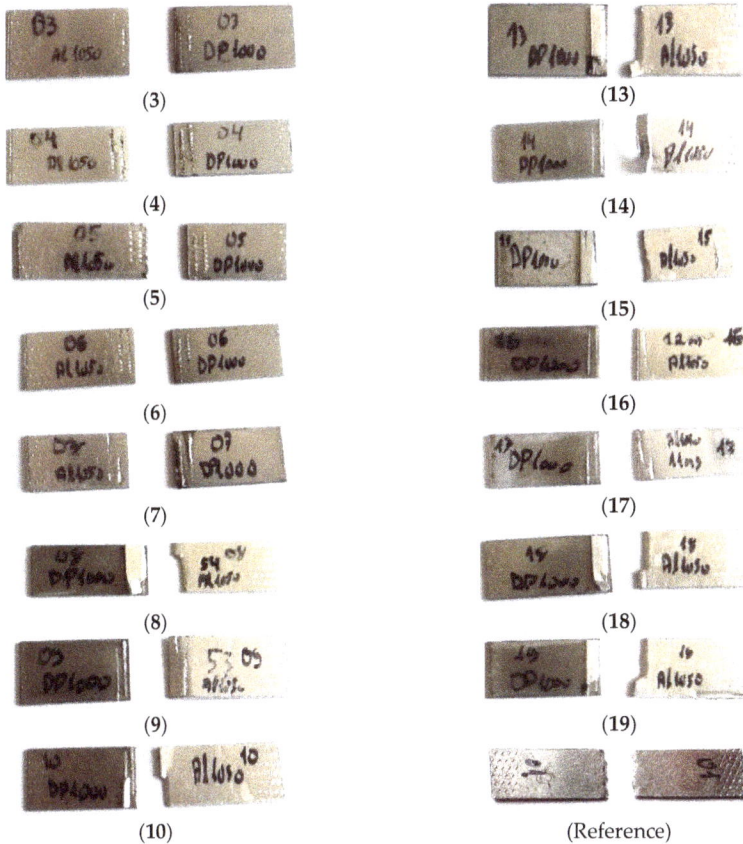

Figure 10. Samples after tensile testing.

Figure 11a shows the stress–strain curve for the samples where fracture occurred partially or completely through the aluminum part (true stress/strain values). On the other hand, Figure 11b shows the samples where fracture occurred through the welding joint. In both, the reference sample stress–strain curve was included for comparison. Since Figure 11 plots the true stress vs. true strain, the curves were only presented up to maximum stress value.

An ideal welded sample would perform similarly to the original (weakest) material in terms of tensile strength and maximum elongation. Regarding tensile strength, the samples that performed better were samples 8 and 18 (Figures 11 and 12). The latter had a slightly lower value but a higher strain value, thus being considered the best of this whole batch, reaching a 20 MPa lower tensile strength and a 1% strain difference compared the reference sample. It is also worth mentioning that for sample 19, although its maximum stress was significantly lower than samples 8 and 18, its maximum strain was closest to the reference sample. In this group, the worst sample was clearly sample 15 which had the lowest tensile strength. It could probably be justified by the excessive laser penetration on the aluminum part, causing its fragilization. Thus, an ideal laser power between 54 and 55, with a pulse duration between 13 and 15 ms can be herein stated.

(a)

(b)

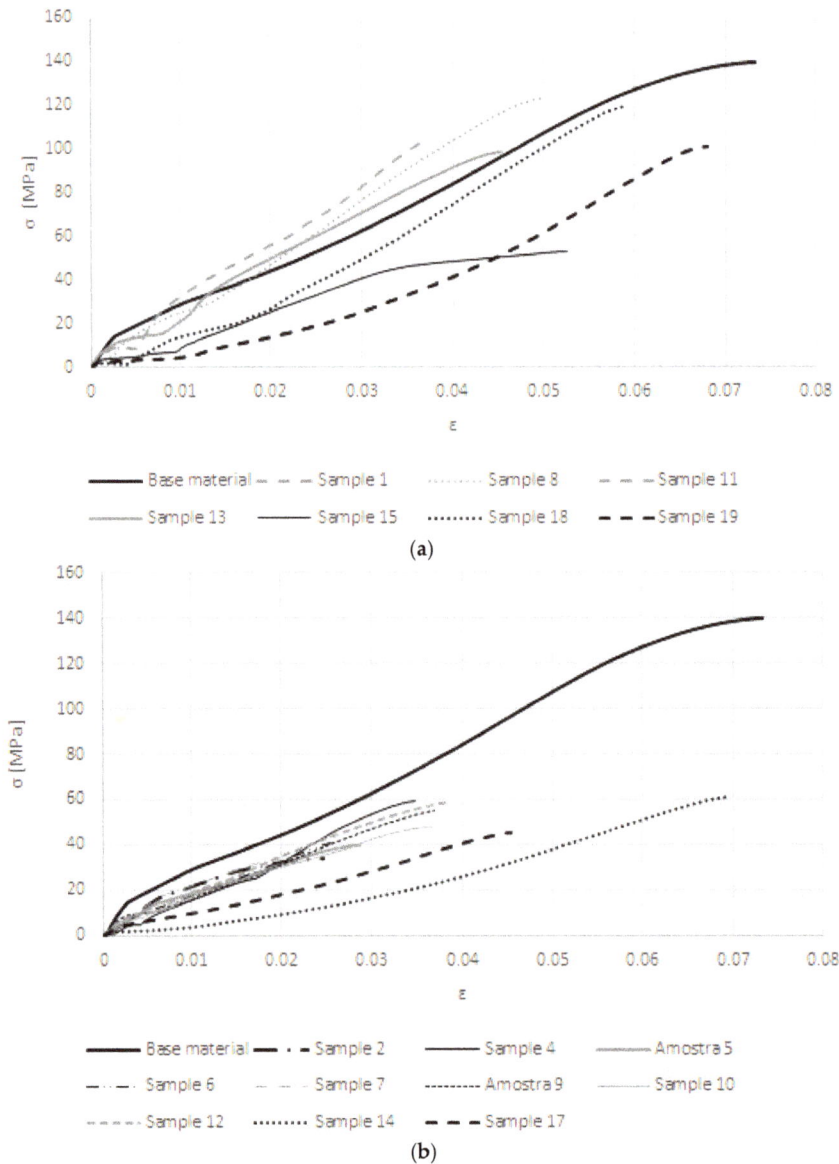

Figure 11. True stress–strain curves for the samples where fracture occurred through: (**a**) the aluminum part; (**b**) the welding joint.

For the samples where the welding joint did not remain intact after tensile testing, the maximum stress recorded was much lower than the one in the reference sample (at best, 80 MPa lower). However, it is important to highlight the sample 14 strain value, which came close to the value of the reference sample. It is worth mentioning that the curves of samples 3 and 16 are not plotted in Figure 11, since the results were irrelevant. For an exact reading of the tensile strength values, as well as their respective strain values, the following bar graphs are seen in Figure 12.

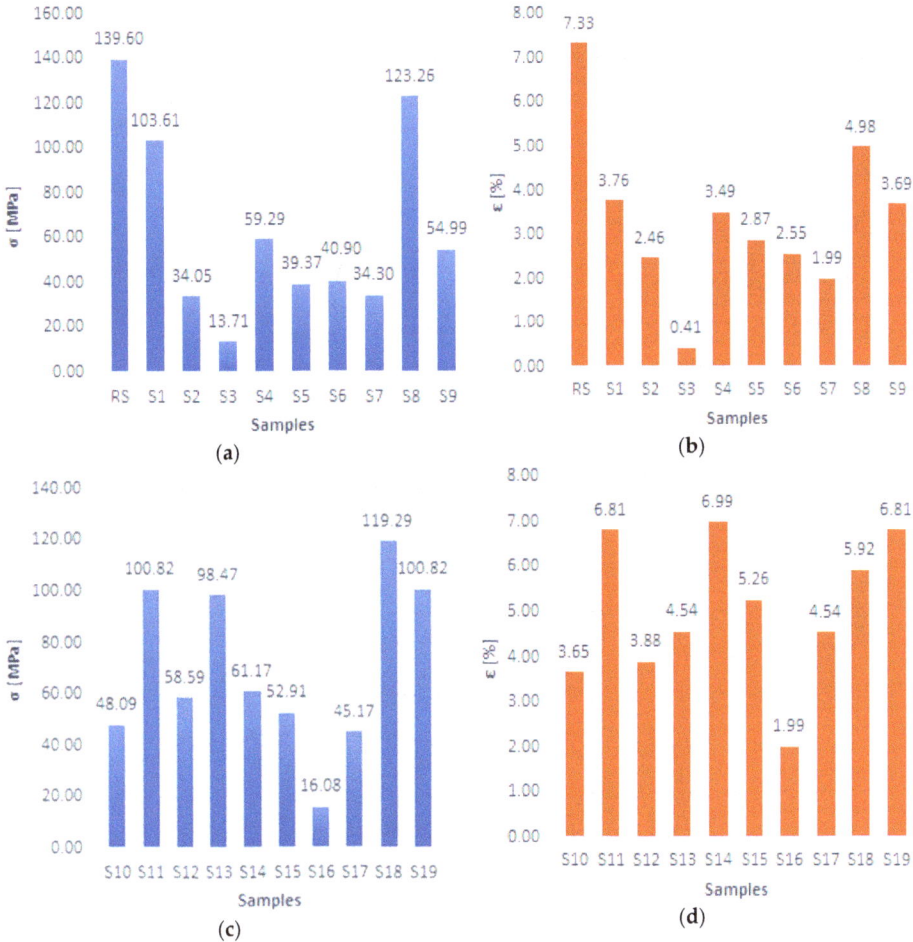

Figure 12. Results from the tensile tests: (**a**) Ultimate tensile strength (samples 1 to 9); (**b**) Strain associated with the ultimate tensile strength (samples 1 to 9); (**c**) Ultimate tensile strength (samples 10 to 19); (**d**) Strain associated with the ultimate tensile strength (samples 10 to 19). (Note: RS means "reference sample" whereas S1 from S19 are the samples from 1 to 19).

In terms of laser power (samples 2 to 12), the best results were obtained as this value decreased. Considering the pulse duration (samples 13 to 17), the contrary could be seen for the same value of the laser power. However, in terms of strain, the results were inconclusive without a linear behavior of any sort of pattern. Sample 18 could be considered the ideal sample, with a tensile strength and strain of only 20 MPa and 1.4%, respectively, which were lower than the reference sample. This ideal sample was repeated to evaluate its microstructure.

3.3. Microstructure

Figures 13 and 14 show the observations using the optical microscope and SEM, respectively. In Figure 13, the welded joint is easily spotted, being observed in the way in which both metals melted and mixed in a successful way. This was also confirmed through the SEM observations in Figure 14, where it shows the interface between the two metals. In Figure 14b, it was possible to spot the Fe particles in the aluminum.

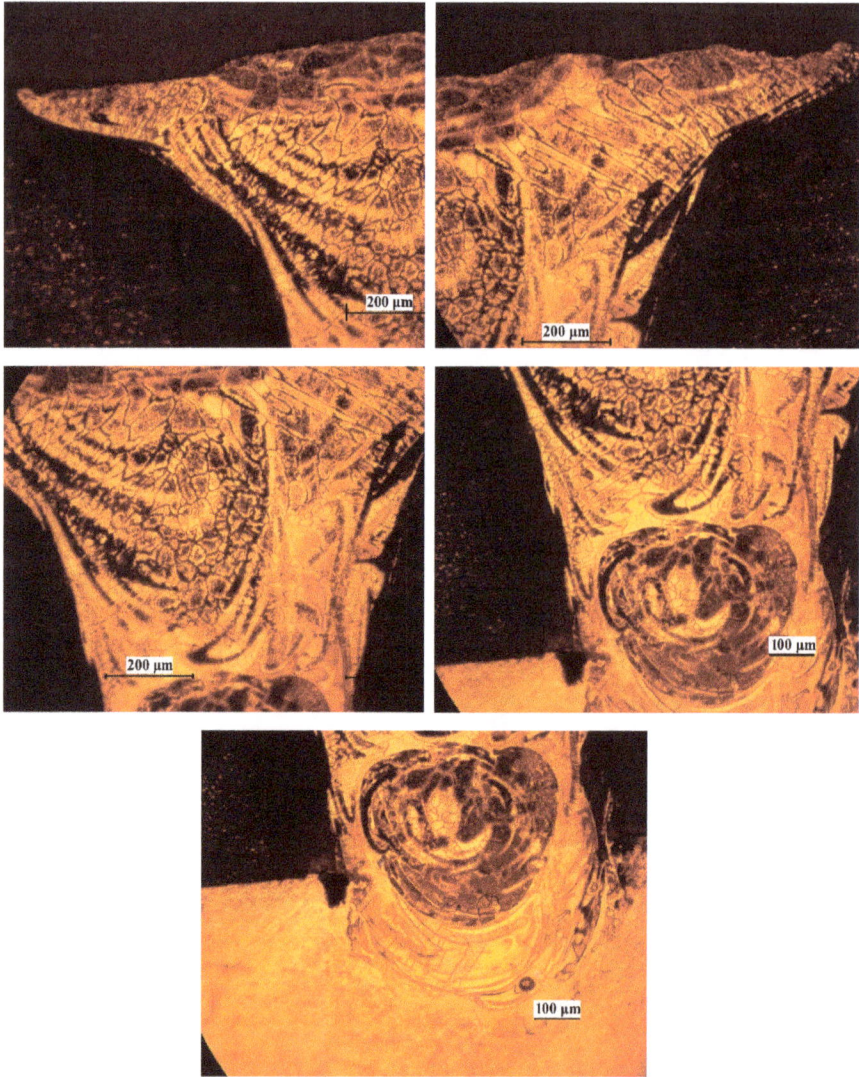

Figure 13. Optical microscope observations of the welded zone.

Figure 15 shows the BSE images collected, as well as the microstructural and compositional analysis of the welded zone. Compositions are detailed in Table 5 for a total of eight different points, six of them in the welded zone and the other two in the base materials used as the reference.

Table 5. Elemental composition mapping (normalized mass concentration [%]).

Spectrum	C	O	Al	Si	Fe
1	8.25	2.10	12.17	-	77.48
2	8.30	2.82	19.13	0.59	69.15
3	13.00	10.87	62.77	9.37	3.99
4	8.18	2.37	4.73	-	84.72
5	9.39	3.33	46.69	-	40.59
6	10.11	3.22	23.85	-	62.82

Table 5. *Cont.*

Spectrum	C	O	Al	Si	Fe
7	7.83	2.11	9.53	-	80.53
8	7.70	2.13	8.86	-	81.31
Mean	9.10	3.62	23.47	4.98	62.58
Sigma	1.77	2.97	20.66	6.21	27.63
SigmaMean	0.63	1.05	7.31	2.20	9.77

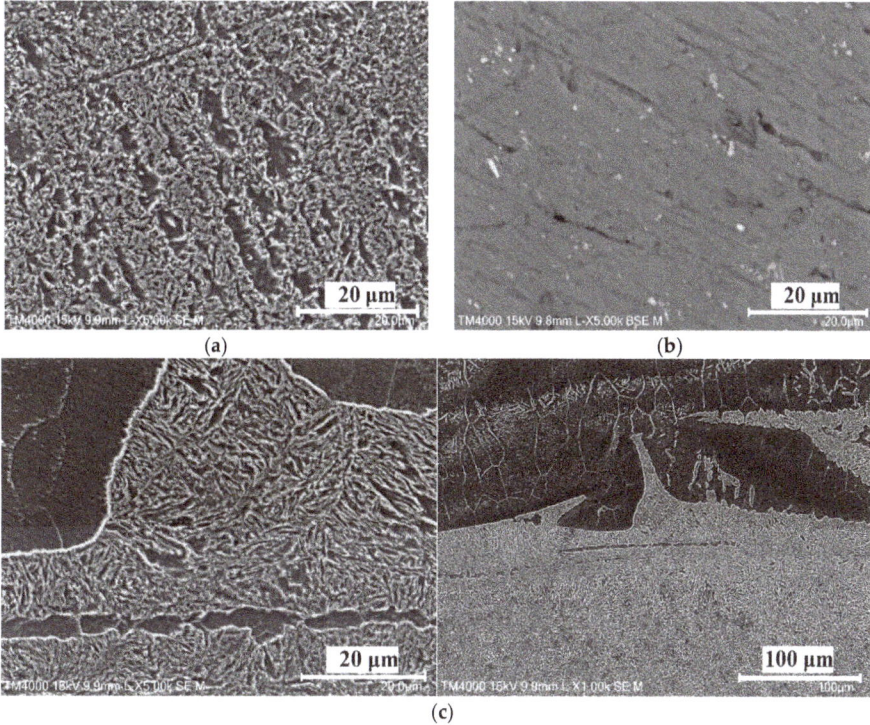

Figure 14. Scanning electron microscope (SEM) observations: (**a**) Steel; (**b**) Aluminum; (**c**) Steel–aluminum joint interfaces.

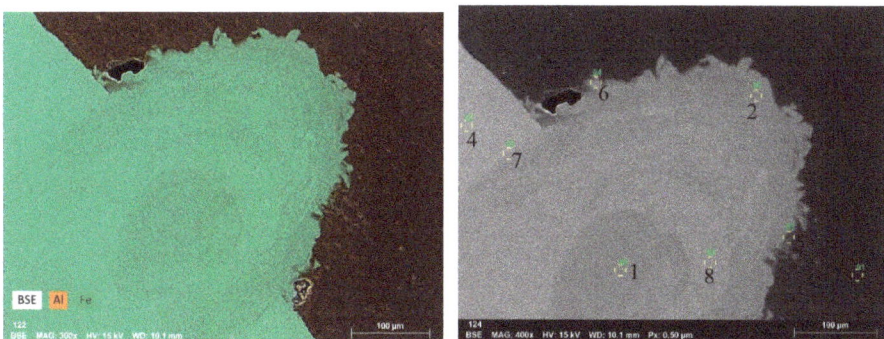

Figure 15. Back-scattered electron detector (BSE) images with elemental composition mapping using Energy-dispersive X-ray spectroscopy (EDS).

4. Conclusions

In this work, the welding of dissimilar metals was performed with success. A pulsed Nd: YAG laser welding machine, the SISMA SWA300, was used to join the DP1000 steel and the AA1050. The adopted strategy was based on the lap welding of two sheet samples of each metal, since butt welding was nearly impossible due to the great disparity in terms of the thermal properties of both materials.

Even if some predicted difficulties in welding these dissimilar metals were found, a good choice of welding parameters was studied resulting in good-quality welded joints. The quality was confirmed by the results obtained in the tensile tests and in the observations performed using the optical microscope and the SEM. It is worth highlighting the sample performance, which reached tensile strengths and maximum elongations close to the weakest base metal, the AA1050.

In conclusion, it was shown that even for highly dissimilar materials, with very distinct material properties (thermal and mechanical), one can declare the success of this work. The possibility to effectively join high strength dual-phase steels with a soft, ductile 1XXX aluminum alloy opens a new range of design possibilities and attests the versatility of laser-type welding operations. The authors hope that this study can serve as a sounding base for any other future work in this area.

Author Contributions: Conceptualization, A.C. and F.R.; Methodology, A.B.P.; Software and Validation, P.M.; Formal Analysis, P.M.; Investigation, A.C. and F.R.; Resources, A.B.P.; Data Curation, F.A.O.F.; Writing-Original Draft Preparation, F.A.O.F.; Writing-Review & Editing, R.J.A.d.S.; Supervision, P.M.; Project Administration, A.B.P.; Funding Acquisition, A.B.P. and R.J.A.d.S.

Funding: Programa Operacional Temático Factores de Competitividade: CENTRO-01-0145-FEDER-022083, Fundação para a Ciência e a Tecnologia, grant UID/EMS/00481/2013-FCT.

Acknowledgments: Center of Mechanical Technology and Automation, Fundação para a Ciência e a Tecnologia, grant CEECIND/01192/2017.

Conflicts of Interest: The authors declare no conflict of interest.

References

1. Sun, Z.; Ion, J.C. Laser welding of dissimilar metal combinations. *J. Mater. Sci.* **1995**, *30*, 4205–4214. [CrossRef]
2. Wang, P.; Chen, X.; Pan, Q.; Madigan, B.; Long, J. Laser welding dissimilar materials of aluminum to steel: An overview. *Int. J. Adv. Manuf. Technol.* **2016**, *87*, 3081–3090. [CrossRef]
3. Katayama, S. Laser welding of aluminium alloys and dissimilar metals. *Weld. Int.* **2004**, *18*, 618–625. [CrossRef]
4. Shah, L.H.; Ishak, M. Review of research progress on aluminum-steel dissimilar welding. *Mater. Manuf. Process.* **2014**, *29*, 928–933. [CrossRef]
5. Shao, L.; Shi, Y.; Huang, J.K.; Wu, S.J. Effect of joining parameters on microstructure of dissimilar metal joints between aluminum and galvanized steel. *Mater. Des.* **2015**, *66*, 453–458. [CrossRef]
6. Song, J.L.; Lin, S.B.; Yang, C.L.; Ma, G.C.; Liu, H. Spreading behavior and microstructure characteristics of dissimilar metals TIG welding–brazing of aluminum alloy to stainless steel. *Mater. Sci. Eng. A* **2009**, *509*, 31–40. [CrossRef]
7. Borrisutthekul, R.; Yachi, T.; Miyashita, Y.; Mutoh, Y. Suppression of intermetallic reaction layer formation by controlling heat flow in dissimilar joining of steel and aluminum alloy. *Mater. Sci. Eng. A* **2007**, *467*, 108–113. [CrossRef]
8. Bouché, K.; Barbier, F.; Coulet, A. Intermetallic compound layer growth between solid iron and molten aluminium. *Mater. Sci. Eng. A* **1998**, *249*, 167–175. [CrossRef]
9. Cui, L.; Chen, H.; Chen, B.; He, D. Welding of Dissimilar Steel/Al Joints Using Dual-Beam Lasers with Side-by-Side Configuration. *Metals* **2018**, *8*, 1017. [CrossRef]
10. Torkamany, M.J.; Tahamtan, S.; Sabbaghzadeh, J. Dissimilar welding of carbon steel to 5754 aluminum alloy by Nd:YAG pulsed laser. *Mater. Des.* **2010**, *31*, 458–465. [CrossRef]
11. Sun, J.; Yan, Q.; Gao, W.; Huang, J. Investigation of laser welding on butt joints of Al/steel dissimilar materials. *Mater. Des.* **2015**, *83*, 120–128. [CrossRef]

12. Meco, S.; Pardal, G.; Ganguly, S.; Williams, S.; McPherson, N. Application of laser in seam welding of dissimilar steel to aluminium joints for thick structural components. *Opt. Lasers Eng.* **2015**, *67*, 22–30. [CrossRef]

13. Corigliano, P.; Crupi, V.; Guglielmino, E. Non linear finite element simulation of explosive welded joints of dissimilar metals for shipbuilding applications. *Ocean Eng.* **2018**, *160*, 346–353. [CrossRef]

14. Corigliano, P.; Crupi, V.; Guglielmino, E.; Mariano Sili, A. Full-field analysis of AL/FE explosive welded joints for shipbuilding applications. *Mar. Struct.* **2018**, *57*, 207–218. [CrossRef]

15. Kaya, Y. Microstructural, Mechanical and Corrosion Investigations of Ship Steel-Aluminum Bimetal Composites Produced by Explosive Welding. *Metals* **2018**, *8*, 544. [CrossRef]

16. Rana, R.S.; Chouksey, R.; Dhakad, K.; Paliwal, D. Optimization of process parameter of Laser beam machining of high strength steels: A review. *Mater. Today Proc.* **2018**, *5*, 19191–19199. [CrossRef]

17. Patidar, D.; Rana, R.S. The effect of CO_2 laser cutting parameter on Mechanical & Microstructural characteristics of high strength steel—A review. *Mater. Today Proc.* **2018**, *5*, 17753–17762. [CrossRef]

18. Chen, S.; Huang, J.; Ma, K.; Zhao, X.; Vivek, A. Microstructures and Mechanical Properties of Laser Penetration Welding Joint With/Without Ni-Foil in an Overlap Steel-on-Aluminum Configuration. *Metall. Mater. Trans. A* **2014**, *45*, 3064–3073. [CrossRef]

19. Zhan, Y.; Zhang, E.; Ge, Y.; Liu, C. Residual stress in laser welding of TC4 titanium alloy based on ultrasonic laser technology. *Appl. Sci.* **2018**, *8*, 1997. [CrossRef]

20. Li, R.; Zhang, F.; Sun, T.; Liu, B.; Chen, S.; Tian, Y. Investigation of strengthening mechanism of commercially pure titanium joints fabricated by autogenously laser beam welding and laser-MIG hybrid welding processes. *Int. J. Adv. Manuf. Technol.* **2018**. [CrossRef]

21. Fernandes, F.A.O.; Oliveira, D.F.; Pereira, A.B. Optimal parameters for laser welding of advanced high-strength steels used in the automotive industry. *Procedia Manuf.* **2017**, *13*, 219–226. [CrossRef]

22. Gale, W.F.; Totemeier, T.C. (Eds.) *Smithells Metals Reference Book*, 8th ed.; Elsevier Butterworth-Heinemann: Oxford, UK, 2004.

23. Esmaeili, A.; Givi, M.K.B.; Rajani, H.R.Z. A metallurgical and mechanical study on dissimilar Friction Stir welding of aluminum 1050 to brass (CuZn30). *Mater. Sci. Eng. A* **2011**, *528*, 7093–7102. [CrossRef]

24. Wang, J.; Yang, L.; Sun, M.; Liu, T.; Li, H. Effect of energy input on the microstructure and properties of butt joints in DP1000 steel laser welding. *Mater. Des.* **2016**, *90*, 642–649. [CrossRef]

metals

MDPI

Article

Interfacial Characteristics of Dissimilar Ti6Al4V/AA6060 Lap Joint by Pulsed Nd:YAG Laser Welding

Xin Xue [1,2], António Pereira [2,*], Gabriela Vincze [2], Xinyong Wu [1] and Juan Liao [1]

[1] School of Mechanical Engineering and Automation, Fuzhou University, Fuzhou 350116, China;
    xin@fzu.edu.cn (X.X.); n170227061@fzu.edu.cn (X.W.); jliao@fzu.edu.cn (J.L.)
[2] Centre for Mechanical Technology and Automation, Department of Mechanical Engineering,
    University of Aveiro, 3810-193 Aveiro, Portugal; gvincze@ua.pt
*   Correspondence: abastos@ua.pt; Tel.: +351-234-370-830

Received: 13 December 2018; Accepted: 9 January 2019; Published: 12 January 2019

Abstract: This paper focuses on the interfacial characteristics of dissimilar Ti6Al4V/AA6060 lap joint produced by pulsed Nd:YAG laser beam welding. The process-sensitivity analysis of welding-induced interface joining quality was performed by using the orthogonal design method. Microstructural tests such as scanning electron microscopy and energy dispersive X-ray spectroscopy were used to observe the interfacial characteristics. The mechanism of interfacial crack initiation, which is an important indicator of joint property and performance, was assessed and analyzed. The preferred propagation paths of welding cracks along the interfaces of different intermetallic layers with high dislocation density were analyzed and discussed in-depth. The results indicate that discontinuous potential phases in the micro-crack tip would mitigate the mechanical resistance or performance of the welded joint, while the continuous intermetallic layer can lead to a sound jointing performance under pulsed Nd:YAG laser welding process.

Keywords: pulsed Nd:YAG laser beam welding; interfacial crack initiation; dissimilar Ti6Al4V/AA6060 lap joint; phase potential

1. Introduction

Dissimilar welding of lightweight alloys is attracting increasing attention in various fields because it can take advantage of specific contributions of each alloy to enhance the properties of a weld joint or bring out new functionalities. For instance, Chen et al. [1] reported that Ti/Al structures have already been applied in the wing structure of airplanes. As an implication, there is a challenge for a way to use the high-level properties of dissimilar alloys such as corrosive properties and strength of titanium and lightweight and low cost of aluminum. Although considerable research has been devoted to the same or similar metal welding techniques that can supply appropriate material and mechanical performances, rather less attention has been paid to dissimilar lightweight metal welding.

To obtain good combinations of titanium with aluminum, many efforts have been made using various technologies. Friction stir welding of dissimilar alloys and the improvement of joint quality was reported by Suhuddin et al. [2]. The AA6061/Ti6Al4V dissimilar laminates by single-shot explosive-welding process was produced by Ege and Inal [3]. By using brazing, Song et al. [4] and Yang et al. [5] pointed out the possibility of dissimilar joining of Ti/Al and steel/Al, respectively. Wang et al. [6] used ultrasonic spot-welding technology to joint dissimilar materials of DP600 steel and AA6022. Tang et al. [7] indicated that the preheating treatment could improve the mechanical properties or fracture load of dissimilar joint in the welding process. The above-mentioned methods have in common that convective mixing and diffusion phenomena are suppressed or attenuated, thus bulk brittle phases in the interfacial layer do not form. Another alternative against brittle phases is the change of the chemistry of fusion zone. Sambasiva Rao et al. [8] reported the dissimilar metal gas tungsten arc welding of aluminum to titanium alloy using Si-containing filler wire. Chang et al. [9] used vacuum brazing method to investigate the role of additional rare-earth elements for aluminum/titanium joining performance.

One of the major problems with dissimilar metal welds is their reduced inter-metallic mechanical property. Vaidya et al. [10] reported that the inter-metallic brittle phase TiAl$_3$ formed at the weld interface between AA6065 and Ti6Al4V sheet by laser in conduction mode and affected mechanical properties: the tensile strength of the junction was found superior to the strength of aluminum alloy and the fracture happened on Al side. Chen et al. [11] tried to use a specific Si-containing filler wire during the laser brazing and indicated the influence of Ti$_7$Al$_5$Si$_{12}$ on decreasing the growth of brittle TiAl$_3$ phase. Majumda et al. [12] reported that the insert of Nb foil barrier allowed suppressing Ti/Al inter-metallic formation and enhanced tensile strength from 57 to 120 MPa. By offsetting the laser beam, different additional inter-metallic phases were observed and affected the mechanical properties. Recently, the micro-hardness, lap shear strength and fracture energy of AA2139-TiAl6V4 joints by ultrasonic welding were investigated by Zhang et al. [13]. They indicated that the peak load and energy of welds increased with an increase in welding time and then reached a plateau.

Recently, laser beam welding opens an attractive perspective for joining strongly dissimilar materials. One of the main advantages of laser beam welding is providing very local energy supply that allows obtaining a good quality weld. The other advantage is that laser beam welding can induce small interaction zone and high welding speed to promote high thermal gradients, which are helpful for local and potential phase content optimization. Thus, the importance of mixing and diffusion phenomena can be mitigated. Furthermore, the mismatch in thermo-physical properties of dissimilar materials is easy to accommodate by changing the laser beam to one of the substrates. Casalino et al. [14] investigated the dissimilar butt joint of AA5754 and T40 by using Yb:YAG laser offset welding. Fabbro [15] reported that, particularly for Nd:YAG pulsed laser welding technology, thickness of Ti/Al intermetallic layers near Ti/Al interface has been controlled to a relatively low level. Ren et al. [16] showed that the interface zone of Ti/Al diffusion bonding included transition zone on Ti substrate, aluminized coating and transition zone on Al substrate. Intermetallic potential phases TiAl and TiAl$_3$ were formed in the transition zone on Ti substrate and aluminized coating, but, in the transition zone on Al substrate, no additional intermetallic phase was found. Tomashchuk et al. [17,18] investigated the intermetallics in dissimilar Ti6Al4V/copper/AISI 316 L in Nd:YAG laser joints through simulation and microstructural testing methods. However, intermetallic layer in the weld joint still had high crack sensitivity, and most of the joints were fractured at Ti/Al interface under mechanical loading conditions. Therefore, it is important to investigate the mechanism and estimation of welding cracks during the thermal welding. However, up to now, few studies on the initiation mechanism and propagation paths of welding cracks produced by Nd:YAG laser welding have been published in literature.

In the present work, the weldability of Ti6Al4V/AA6060 dissimilar metal alloys using Nd:YAG pulsed laser welding and the interface characteristics of lap joint were addressed. The influence of key laser welding process parameters on weld morphology, microstructure and mechanical properties were investigated by means of microstructure characterizations and experimental design approach. Initiation mechanism and estimation of interfacial crack was analyzed as well as the exploration of interfacial phase constituents.

2. Base Materials

In this study, commercially available Ti6Al4V and AA6060 with dimensions of 35 mm × 14 mm × 0.8 mm and 35 mm × 14 mm × 1.5 mm, respectively, were used as the studied base metals. The chemical compositions of two studied materials are listed in Table 1. Element Si has been demonstrated to be effective for the growth of Ti/Al brittle intermetallic [19]. The thermal physical properties of the received Ti6Al4V and AA6060 were provided by Titanium International Group SRL in Italy and EXTRUSAL S.A in Portugal, as listed in Table 2. Both base materials are active light metals compared to traditional steels. The detailed data information about shear strength of AA6060 can be referred to in the publication of the previous work [20], and the shear strength of Ti6Al4V can be referred to in the work of He et al. [21]. It also can be seen that the thermal physical properties, particularly the melting point and specific heat and thermal expansion coefficient, have wide discrepancies for the two studied base metals. This means that diffusion bonding or joining of dissimilar titanium and aluminum alloys is a challenging task, and the welding parameters should play a considerable role in the weld quality.

Table 1. Chemical composition of base materials used in the present study in wt%.

Ti6Al4V	Al	V	Fe	O	Si	C	N	H	Other Elements	Ti
	5.5–6.8	3.5–4.5	0.3	0.2	0.15	0.10	0.05	0.01	0.5	Balance
AA6060	Al	Si	Fe	Cu	Mn	Mg	Cr	Zn	Ti	
	Balance	0.3–0.6	0.1–0.3	≤0.10	≤0.10	0.35–0.6	≤0.05	≤0.15	≤0.10	

Table 2. Thermal physical properties of the as-received Ti6Al4V and AA6060.

Material	Density (Kg/m^3)	Melting Point (°C)	Specific Heat (J/Kg·°C)	Thermal Conductivity W/(M·K)	Thermal Expansion Coefficient (°C)	Shear Strength (Mpa)
Ti6Al4V	4.44×10^3	1660	610	7.955	8.6×10^{-6}	760
AA6060	2.70×10^3	657	934.8	167	23.4×10^{-6}	56

Microstructure morphologies of the as-received Ti6Al4V and AA6060 are shown in Figure 1. For Ti6Al4V, the α-Ti phase is shown in dark and in the form of equiaxed grains and the β-Ti phase is represented by the bright regions [22,23]. The microstructure contains a volume fraction of 93.9% of the α-Ti phase and 6.1% of the β-Ti phase. Aluminum alloy AA6060 heat treatable series alloys are the most widely used for the industrial applications. Louvis et al. [24] reported that the amount of solid solutions such as Mg and Si is very important for the strength of the welds as well as the size and precipitate particle distributions.

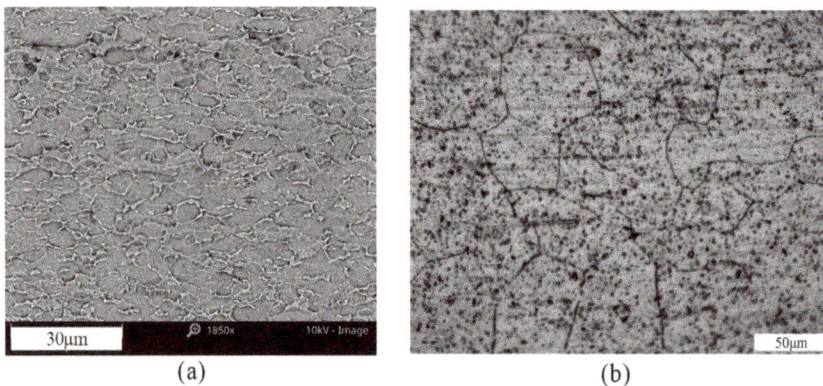

(a)

(b)

Figure 1. Microstructure observations of base materials: (**a**) Ti6Al4V; and (**b**) AA6060.

3. Experimental Procedures

3.1. Pulsed Nd:YAG Laser Welding

In this work, pulsed Nd:YAG laser system at Department of Mechanical Engineering, University of Aveiro, i.e., SISMA SWA300, (SISMA, Vicenza, Italy) was used to perform welding experiments, as illustrated in Figure 2. The process configurations for this apparatus are listed in Table 3.

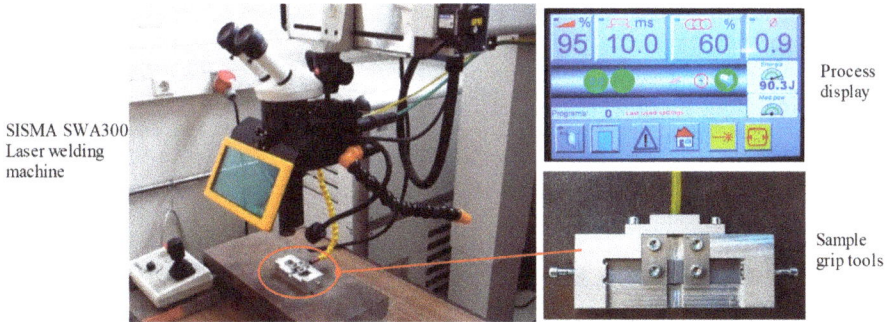

Figure 2. Experimental apparatus of pulsed Nd:YAG laser welding.

Table 3. Process configurations of pulsed Nd:YAG laser welding for SISMA SWA300.

No.	Process Configurations	Parameters
1	Source type	Nd:YAG (flash lamp)
2	Laser beam transport	Fiber-coupled
3	Average laser power	300 W
4	Peak laser power	12 kW
5	Wave length	1064 nm
6	Spot diameter	0.6–2.0 mm
7	Maximum pulse energy	100 J
8	Pulse range	0.2–25 ms
9	Argon shielding gas purity	99.99%
10	Maximum focusing optics	120 mm

To avoid an accidental movement of the welding sample, a characteristic sample fixture was developed for the stable and consistent laser welding tests. It is relatively simple and user-friendly for the fixture of sheet samples. The welding samples can be snapped into place and clamped with screws. A porthole inlet structure of assisted gas flow was designed to ensure welding samples have full gas shielding protection during the whole laser welding process. This can lead to the possibility to reduce the oxidation on the surfaces of the welding sample.

3.2. Interface Characterizations

In this work, typical lap joints with dissimilar lightweight Ti6Al4V/AA6060 were produced by pulsed Nd:YAG laser welding, which is basically categorized into fusion and solid state, respectively. At the stage of the fusion bonding, the heat source induces an inhomogeneous temperature field or a nonlinear variation of thermal field. At the stage of non-uniform solidification, the microstructural variations in the fusion zone, melted area and heat-affected zone dominate the evolutions in the of interfacial joints. The present task was focused on revealing the potential interface crack initiation and micro-structural characterizations, e.g., the changes with Ti/Al intermetallic layer in the dissimilar joint. Developments of possible metallic phases in the interface as well as crack initiation were also observed by means of scanning electron microscopy (SEM), while the distribution of chemical composition across the welded joint was determined by using energy dispersive X-ray spectroscopy (EDS). A comprehensive understanding of the microstructure mechanics of adhesion at the dissimilar metal interface represents the first step towards improvement of inter-metallic mechanical properties and weld quality.

Both Ti6Al4V and AA6060 are active lightweight alloys, and all sheet surfaces should be cleaned before laser welding. Oxidation films as-received on the surfaces of base materials were eliminated by polishing process. The sample was firstly cut to a suitable size using a diamond wire, and then progressive wet ground with 600–5000 grit sandpaper. Later, 2.5 μm diamond grinding cream and glue-free fabric cloth and Al_2O_3 polishing liquid were used for the final polish. Finally, the microstructural testing samples were cleaned by 75% alcohol solution and ultrasonic cleaning machine. After welding, typical cross-sections of the lap joints perpendicular to the welding direction were prepared for microstructure characterization near Ti/Al interface. Nova NanoSEM 230 field emission scanning electron microscopy (FE-SEM, FEI company, Hillsboro, OR, USA) and Quanta 250 energy-dispersive X-ray spectroscopy (EDS, Malvern Panalytical, Eindhoven, The Netherland) were used for microstructure analyses. The samples were ground and polished according to the standard metallographic methods. The surfaces of Ti6Al4V plates were cleaned in acidic solution, namely Keller reagents (1 mL HF + 1.5 mL HCl + 2.5 mL HNO_3 + 95 mL H_2O). Each erosion time was about 3–4 min. Then, 10% HF was used for further erosion of the sample. This persisted time is short about 5–7 s. The surfaces of AA6060 plates were cleaned in alkali liquor (NaOH 8 vol.%, H_2O 92 vol.%). Finally, all plates were cleaned using an ultrasonic cleaner, and then dried through air.

3.3. Mechanical Test

To investigate the pulsed Nd:YAG laser parameters on mechanical resistance of the Ti/Al lap joints, tensile shear testing of at least three samples for each process configuration was performed, as shown in Figure 3, using a Shimadzu AG 10 kN tensile test machine (Shimadzu, Japan). The sizes of weld pieces for Ti6Al4V and AA6060 were 30 mm × 14 mm × 0.8 mm and 35 mm × 14 mm × 1.5 mm, respectively.

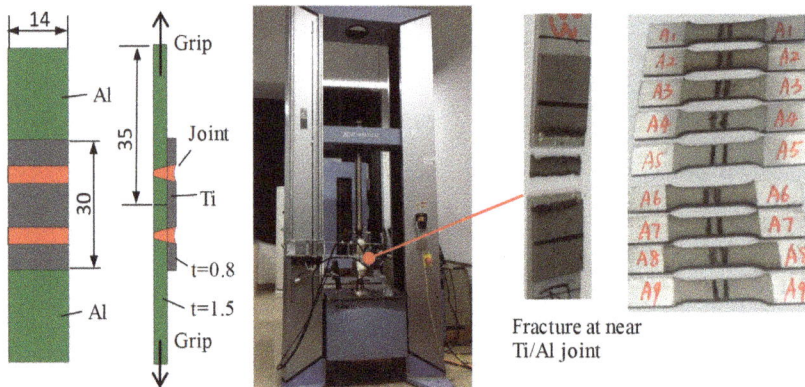

Figure 3. Experimental apparatus and fracture sample of tensile shear test.

4. Results and Discussions

4.1. Influence of Pulsed Nd:YAG Laser Parameters on Tensile Shear Strength

Since pulsed Nd:YAG laser welding is a complex multi-physical process coupling multi-factor interactive effects, the orthogonal experimental design method was introduced to determine the roles for deciding process parameters. The design of experiment was set up as L9_4_3 orthogonal tests, considering the power percent, duration, overlap and laser beam diameter. The selected factors of the orthogonal design and the corresponding results ARE shown in Table 4. K1, K2 and K3, respectively, indicate the average of factors in each level. It should be noted that the shear strength of Ti/Al lap joint is about up to 56–78% compared to that of the single AA6060. One of the reasons may be that the tested samples with Ti/Al lap joint were not heat treated. Furthermore, there were some potential microcracks in the interface, which led to the reduction of shear strength.

Table 4. Factors and results of L9_4_3 orthogonal tests.

Sample No.	Power Percent (%)	Duration (ms)	Overlap (%)	Laser Beam Diameter (mm)	Peak Shear Strength (MPa)
TA1	80	8	50	0.8	35.87
TA2	80	9	60	0.9	37.25
TA3	80	10	70	1.0	32.06
TA4	90	8	60	1.0	37.81
TA5	90	9	70	0.8	35.27
TA6	90	10	50	0.9	37.44
TA7	95	8	70	0.9	29.74
TA8	95	9	50	1.0	43.43
TA9	95	10	60	0.8	34.52
K1	35.060	34.473	38.913	35.220	
K2	36.840	38.650	36.527	34.810	
K3	35.897	34.673	32.357	37.767	
Variation Range (R)	1.780	4.177	6.556	2.957	
Sensitivity Order	4	2	1	3	
Optimal Value	90	9	50	1.0	

To achieve exhaustive high-quality lap joint after laser beam welding, it was necessary to investigate the welding process optimization. After some "trial and error" experimental tests, it was evaluated that the suitable power for the laser welding in this case should be between 80% and 95%. Then, further optimization was carried out to analyze the influence of Ti/Al interface characteristics. According to the results of orthogonal experiments, the sensitivity of laser beam power on peak shear strength of Ti/Al lap joint was the least significant. The overlap was related to the selection of pulse duration, spot size (diameter), and traverse velocity for a specific mean power. The suitable overlap should be from 50% to 70% for the studied Ti/Al lap joint by pulsed Nd:YAG laser beam welding. Although the selected overlap hardly affected penetration, it had a significant effect on peak shear strength of Ti/Al lap joint.

The tensile shear strength of the above tested samples was found obviously inferior to the resistance of aluminum alloy: the fracture happened at near Ti/Al welded joint. Many previous efforts [10,25] tried to increase the tensile strength of Ti/Al dissimilar butt joints, but the mechanical strength was still at the level of about 60–70% of aluminum alloy. The other key source of mechanical strength decrease was the thickness reduction of the intermetallic layer caused by the Si loss during the laser welding. That is why some researchers [1,26] attempted to use filler wire containing high silicon content for changing the intermetallic type and obtaining a beneficial effect on depressing the growth of intermetallic layer. The sensitivity order of the selected key process parameters on peak shear strength was: overlap, duration, laser beam diameter and power percent. One group of relative optimal process parameters was 90% power, 9 ms duration, 50% overlap and 1.0 mm laser beam diameter. However, under this optimal process configuration, there were still some welding defects, e.g., crack initiation, surface oxidation and splash, as shown in Figure 4. This is because the impact of laser welding affects the physical properties of the welded dissimilar materials. In other words, the aluminum base material encompasses inherent characteristics including oxide surface films, low absorptivity to laser beam, low boiling point elements and a tendency to form low melting constituents [27]. Therefore, it was essential to further investigate the interface characteristics associated with the microstructural evolution.

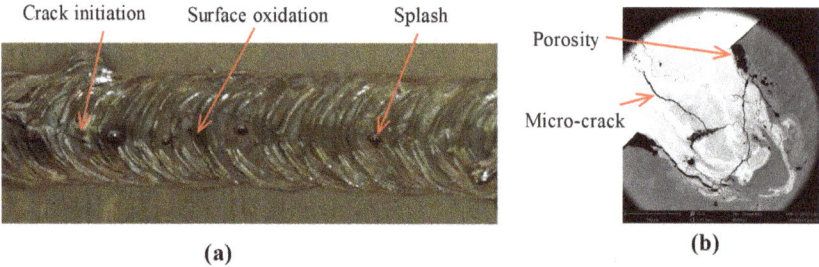

Figure 4. Welding-induced defects on Ti/Al lap joint by pulsed Nd: YAG laser welding: (**a**) macro-defects; and (**b**) micro-defects.

4.2. Estimation of Interfacial Crack Initiation of Dissimilar Ti6Al4V/AA6060 Lap Joint

The major problem with dissimilar welds originated from the different thermal expansion and contraction, leading to the reduction of the jointing properties after welding. The microstructure mechanics of adhesion (e.g., chemical potential) at the dissimilar metal interface and the non-homogeneous crack initiation under different process conditions were not ensured. Ti/Al hybrid structures have advantage in comparison to single material for both performance and lightweight requirements. Since titanium and aluminum have low inter-solubility and brittle intermetallic, they easily form interface crack during thermal welding, which would seriously degrade properties of the Ti/Al joint. Thus, it is vital to produce sound Ti/Al weld joints without defects.

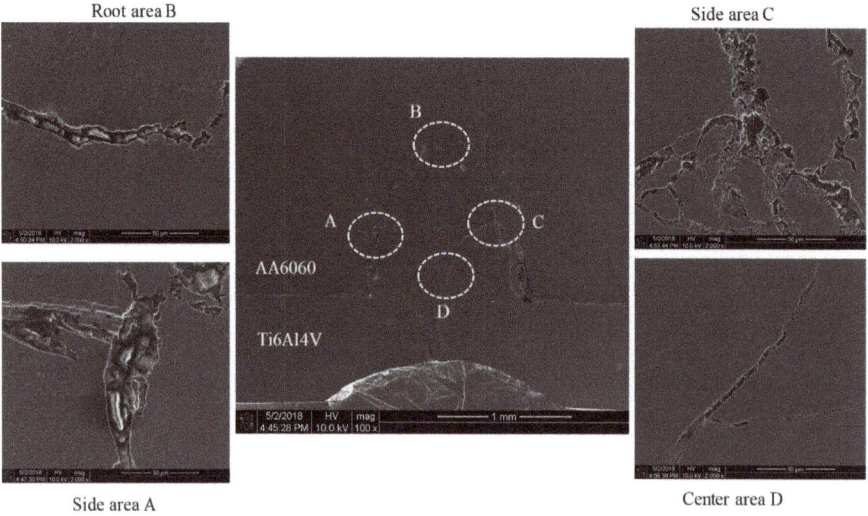

Figure 5. Morphology and microstructure of dissimilar Ti/Al lap joint: interface at different areas.

Typical cross-section of the lap joint with almost full penetration is illustrated in Figure 5. Numerous cracks crossing interface are visible. It can be supposed that they generated or formed immediately after the solidification of the weld as a result of local accumulation of residual stress. In other words, the shrinkage of the melt served as the initiator of interface cracks. Microstructures of weld zone near Ti/Al interface at different locations are marked as A–D zones. Because pulsed Nd:YAG laser welding process has high temperature gradient in the thickness direction, weld thermal cycle suffered at Ti/Al interface as top/side and bottom/root of the lap joint were different. Due to direct heating by laser beam, titanium material was fully melted and partially mixed with base aluminum in the interface. Since the Ti/Al intermetallic layers have intrinsic brittleness, cracks could initiate easily with welding-induced residual stress. Eventually, formation of brittle interfacial phases would affect properties of the lap joint for having an evident crack sensitivity. In the side areas, Zones A and C, peak temperature at Ti/Al interface was too high, and titanium as fully melted. A thick serrate reaction layer was generated. In the root area, Zone B, due to insufficient reactions between titanium and mixed aluminum, a thin serrate reaction layer was observed. In the center area, Zone D, because titanium was not directly combined with aluminum, there are almost no obvious reaction layers.

The lap joint was cut from the Ti/Al interface, and then analyzed the interface microstructure by using SEM and XRD. The test results are shown in Figure 6. The potential phases TiAl, TiAl$_2$, and TiAl$_3$ were observed near the Ti/Al interface. Near the layer interface, there are two main zones: Al-rich and Ti-rich. Figure 6 shows that Al dominates on interface of Ti/Al lap joint. This means that phase change is situated mainly in Al-rich melted zone. However, sometimes it crosses interface and even touches Ti-rich melted zone as some quantities of intermetallic phase also present. For example, brittle TiAl$_3$ phase was found in low quantities, as shown in Figure 7. The combination of relatively fast flash speed and large laser power enables performing laser welding in capillary mode with short time.

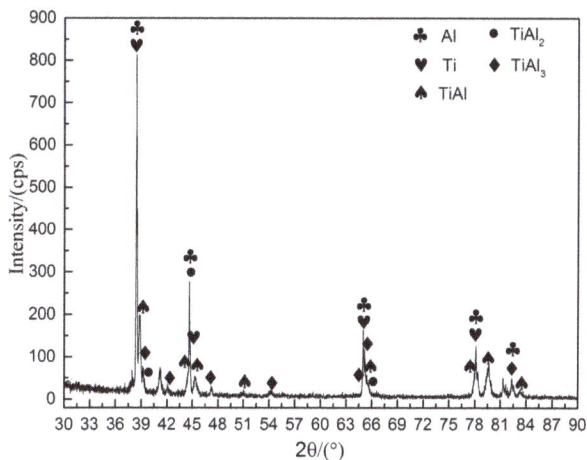

Figure 6. X-ray diffraction profiles of the Ti/Al lap joint interface.

The propagation path of phase change in the Ti/Al interface can be explained by mismatch in thermo-physical properties and by thermodynamic factor. As the Ti6Al4V has much higher fusion temperature than AA6060 (Table 2), the volume of melted aluminum was much larger than volume of melted titanium. Due to the high solidification rates proper to laser welding, the main solidification process was the local equilibrium at solid–liquid interface and the convective mixture between melted materials is poor. The previous study indicates that the solidification associated with thermodynamic properties of Al-Ti system begins from solid solutions and β-Ti phase. Besides the coexisting Al-rich liquids, the key change of Al in Ti-rich melted zone was dependent on lower Gibbs energy of (β-Ti) formation. In addition, diffusion coefficient of aluminum in liquid titanium overestimates that of titanium in liquid aluminum. Thus, compared to the concentration of Ti in Al-rich zone, Al in Ti-rich zone was more active during the separation of the materials through the contact interface. As for the stage of solidification, Ti-rich melted zone was relatively slowly depleted in Ti. Then, some (β-Ti) solid solutions having high Al content were formed. Consequently, in Ti-rich zone, these (β-Ti) solutions could be transformed into α-Ti and/or Ti3Al at the further cooling stage, which should be depending on Al content.

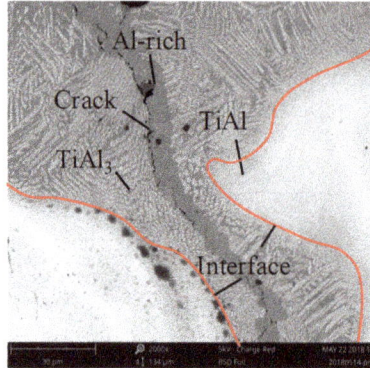

Figure 7. Propagation paths of potential phase distribution in the Ti/Al interface.

Based on the above analysis, it can be indicated that the rapid rate with respect to cooling or solidification rate seriously affected the crack initiation. It should be noticed that the time available for the residual liquid to refill and heal the initiated cracks may be mitigated by high cooling rates. Kanazawa [28] showed that the higher susceptibility of hot tearing and porosity could be obtained at lower duty cycles through pulsed Nd:YAG laser welding. In other words, the shorter is the average beam interaction time, the higher is the average temperature gradient of the interface and the faster is the cooling rate. Thus, it is generally recommended to shorten the off-time of the pulse or to use high duty cycle to reduce the solidification time of the molten pool. For the pulsed laser welding, some defects such as crack may be prevented if the next pulse occurs before the initiation of solidification cracking caused by the previous pulse. Another control strategy to reduce interfacial crack is to apply preheating to the work-piece because of an influence on cooling rate. For example, Ion [29] indicated that preheating to 500 °C could reduce solidification cracking in a series of aluminum alloys by reducing the cooling rate. It may be explained as it allows voids to heal before solidification is complete. Nowadays, the preheating control strategy is widely adopted during industrial applications.

4.3. Interface Microstructural Characterizations of Dissimilar Ti6Al4V/AA6060 Lap Joint

The distribution order of intermetallic in the selected reaction layers near Ti/Al interface was determined using EDS. The selected interface without micro-cracks, as illustrated in Figure 8, is helpful for the analyses of phase change and chemical potential. Chemical compositions of the selected locations marked P1–P5 on the interface were analyszd. As shown in Figure 9, the atomic ratio of Ti to Al at Point 1 was about 2:1. The layer was supposed to contain α-Ti solution and to some extent TiAl. Atomic percent of Ti at Point 2 was nearly equal to Al, thus the layer was mainly composed of intermetallic TiAl. At Point 3, atomic ratio of Ti to Al was about 1:2. The layer was mainly composed of intermetallic $TiAl_2$. At Point 4, atomic ratio of Ti to Al was about 1:4. The layer was composed of intermetallic $TiAl_3$ and α-Ti. The atomic percentages of Ti and Al at Point 5 were 94.71% and 5.01%, respectively. The results suggest the layer is α-Ti solution. Some other potential phases, e.g., Ti_5Si_3, might exist in the interface. Because $TiAl/TiAl_3$ is prone to be eutectic mixture structure with the potential Ti_5Si_3, it should be further investigated using nanoparticle observation. However, the authors think the amount of Ti_5Si_3 would be scant if it existed in the interface.

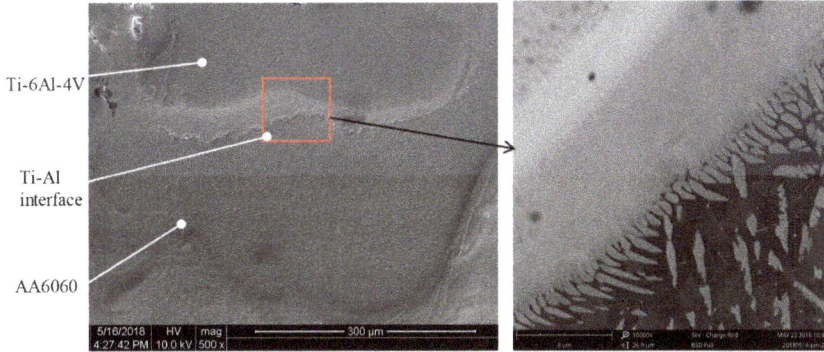

Figure 8. The selected Ti/Al interface without micro-crack for EDS observation.

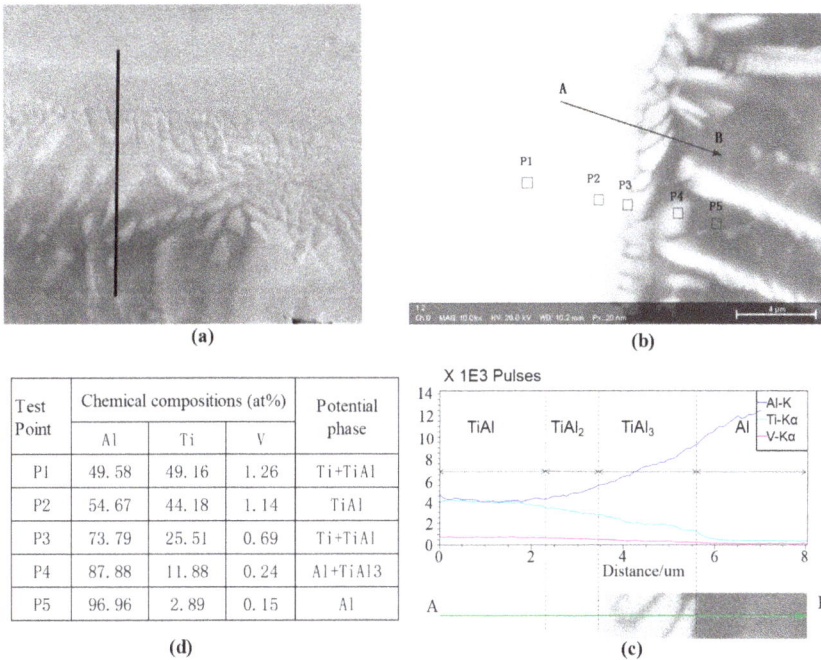

(a)

(b)

Test Point	Chemical compositions (at%)			Potential phase
	Al	Ti	V	
P1	49. 58	49. 16	1. 26	Ti+TiAl
P2	54. 67	44. 18	1. 14	TiAl
P3	73. 79	25. 51	0. 69	Ti+TiAl
P4	87. 88	11. 88	0. 24	Al+TiAl3
P5	96. 96	2. 89	0. 15	Al

(d)

(c)

Figure 9. EDS element analysis at selected points in the interface: (**a**) the selected interfacial area; (**b**) the selected points in the interface; (**c**) chemical compositions of the selected points; and (**d**) EDS element distribution along the selected interface.

Owing to different physical performance, including the coefficient of linear expansion and lattice structure, the Ti/Al intermetallic layers have different solidification processes. The divergence of two dissimilar base materials resulted in high edge dislocation density near interfaces of the Ti/Al multi-layers. Due to welding-induced residual stress [30,31], dislocations migrated and accumulated to the interface, and some micro-crack sources formed along the interface, as shown in Figure 10. Once welding micro-crack initiates in the melted zone, propagation tendency of welding micro-cracks would be along the interface with high dislocations density. Welding cracks mainly propagate along the longitudinal direction of the interface between TiAl layer and TiAl₃ layer. It should be noted that there are still some potential phase and discontinuous distribution in the micro-crack tip. It might be

Ti$_5$Si$_3$ or other brittle phases, which would mitigate the mechanical strength or performance of the welded joint.

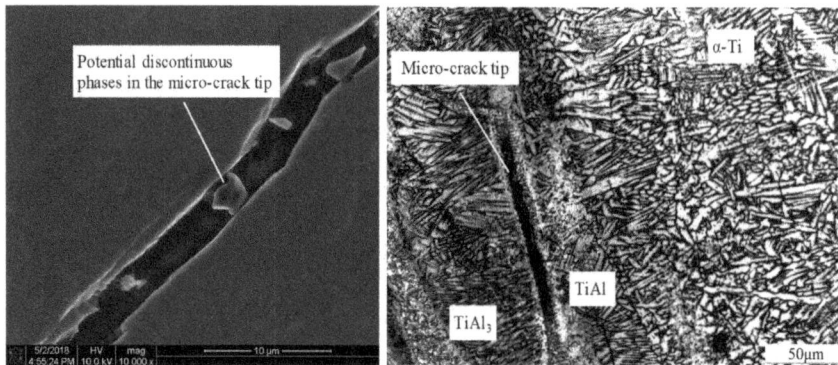

Figure 10. Welding micro-crack initiation in the Ti/Al interface.

It should be noted that the effect of the thickness of intermetallic layer on the mechanical strength or other properties of the Ti/Al joint is considerable. Although microstructures in the laser beam Ti/Al dissimilar joint by using silicon filler wire and the effect of intermetallic layer morphology in the fracture behavior of the joint were demonstrated by Chen et al. [1,8], the authors preferred to focus on the laser welding of dissimilar Ti/Al alloys without the addition of filler wire. This is because the formation of dissimilar Ti/Al alloys without additional fillers is prone to industrial applications. Of particular interest was continuous distribution of potential phase in the Ti/Al interface, as shown in Figure 11. It is not common but generated in the TA8 sample in this study. The sample was obtained under the following laser welding parameters: 95% power percent, 9 ms duration, 50% overlap and 1.0 mm laser beam diameter. It was found that the mechanical resistance of the Ti/Al lap joint is prominent compared to the other welding configurations. The continuous intermetallic layer can lead to a sound jointing performance under Nd:YAG laser welding process.

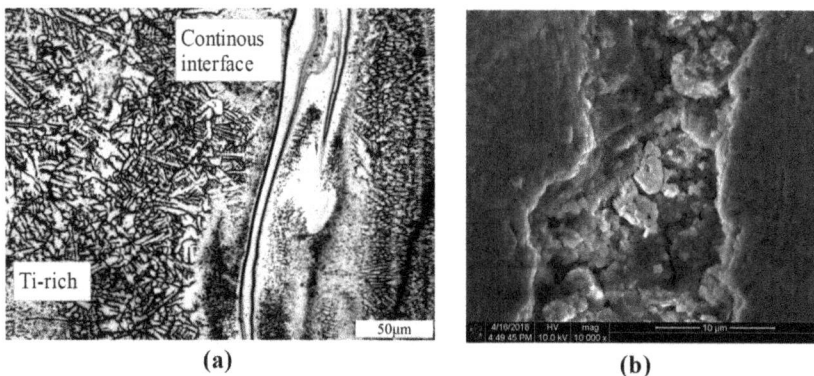

(a)

(b)

Figure 11. Continuous distribution of potential phase in the Ti/Al interface: (**a**) continuous intermetallic layer; and (**b**) potential continuous phases in the interfacial layer.

5. Conclusions

The interfacial characteristics of dissimilar Ti6Al4V/AA6060 lap joint produced by pulsed Nd:YAG laser beam welding was addressed. To better understand the interaction relationship of interfacial characteristics and process parameters, the corresponding experimental optimization design and microstructural observations were performed as well as the sensitivity analysis of welding-induced interfacial joining quality. The main conclusions can be drawn as follows:

(1) By using the orthogonal experimental design method, the sensitivity order of the selected key process parameters on peak shear strength was: overlap, duration, laser beam diameter and power percent. One group of relatively optimal process parameters was: 90% power, 9 ms duration, 50% overlap and 1.0 mm laser beam diameter.

(2) The potential phases TiAl, $TiAl_2$, and $TiAl_3$ were observed near the Ti/Al interface. The phase change was situated mainly in Al-rich melted zone. However, sometimes it crosses interface and even touches Ti-rich melted zone as some quantities of intermetallic phase were also present. The estimation of interfacial crack initiation was analyzed and discussed in detail.

(3) The discontinuous potential phases in the micro-crack tip can lead to mitigating the mechanical strength or performance of the welded joint, and a better jointing performance under pulsed Nd:YAG laser welding process may be obtained with the formation of continuous intermetallic layer.

Author Contributions: Methodology. X.X., A.P. and J.L.; investigation, X.X., A.P., and G.V.; data curation, X.X. and X.W.; writing—original draft preparation, X.X. and J.L.; writing—review and editing, A.P. and G.V.; supervision, A.P. and G.V.; project administration, A.P. and G.V.; and funding acquisition, X.X., A.P., G.V. and J.L.

Funding: This research was funded by the Portuguese Foundation of Science and Technology (SFRH/BPD/114823/2016 and UID/EMS/00481/2013-FCT), project CENTRO-01-0145-FEDER-022083, National Natural Science Foundation of China (No. 51705080 and No. 51805087), and Natural Science Foundation of Fujian Province (No. 2018J01761 and No. 2018J01764).

Acknowledgments: The authors wish to thank Rafael Gomes at the Department of Mechanical Engineering, University of Aveiro for the assistance with some dissimilar welding experiments.

Conflicts of Interest: The authors declare no conflict of interest.

References

1. Chen, S.; Li, L.; Chen, Y.; Huang, J. Improving interfacial reaction nonhomogeneity during laser welding-brazing aluminum to titanium. *Mater. Des.* **2011**, *32*, 3913–3919. [CrossRef]
2. Suhuddin, U.F.H.; Fischer, V.; Kostka, A.; Santos, J.F. Microstructure evolution in refill friction stir spot weld of a dissimilar Al–Mg alloy to Zn-coated steel. *Sci. Technol. Weld. Join.* **2017**, *22*, 658–665. [CrossRef]
3. Ege, E.S.; Inal, O.T. Stability of interfaces in explosively welded aluminium titanium laminates. *J. Mater. Sci. Lett.* **2000**, *19*, 1533–1535. [CrossRef]
4. Song, Z.; Nakata, K.; Wu, A.; Liao, J. Interfacial microstructure and mechanical property of Ti6Al4V/AA6061 dissimilar joint by direct laser brazing without filler metal and groove. *Mater. Sci. Eng. A* **2013**, *560*, 111–120. [CrossRef]
5. Yang, J.; Yu, Z.S.; Li, Y.L.; Zhang, H.; Guo, W.; Zhou, N. Influence of alloy elements on microstructure and mechanical properties of Al/steel dissimilar joint by laser welding/brazing. *Weld. World* **2018**, *62*, 427–433. [CrossRef]
6. Wang, T.H.; Shivakant, S.; Frank, M.; Mishra, R.S. Evolution of bond formation and fracture process of ultrasonic spot welded dissimilar materials. *Sci. Technol. Weld. J.* **2019**, *24*, 171–177. [CrossRef]
7. Tang, J.M.; Shen, Y.F. Effects of preheating treatment on temperature distribution and material flow of aluminum alloy and steel friction stir welds. *J. Manuf. Process.* **2017**, *29*, 29–40. [CrossRef]
8. Sambasiva Rao, A.; Madhusudhan Reddy, G.; Satya Prasad, K. Microstructure and tensile properties of dissimilar metal gas tungsten arc welding of aluminium to titanium alloy. *Mater. Sci. Technol.* **2011**, *27*, 65–70. [CrossRef]
9. Chang, S.Y.; Tsao, L.C.; Lei, Y.H.; Mao, S.M.; Huang, C.H. Brazing of 6061aluminum alloy/Ti6Al4V using Al–Si–Cu–Ge filler metals. *J. Mater. Process. Technol.* **2012**, *212*, 8–14. [CrossRef]
10. Vaidya, W.V.; Horstmann, M.; Ventzke, V.; Pertovski, B.; Kocak, M.; Kocik, R.; Tempus, G. Improving interfacial properties of a laser beam welded dissimilar joint of aluminum AA6056 and titanium Ti6Al4V for aeronautical applications. *J. Mater. Sci.* **2010**, *45*, 6242–6254. [CrossRef]

11. Chen, S.; Li, L.; Chen, Y.; Huang, J. Joining mechanism of Ti/Al dissimilar alloys during laser welding-brazing process. *J. Alloys Compd.* **2011**, *509*, 891–898. [CrossRef]
12. Majumda, B.; Galun, R.; Weisheit, A.; Mordike, B.L. Formation of crack-free joint between Ti-alloy and Al alloy by using a high-power CO_2 laser. *J. Mater. Sci.* **1997**, *32*, 6191–6200. [CrossRef]
13. Zhang, C.Q.; Robson, J.D.; Prangnell, P.B. Dissimilar ultrasonic spot welding of aerospace aluminum alloy AA2139 to titanium alloy TiAl6V4. *J. Mater. Process. Technol.* **2016**, *231*, 382–399. [CrossRef]
14. Casalino, G.; Mortello, M.; Peyre, P. Yb–YAG laser offset welding of AA5754 and T40 butt joint. *J. Mater. Process. Technol.* **2015**, *223*, 139–149. [CrossRef]
15. Fabbro, R. Developments in Nd–Yag laser welding. In *Handbook of Laser Welding Technologies*; Katayama, S., Ed.; Woodhead Publishing Limited: Oxford, UK, 2013; pp. 47–72.
16. Ren, J.W.; Li, Y.J.; Feng, T. Microstructure characteristics in the interface zone of Ti/Al diffusion bonding. *Mater. Lett.* **2002**, *56*, 647–652.
17. Tomashchuk, I.; Sallamand, P.; Jouvard, J.M.; Grevey, D. The simulation of morphology of dissimilar copper–steel electron beam welds using level set method. *Comput. Mater. Sci.* **2010**, *48*, 827–836. [CrossRef]
18. Tomashchuk, I.; Sallamand, P.; Andrzejewski, H.; Grevey, D. The formation of intermetallics in dissimilar Ti6Al4V/copper/AISI 316 L electron beam and Nd:YAG laser joints. *Intermetallics* **2011**, *19*, 1466–1473. [CrossRef]
19. Oliveira, A.C.; Moreira, A.F.R.; Mello, C.B.; Riva, R.; Oliveira, R.M. Influence of Si coating on interfacial microstructure of laser joining of titanium and aluminium alloys. *Mater. Res.* **2018**, *21*, e20161109. [CrossRef]
20. Xue, X.; Liao, J.; Vincze, G.; Pereira, A.B. Control strategy of twist springback for aluminium alloy hybrid thin-walled tube after mandrel-rotary draw bending. *Int. J. Mater. Form.* **2018**, *11*, 311–323. [CrossRef]
21. He, P.G.; Chen, K.; Yu, B.; Yue, C.Y.; Yang, J.L. Surface microstructures and epoxy bonded shear strength of Ti6Al4V alloy anodized at various temperatures. *Compos. Sci. Technol.* **2013**, *82*, 15–22. [CrossRef]
22. Squillace, A.; Prisco, U.; Ciliberto, S.; Astarita, A. Effect of welding parameters on morphology and mechanical properties of Ti6Al4V laser beam welded butt joints. *J. Mater. Process. Technol.* **2012**, *212*, 427–436. [CrossRef]
23. Vaithilingam, J.; Goodridge, R.D.; Richard, J.M.; Hague, R.J.M.; Christie, S.D.R.; Edmondson, S. The effect of laser remelting on the surface chemistry of Ti6Al4V components fabricated by selective laser melting. *J. Mater. Process. Technol.* **2016**, *232*, 1–8. [CrossRef]
24. Louvis, E.; Fox, P.; Christopher, J.; Sutcliffe, C.J. Selective laser melting of aluminium components. *J. Mater. Process. Technol.* **2011**, *211*, 275–284. [CrossRef]
25. Moller, F.; Grden, M.; Thomy, C.; Vollertsen, F. Combined laser beam welding and brazing process for aluminium titanium hybrid structures. *Phys. Procedia* **2011**, *12*, 215–223. [CrossRef]
26. Saida, K.; Ohnishi, H.; Nishimoto, K. Laser brazing of TiAl intermetallic compound using precious brazing filler metals. *Weld. World* **2015**, *59*, 9–22. [CrossRef]
27. Cam, G.; Koçak, M. Progress in joining of advanced materials Part 2: Joining of metal matrix composites and joining of other advanced materials. *Sci. Technol. Weld. Join.* **1998**, *3*, 159–175. [CrossRef]
28. Kanazawa, H. Welding performance of high power YAG lasers in aluminium alloys. *J. Light Met. Weld. Constr.* **1997**, *35*, 10–15. [CrossRef]
29. Ion, J.C. Laser beam welding of wrought aluminium alloys. *Sci. Technol. Weld. Join.* **2013**, *5*, 265–276. [CrossRef]
30. Eisazadeh, H.; Bunn, J.; Coules, H.E.; Achuthan, A.; Goldak, J.; Aidun, D.K. A residual stress study in similar and dissimilar welds. *Weld. J.* **2016**, *95*, 111–119.
31. Liu, F.; Liu, Y.; Wu, Y.C. Effect of lattice matching degree and intermetallic compound on the properties of Mg/Al dissimilar material welded joints. *Sci. Technol. Weld. Join.* **2017**, *22*, 719–725. [CrossRef]

metals

MDPI

Article

Welding of Dissimilar Steel/Al Joints Using Dual-Beam Lasers with Side-by-Side Configuration

Li Cui *, Hongxi Chen, Boxu Chen and Dingyong He *

College of Materials Science and Engineering, Beijing University of Technology, 100# Pingleyuan, Chaoyang District, Beijing 100124, China; chx@emails.bjut.edu.cn (H.C.); thecbx92@emails.bjut.edu.cn (B.C.)
* Correspondence: cuili@bjut.edu.cn (L.C.); dyhe@bjut.edu.cn (D.H.);
  Tel.: +86-10-67932523 (L.C.); +86-10-67932168 (D.H.)

Received: 13 November 2018; Accepted: 1 December 2018; Published: 4 December 2018

Abstract: Welding of dissimilar steel/Al lapped joints of 1.5 mm in thickness was carried out by using dual-beam laser welding with side-by-side configuration. The effect of the major process parameters including the dual-beam power ratio of (Rs) and dual-beam distance (d_1) on the steel/Al joint characteristics was investigated concerning the weld shape, interface microstructures, tensile resistance and fracture behavior. The results show that dual-beam laser welding with side-by-side configuration produces soundly welded steel/Al lapped joints free of welding defects. The processing parameters of Rs and d_1 have a great influence on the weld appearance, the weld penetration in the Al alloy side (P2) and the welding defects. Variation in the depth of the P2 and the locations at the Al/weld interface cause heterogeneous microstructures in the morphology and the thickness of the intermetallic compound (IMC) layers. In addition, electron back scattered diffraction (EBSD) phase mapping reveals that the IMC layer microstructures formed at the Al/weld interface include the needle-like θ-Fe_4Al_{13} phases and compact lath η-Fe_2Al_5 layers. Some very fine θ-Fe_4Al_{13} and η-Fe_2Al_5 phases generated along the weld grain boundaries of the steel/Al joints are also confirmed. Finally, there is a matching relationship between the P2 and the tensile resistance of steel/Al joints, and the maximum tensile resistance of 109.2 N/mm is obtained by the steel/Al joints produced at the Rs of 1.50 during dual-beam laser welding with side-by-side configuration. Two fracture path modes have taken place depending on the P2, and relatively high resistance has been achieved for the steel/Al joints with an optimum P2.

Keywords: dual-beam laser welding; steel/Al joint; side-by-side configuration; tensile resistance; EBSD phase mapping

1. Introduction

Joints between dissimilar metals are particularly common in components used in the power generation, chemical, petrochemical, nuclear, and electronics industries [1]. Current and potential dissimilar welding applications in the automotive industry include wind shield frame, center pillar, bumper reinforcement, and floor pan, among others [2]. With the increased use of dissimilar metal parts in the industry applications, joining of dissimilar joints has become increasingly important. The total weight reduction of the composite components fabricated by dissimilar joining of steels to Al alloys makes it attractive to various industries, such as the automotive [2–5] and shipbuilding industries [6–8]. Several welding techniques for joining dissimilar steel and Al metals have been reported. For example, Explosion Welding (EW) technique is used for successful joining the bimetallic plates in shipbuilding applications [7–9]. However, EW can produce high quality joints [10–12], but the cost effectiveness and mass efficiency are thus reduced in such structures [6]. Fusion-based welding processes have been recently investigated due to the flexibility compared with the EW. It has been reported that it is still a large challenge in the field of fusion-based welding due to the huge disparity in thermal-physical properties between steels and Al alloys [2–4]. The main issue associated with welding of steel/Al joints is the formation of the brittle intermetallic compound (IMC) phases as a result of the reaction between iron (Fe) and aluminum (Al), which is detrimental to the mechanical properties of the steel/Al joints [5,13]. The IMC layer thickness at the interface is generally accepted with less than 10 μm to avoid the degradation of the strength of the steel/Al joints [5,13,14].

As the formation of the IMC layers is mainly controlled by the temperature and the time [15], the welding process should have simultaneously low heat input and high cooling rate [13] for successfully joining of the steel/Al joints. In this respect, laser welding offers some distinct advantages over the convention arc welding, such as high energy density, controllable heat input, accurate laser beam location, small heat-affected zones, high welding speed, to meet the increasing demands for high performance of welding of steel/Al joints [16,17]. During laser welding of steel/Al joints, three welding processes are reported involving reactive wetting, welding-brazing and keyhole welding [18]. Among these three methods, current state-of-the-art of the welding process has the main focus on welding-brazing for dissimilar metal joints, and it has been proved as one of the most effective welding methods for dissimilar metal parts [19,20]. However, the good wettability of molten Al on the solid steel [21] has been difficult to obtain, because factors such as unequal temperature distribution, rapid cooling rate, and deficient heating during laser welding-brazing deteriorated the wetting and spreading of liquid Al alloy on the steel surface [22]. To improve the wettability, a variety of filler materials or specific techniques have normally been required, such as the use of chemical flux, brazing in vacuum or inert gas atmosphere, surface preparation [21,23], which resulted in additional costs generated by the flux application and the subsequent cleaning necessary to remove all residues detrimental to corrosion resistance [21]. To avoid the wettability problems, laser keyhole welding could be utilized [5,17,21], which provides very short interaction times between liquid steel and liquid Al to limit IMC phases without any filler materials [24]. This method has been of important technological interest, because it allows a reduction in joint preparation time and has fewer parameters to control, making it easier to obtain reproducible and stable results [24]. However, the laser keyhole welding presents issues related to the instability of the keyhole, which lead to excessive welding defects, such as pores and spatters [25].

To overcome the disadvantage of single-beam laser welding, dual-beam laser welding has been proposed as one of the alternative methods in some earlier research works for welding of aluminum alloys, titanium alloys, steels [26–28], etc. The original idea behind this approach, namely utilizing two individual focal spots at the workpiece instead of single beam, is to shape the keyhole by appropriate double-focus geometry such that the tendency of a keyhole collapse is considerably reduced [29]. During the dual-beam laser welding, the dual beams are arranged in either tandem or side-by-side (as shown in Figure 1a) in dual-beam laser processing [28,30]. The two beams are arranged preferably in a way that they can compensate and benefit from each other. The dual-beam configuration has been demonstrated as a key factor affecting strongly the temperature and the time at the interface of the steel/Al joints [28,31]. Dual laser beams arranged in tandem have been reported to provide benefits over conventional single-beam laser welding such as improved weld quality [26–28,32]. A study on using side-by-side laser beams for improved fit-up tolerance has been reported in welding tailored blanks [33]. The previous studies demonstrated that there was a significant difference in process efficiency between the two configurations regarding the molten seam volume per energy unit (mm^3/kJ): the molten area of the cross section in tandem configuration—meaning the second laser was following the first one—was considerably smaller than that in the side-by-side configuration [27,29].

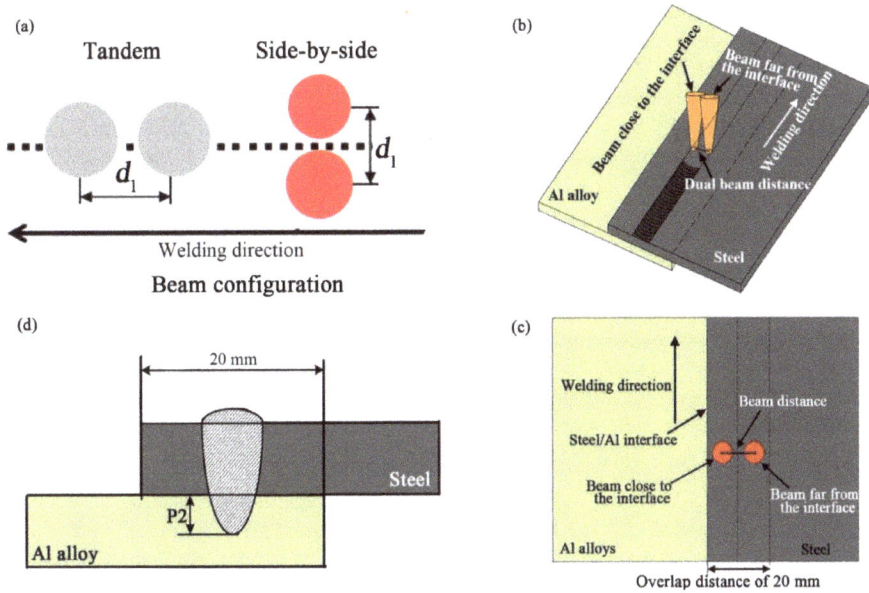

Figure 1. Schematic diagram of dual-beam laser welding of steel/Al joints: (**a**) Tandem beam and side-by-side configuration [28,30]; (**b**) Setup of the side-by-side configuration; (**c**) Main processing parameters with side-by-side configuration; (**d**) Dimension of the cross section of the welds.

Most recently, dual-beam laser welding has been introduced for dissimilar welding of steel/Al joints. Laukant et al. [34] reported that the melt flow could be controlled more effectively and the wetting behavior could be improved by preheating the zinc-covered steel sheet surface with a second laser beam when joined zinc-coated steel and Al sheets in overlap geometry. Shi et al. [35] demonstrated that the application of a dual-beam laser keyhole welding could generate an acceptable steel/Al joint and effectively reduced the presence of the welding defects. Chen et al. [36] found that the dual-beam laser welding exhibited better process stability leading to better weld appearance, and bigger effective joining width, which enhanced tensile capacity. Our previous study on dual-beam laser welding of steel/Al joints with tandem configuration demonstrated that soundly welded steel/Al joints had been achieved by using dual-beam laser keyhole welding at optimum welding conditions [37]. The above studies show that dual-beam laser welding of steel/Al joints maintained the key advantages of laser keyhole welding and even improved efficiently the mechanical properties of the steel/Al joints. However, these studies reported on dual-beam laser welding of steel/Al joints were limited to the tandem configuration. In addition, one new piece of research investigated the effect of dual-beam modes between tandem and side-by-side configuration on welding of steel/Al joints [38], indicating that the side-by-side configuration offered some superiority over the tandem one. This cited study has yielded some important knowledge on the possible use of dual-beam laser welding of steel/Al joints with side-by-side configuration, but the relationship between the welding parameters and weld shape have not been yet examined. However, it is important that details of processing parameters on the weld shape during the dual-beam laser welding with side-by-side configuration be well defined.

Compared to the single-beam laser welding, the dual-beam welding process provides additional process parameters, which allows for affecting the weld shape and welding quality. The dual-beam power ratio (Rs) and dual-beam distance (d_1) are the most important parameters. Thus, the effect of the process parameters on the steel/Al joint characteristics is essential to well understand the dual-beam laser welding of steel/Al joints. However, studies on the effect of process parameters on the weld shape of the steel/Al joints are scarce. Details of the effect of process parameters on the steel/Al joint characteristics are still far from being completely understood.

In the present study, dual-beam laser welding of steel/Al lapped joints was conducted by means of side-by-side configuration. Effect of the Rs and d_1 on the weld shape, Al/weld interfacial microstructures, and tensile resistance and fracture behaviors of the steel/Al joints were investigated. Using EBSD analysis, the phase composition, grain shape and grain size of the Al/weld interface microstructures with different locations were studied.

2. Materials and Methods

Q235 low carbon steel and 6061 Al alloy sheets in the dimension of 150 mm × 100 mm × 1.5 mm were used in the present study. The nominal compositions of the Q235 steels and 6061 Al alloys are listed in Table 1. Before welding, the specimen surface was polished by angle grinder and was chemically cleaned with acetone to eliminate surface contamination. The specimens were arranged in a lapped configuration with the overlap distance of 20 mm, as shown in Figure 1b. Throughout the experiments, the welding operation was shielded by the trailing and back shielding gas supplied by purity argon at a flow rate of 20 L/min and 15 L/min.

Table 1. Nominal chemical composition of 6061 alloys and Q235 steels (wt %).

Materials	Mg	C	P	Ni	S	Mn	Cr	Fe	Si	Zn	Ti	Cu
6061	0.8–1.2	-	-	-	-	≤0.15	0.04–0.3	≤0.7	0.4–0.8	≤0.25	≤0.15	0.15–0.4
Q235	-	≤0.2	≤0.04	≤0.3	≤0.04	0.3–0.7	≤0.30	Bal.	≤0.35	-	-	≤0.30

Dual-beam laser welding was performed on steel/Al lapped joint using a 6 kW YLR-6000 Yb fiber laser (IPG Photonics, Oxford, MA, USA). This fiber lasers with an emission wave length of 1070 nm can deliver in continuous wave mode through an output fiber core diameter of 100 μm. The welding system was combined by a focal length of 200 mm collimating lens and a focal length of 200 mm focusing lens. The beam diameter of the laser beam at the focal point was 0.3 mm. The dual beams were obtained by using an optical prism put between the collimating lens and the focusing lens, and through it the single beam was split into two ones. The setup of the dual beams arranged in side-by-side configuration is illustrated in Figure 1b. The Rs was calculated by the ratio of the beam power close to the Al/weld interface to the far one, and the Rs of 0.25, 0.5, 0.67, 1.50, 2.00 and 4.00 was tested for the welding experiments, which was determined by the dual-beam laser power fraction of 20/80 33/66, 40/60, 60/40, 66/33, 80/20, respectively. As a result, the dual-beam lasers with whether the smaller or the larger energy fraction will have the relatively strong asymmetric energy distribution [29]. The d_1 was defined as the distance between the two focusing spot of the beams on the surface of the welding specimens. Based on the preliminary parametric study, the d_1 of 0.5 mm, 1.0 mm, 1.5 mm and 2.0 mm were examined in details. The other welding parameters, such as, the total laser power (P_L) of 3000 W, the travel speed (v) of 33 mm/s, and the beam defocusing distance (ΔZ) of 0 mm were kept constant.

After welding, the welded joints were cut transversely from the welds and prepared for metallographic observation through mounting, polishing, and etching with Keller's reagent (HF:HCl:HNO$_3$:H$_2$O = 1:1.5:2.5:95) for Al alloys and 4% Nital acid for the Q235 steels to display the weld shape. The P2 was the depth of the weld in the Al alloy side of the steel/Al joints as illustrated in Figure 1d. Microstructural analysis was performed using an optical microscope, scanning electron microscopy (SEM) and EBSD. The morphology and thickness of the IMC layers at the Al/weld interface were examined using SEM of QUANTA FEG 650 (FEI, Hillsboro, OR, USA). The EBSD analysis was conducted by a field emission gun Quanta FEI 650 (FEI, Hillsboro, OR, USA) SEM operated at 30 kV and 1.0 nm. The EBSD sample was oriented at 70° tilt in the chamber at a working distance in the range from 5 mm to 30 mm. Samples for EBSD with the size of 10 × 10 × 10 mm^3 dependent on the ability of the SEM's chamber were cut, mounted, grounded, mechanically polished and subsequently, electropolished by immersion in a 30% nitric acid in methanol solution cooled to −25 °C at a voltage of 20 V for 30 s. Orientation image map (OIM) and phase mapping were measured in a rectangular zone using a step size of 0.3 μm between two measurements.

The tensile tests of the prepared specimens at the *Rs* of 0.25~4.00 and the d_1 of 0.5~2.0 mm were performed in accordance with the standard of GB/T 2651-2001 using a MTS810 (MTS, Eden Prairie, MN, USA) testing machine having the maximum capacity of 100 kN operating in a stress control mode with a load rate of 1 mm/min at room temperatures. For the overlap configuration, the joint zone of the tensile specimens were rotated during mechanical test because two forces in the specimen were not in line, and thus a torque was generated [17,36]. In this case, the researchers did not evaluate whether the forces were shear or tensile, and the normalized stress at failure (based on bonded area or fracture area) was not calculated. The tensile property was supposed by expressing as the maximal load per millimeter, i.e., linear failure strength (N/mm) [17,24,36]. Therefore, in the present study, the tensile resistance was evaluated by means of the linear failure strength (N/mm) as the tensile property of the steel/Al joints. The average tensile resistance value was determined by means of tensile tests carried out on 3 specimens. Fracture path and the fracture surface morphology of the fractured joints were observed using SEM.

3. Results and Discussion

3.1. Weld Shape of the Steel/Al Joints

3.1.1. Effect of the Dual-Beam Power Ratio (*Rs*)

Figure 2 shows the weld appearances of the steel/Al joints produced by dual-beam laser welding with side-by-side configuration at different *Rs* varied from 0.25 to 4.00 while the d_1 of 1.0 mm, the P_L of 3000 W, the *v* of 33 mm/s, and the ΔZ of 0 mm were fixed. All the welds show regular ripples without any visible welding defects, such as cracks, pores, and so on. This was a benefit of the improved keyhole stability of dual-beam laser welding [34]. The welds exhibit uniform and smooth bead surface at the *Rs* of 0.25, 0.67, 1.50 and 4.00. However, at the *Rs* of 0.50 (33/66) and 2.00 (66/33), the regularity of weld and smoothness are slightly reduced, meaning the energy distribution of the dual beams at the *Rs* of 33/66 or 66/33 is not suitable for the weld appearance quality. Therefore, the *Rs* of 0.25 (20/80), 0.67 (40/60), 1.50 (60/40) and 4.00 (80/20) are appropriate for obtaining smooth welds of the steel/Al joints in dual-beam laser welding with side-by-side configuration.

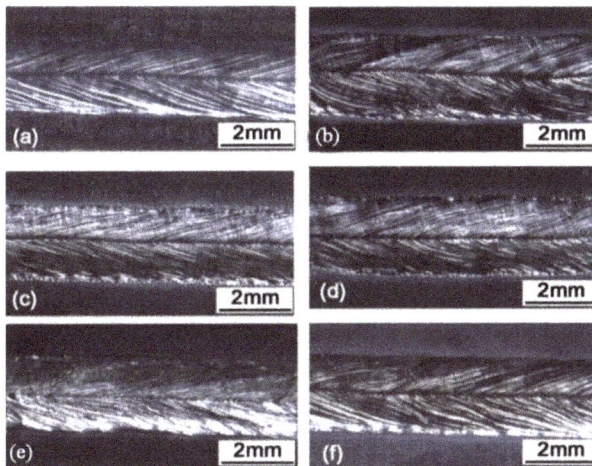

Figure 2. Effect of *Rs* on the weld appearances of the steel/Al joints produced by dual-beam laser welding with side-by-side configuration while the d_1 of 1.0 mm, the P_L of 3000 W, the *v* of 33 mm/s, and the ΔZ of 0 mm were fixed: (**a**) 0.25; (**b**) 0.50; (**c**) 0.67; (**d**) 1.50; (**e**) 2.00; (**f**) 4.00.

Figure 3 shows the effect of Rs on the cross sections of the steel/Al joints produced by dual-beam laser welding with side-by-side configuration while the d_1 of 1.0 mm, the P_L of 3000 W, the v of 33 mm/s, and the ΔZ of 0 mm were fixed. All the welds obtained at different Rs show a funnel weld shape, but produce obviously different P2. At the Rs of 0.25, 2.00 and 4.00, the welds completely penetrate the Al alloy side with an over-deep P2 of 1807 μm, 1741 μm and 1761 μm, respectively, as observed in Figure 3a,e,f. This means that the high penetration of P2 is obtained when the energy fraction of the dual beams is much more unequal, i.e., at the Rs of 0.25 (20/80), 2.00 (66/33) and 4.00 (80/20). However, at the Rs of 0.50 (33/66), 0.67 (40/60), 1.50 (60/40), the welds exhibit a relatively suitable P2 in the range of 93.2~554.2 μm due to the relatively equal energy distribution, as seen in Figure 3b–d. As a result, the P2 formed in the steel/Al joints is strongly affected by the Rs.

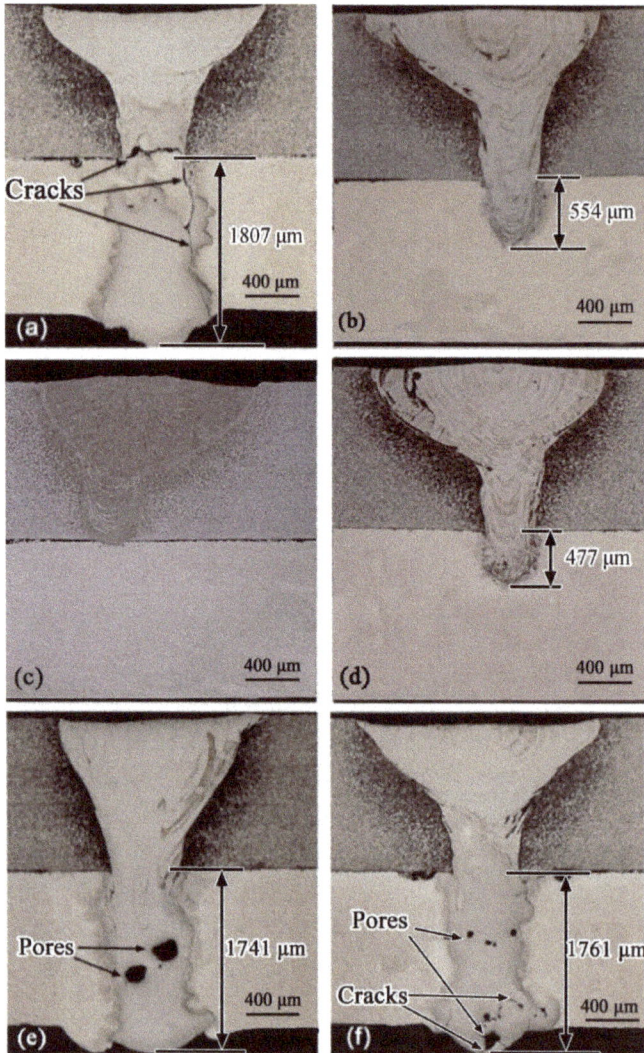

Figure 3. Effects of Rs on the cross sections of the steel/Al joints produced by dual-beam laser welding with side-by-side configuration while the d_1 of 1.0 mm, the P_L of 3000 W, the v of 33 mm/s, and the ΔZ of 0 mm were fixed: (**a**) 0.25; (**b**) 0.50; (**c**) 0.67; (**d**) 1.50; (**e**) 2.00; (**f**) 4.00.

Moreover, regarding the welding quality, it is interesting to find that increasing the P2 reduces the weld quality of the steel/Al joints. It is found that the local welding defects generated in the welds, such as pores and cracks, are also highly related to the over-deep P2 for the relatively unequal energy distribution of the Rs. No welding defects are observed in the welds with a suitable P2 at the Rs of 0.50, 0.67 and 1.50. This means that the steel/Al joints obtained at the relatively equal energy distribution exhibit no welding defects, and dividing the laser beam into unequal parts is not beneficial. Similar correlations between the energy ratio and welding defects were confirmed by Gref W et al. [29], who demonstrated that the cross sections obtained at the equal energy distribution of 50/50 show no porosity, while with a change in the distribution to 20/80, for example, large process pores are visible. Consequently, it can be concluded that the good weld shape should be obtained at the relatively equal power distribution Rs of 0.50, 0.67 and 1.50 during side-by-side dual-beam laser welding of the steel/Al joints.

3.1.2. Effect of the Dual-Beam Distance (d_1)

Figure 4 shows the weld appearance of the steel/Al joints produced by dual-beam laser welding with side-by-side configuration at different d_1 varied from 0.5 mm to 1.5 mm while the Rs of 1.5, the P_L of 3000 W, the v of 33 mm/s, and the ΔZ of 0 mm were fixed. At the small distance of 0.5 mm, as shown in Figure 4a, the dual-beam laser welding of the steel/Al joints provides an irregular-looking weld, but no visible welding defects of cracks and pores are observed. This is because the welding process became unstable when a single beam or a small beam distance was used [27]. Due to the small dual-beam distance, the keyhole-opening is similar to that of the single-beam welding, resulting in the instability of the keyhole [29]. With larger dual-beam distance, incasing d_1 from 0.5 mm to 1.5 mm leads to an improvement of the bead appearance, especially in view of the regularity, as shown in Figure 4b,c, namely, the weld appearances exhibit uniform and regular ripples without any visible cracks and pores, due to the good keyhole stability of the dual-beam laser welding with side-by-side configuration [27]. Figure 4d shows two separated welds are clearly observed at the maximum d_1 of 2 mm, because the keyhole begins to separate into two individual ones. As a result, the optimum d_1 for obtaining good bead appearance with side-by-side configuration should be in the range of 0.5~1.5 mm.

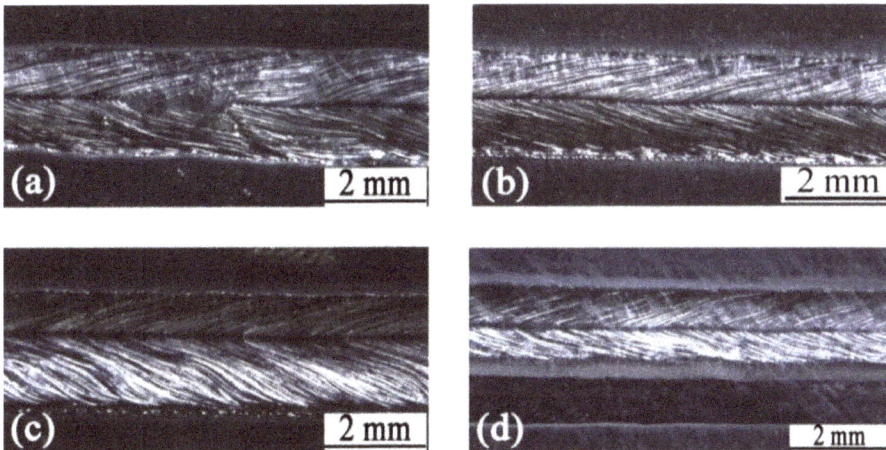

Figure 4. Effects of d_1 on the weld appearances of steel/Al joints produced by using dual-beam laser welding with side-by-side configuration while the Rs of 1.5, the P_L of 3000 W, the v of 33 mm/s, and the ΔZ of 0 mm were fixed: (**a**) 0.5 mm; (**b**) 1.0 mm; (**c**) 1.5 mm; (**d**) 2.0 mm.

Figure 5 shows the cross sections of the steel/Al joints produced by dual-beam laser welding with side-by-side configuration at different d_1 varied from 0.5 mm to 1.5 mm while the Rs of 1.5, the P_L

of 3000 W, the v of 33 mm/s, and the ΔZ of 0 mm were fixed. As seen in Figure 5a–c, each weld of the steel/Al joints has a funnel weld shape free of pores and cracks. At the maximum d_1 of 2.0 mm, as seen in Figure 5d, two segregated welds, including one with semicircle shape formed by conduction mode and one with funnel shape generated by a keyhole mode, are clearly observed. The two independent welds formed may be probably attributed to the formation of separated keyhole from one welding pool into two different parts [27]. In addition, the d_1 also has a significant influence on the P2. At the d_1 of 0.5 mm, the maximum P2 of 712.9 μm is achieved. Increasing d_1 from 0.5 mm to 1.0 mm decreases the P2 from 712.9 μm to 94.2 μm. At the d_1 of 2.0 mm, the minimum depth of P2 is zero, because of no penetration in the Al alloy side. As a result, the maximum P2 is achieved in the small d_1 of 0.5 mm, and enlarging d_1 from 0.5 mm to 2.0 mm significantly decreases the P2 of the welds of steel/Al joints produced by dual-beam laser welding with side-by-side configuration. This is because the interaction between the dual beams was weakening with the increase of dual-beam distances [27], and the laser welding efficiency was decreasing with larger dual-beam distance [29]. Therefore, the optimum d_1 for obtaining good weld shape is limited to 0.5 mm and 1.0 mm for side-by-side dual-beam laser welding of the steel/Al joints.

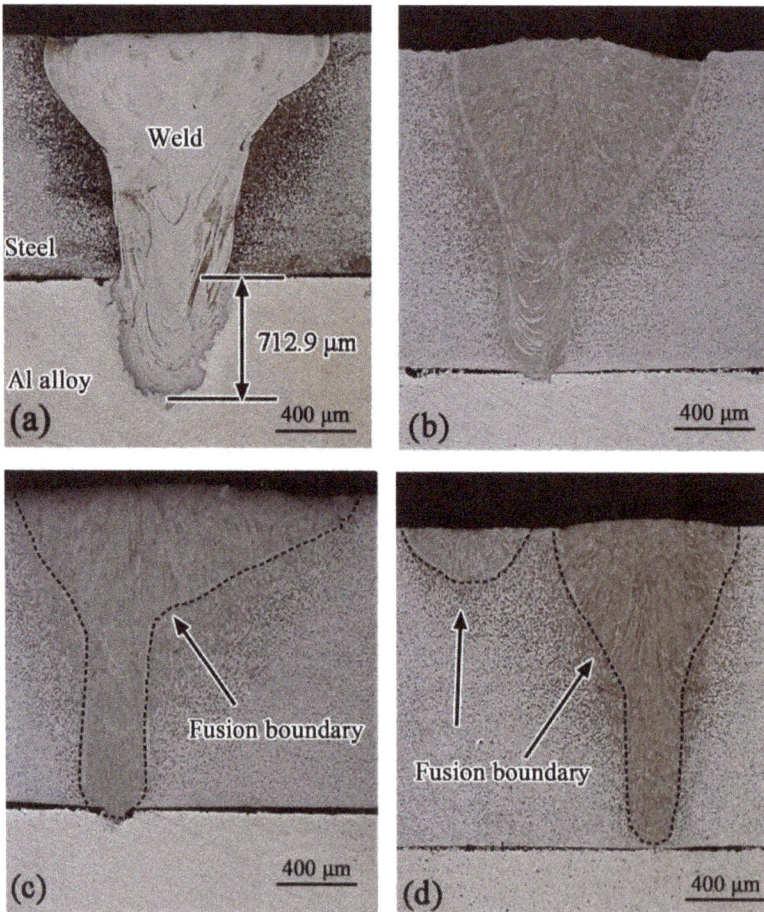

Figure 5. Effects of d_1 on cross sections of steel/Al joints produced by dual-beam laser welding with side-by-side configuration while the Rs of 1.5, the P_L of 3000 W, the v of 33 mm/s, and the ΔZ of 0 mm were fixed: (**a**) 0.5 mm; (**b**) 1.0 mm; (**c**) 1.5 mm; (**d**) 2.0 mm.

3.2. Microstructures of the Al/Weld Interface

3.2.1. Morphology and Thickness

Backscatter electron (BSE) analysis was performed to exhibit the Al/weld interface microstructures of the steel/Al joints produced by dual-beam laser welding with side-by-side configuration. Figure 6a shows the typical weld shape with a P2 of 1741 μm. In this case, the melted steel caused by dual-beam lasers is penetrated completely into the Al alloy side, leading to the pores and cracks generated inside the weld. High magnification of the Al/weld interface microstructures of different Zones marked by rectangles "b", "c" and "d" in Figure 6a are shown in Figure 6b–d. It is found that the irregular IMC layers formed at the Al/weld interface display some needle-like phases and island-shape structures surrounded by lath-like layer. The formation of the island-shape structures were probably due to the stirring effect of keyhole resulting from strong convection in the welding pool [25]. This Al/weld interface characteristics of the steel/Al joints is in agreement with that of the tandem configuration [36,37].

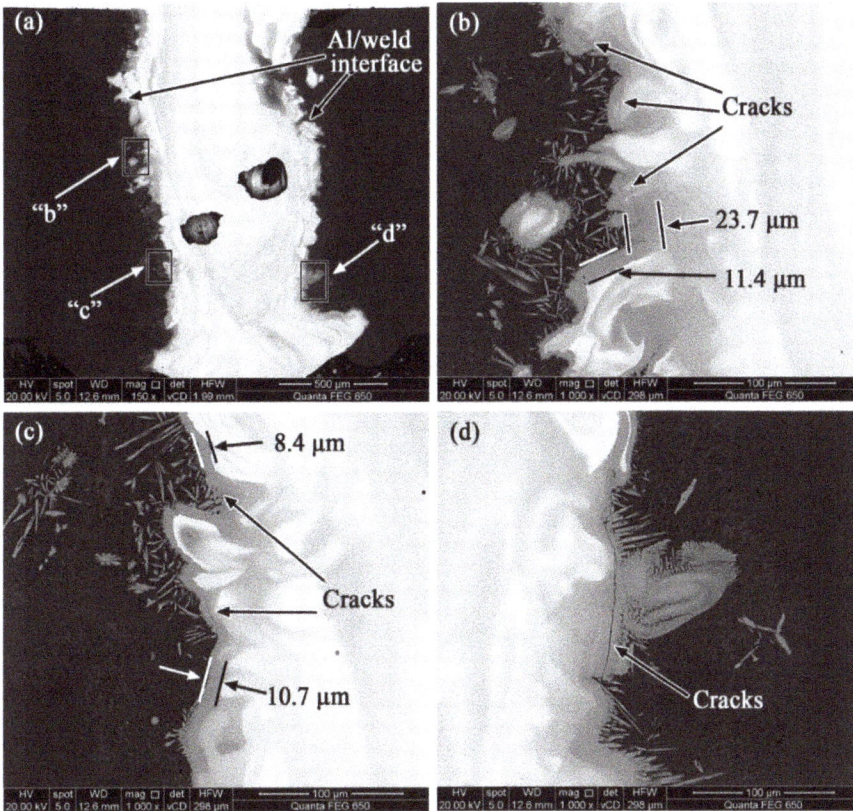

Figure 6. Backscatter electron (BSE) images of the Al/weld interface of the steel/Al joint having a P2 of 1741 μm produced by dual-beam laser welding with side-by-side configuration: (**a**) Cross section of the weld and the location of rectangle of Zone "b"–"d"; (**b**) Enlarged view of Zone "b"; (**c**) Enlarged view of Zone "c"; (**d**) Enlarged view of Zone "d".

The Al/weld interface microstructure in different locations consists of the island-shape structures and lath-like layer; however, the morphology and the lath-like layer thickness are different in each

Zone. Concerning the thickness of the IMC layers, the lath-like layer thickness was utilized as an evaluation indicator due to its compact and continuous morphology for a comparative study. At the upper part of Zone "b", as shown in Figure 6b, the lath-like layers are rather thick with a thickness of 11.4~23.7 μm, and some cracks are evidently observed. At the lower part of Zone "c", as presented in Figure 6c, the island-shape structures have relatively regular morphology surrounded by lath-like layers with thinner thickness of 8.4~10.7 μm. Figure 6d shows Zone "d" located at the other side of the Al/weld interface microstructures. This location has the same distance to the steel/Al interface as the Zone "c". It is seen that the thickness of the lath-like layers is similar to that of Zone "c", but the needle-like phases grown from the lath-like layers are much coarser. In addition, a long crack along the lath-like layers and cross the needle-like phases is clearly observed. Therefore, the upper part primarily displays irregular island-shape structures and thicker lath-like layers, whereas the lower Zones have fine needle-like phases, thinner lath-like layers directly influenced by the relatively small heat input as a result of the far distance from the steel/Al interface [39]. The variation in the morphology and thickness with the different locations at the Al/weld interface microstructure can be finally attributed to the temperature gradients and cooling rates.

BSE images of the Al/weld interface microstructures with a suitable P2 of 477 μm are presented in Figure 7. Figure 7a shows the typical weld shape formed in the Al alloy side free of any welding defects. The enlarged views of the Al/weld interface microstructures of the different Zones marked by rectangles "b", "c" and "d" in Figure 7a are shown in Figure 7b–d. At the upper part of Zone "b", a large number of needle-like phases and irregular island-shape structures surrounded by lath-like layers are found, whereas the thickness of the lath-like layers is relatively smaller in the range of 6.2~11.9 μm. This thickness is obviously thinner than that of the over-deep P2 shown in Figure 6. Figure 7c exhibits the Al/weld interface microstructures of the Zone "c" having the same distance to the steel/Al interface as Zone "b", but locates at the other side of the Al/weld interface. Not only is the quantity of the needle-like phases decreased, but also the thickness of the lath-like layers is significantly decreased to 2.7~4.1 μm. At the bottom part of Zone "d", as shown in Figure 7d, both the thickness of the lath-like layers and the needle-like phases are also decreased significantly. Therefore, it can be concluded that the P2 has a more significant influence on the morphology and the lath-like layer thickness, and controlling the P2 is effective to inhibit the formation of IMC layers at the Al/weld interface.

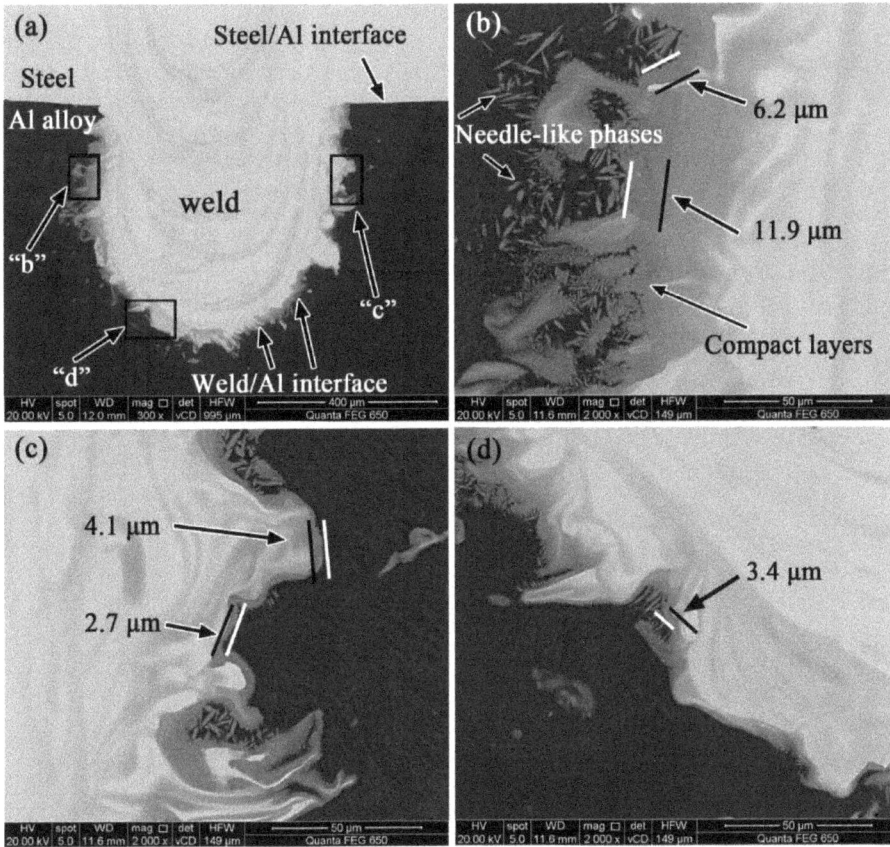

Figure 7. BSE images of the Al/weld interface microstructures of the steel/Al joint having a P2 of 477 μm produced by dual-beam laser welding with side-by-side configuration: (**a**) Cross section of the steel/Al joint and the location of Zones "b"–"d"; (**b**) Enlarged view of Zone "b"; (**c**) Enlarged view of Zone "c"; (**d**) Enlarged view of Zone "d".

3.2.2. Phase Identification

To identify the phase type of the needle-like phases and the lath-like layers formed at the Al/weld interface, EBSD phase mapping was performed at the different locations marked by rectangle Zone "a–d" in Figure 8e, in which the grain color specifies the phase type distribution according to the color indicated in the phase legend for the cubic symmetry. The phase mapping of Zones "a"–"c" include the Al alloy, the Al/weld interface and the weld region, and Zone "d" is the full weld exactly right across the steel/Al interface.

At the Al/weld interface, it is found that the blue island-shape structures are surrounded by the green lath-like layers adjacent to the blue welds, as presented in Figure 8a–c, and the red needle-like phases are dispersed in the Al alloy or grown from the green lath-like layers. According to the phase legend shown in the Figure 8, the green lath-like layers and the red needle-like phases are determined to be η-Fe_2Al_5 and θ-Fe_4Al_{13} phases. In addition, the phase mapping confirms that the variation in the location at the Al/weld interface results in the various morphology of the η-Fe_2Al_5 layers and θ-Fe_4Al_{13} phases and the different thickness of the η-Fe_2Al_5 layer in each zone. At the upper part of the Al/weld interface of Zone "a", the η-Fe_2Al_5 layers are rather thick, and a large number of the θ-Fe_4Al_{13} phases are confirmed. At the lower part of Zone "b", the weld region is found to be dominated by α-Fe phase with curved columnar grains and the less θ-Fe$_4$Al$_{13}$ phases and the thinner η-Fe$_2$Al$_5$ layers formed at the Al/weld interface have been proved. In particular, it is noted that some very fine θ-Fe$_4$Al$_{13}$ and η-Fe$_2$Al$_5$ phases distributed along α-Fe grain boundaries are observed inside the weld region, which is proved by further examining the Kikuchi diffraction patterns and the lattice constants. Figure 8c proves the highly reduced quantity of the θ-Fe$_4$Al$_{13}$ phases and the η-Fe$_2$Al$_5$ layer thickness are generated at the Al/weld interface at the bottom part of Zone "c". In addition, the fine θ-Fe$_4$Al$_{13}$ and Fe$_2$Al$_5$ phases are confirmed to generate along α-Fe grain boundaries inside the weld zone close to the Al/weld interface. Figure 8d presents the phase distribution of the weld Zone "d" across the steel/Al interface, exhibiting that the very fine θ-Fe$_4$Al$_{13}$ and η-Fe$_2$Al$_5$ phases are formed along α-Fe grain boundaries inside the weld zone. With these EBSD observations, it can be concluded that the θ-Fe$_4$Al$_{13}$ and η-Fe$_2$Al$_5$ phases are formed at the Al/weld interface, and the very fine θ-Fe$_4$Al$_{13}$ and η-Fe$_2$Al$_5$ phases are generated along α-Fe grain boundary inside the weld zone of the steel/Al joints.

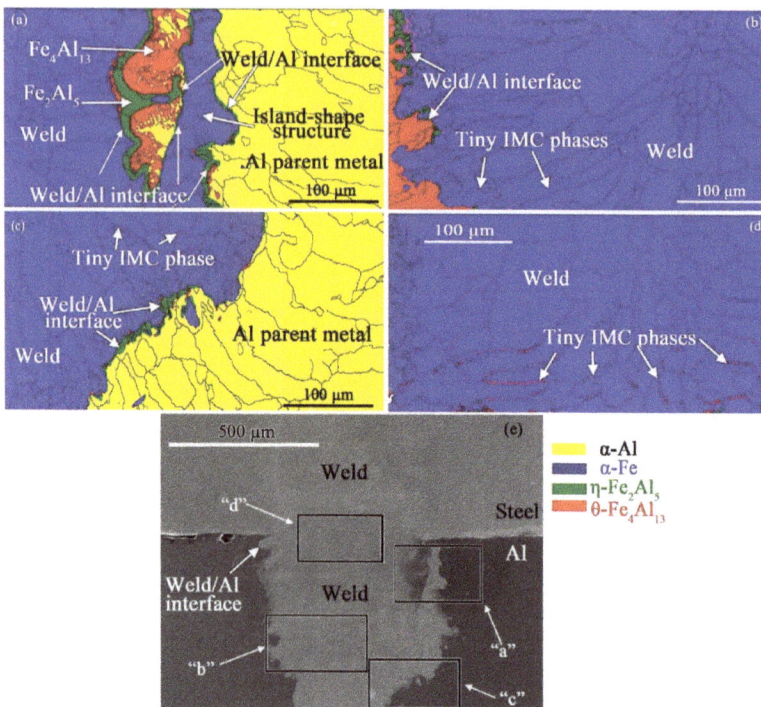

Figure 8. EBSD phase mapping images of the different zones in the steel/Al joints produced by dual-beam laser welding with side-by-side configuration: (**a**) Zone "a" marked in (**e**); (**b**) Zone "b" marked in (**e**); (**c**) Zone "c" marked in (**e**); (**d**) Zone "d" marked in (**e**); (**e**) Cross section of steel/Al joint and the locations of Zones "a–d".

This present study indicates that the θ-Fe$_4$Al$_{13}$ and η-Fe$_2$Al$_5$ phases are formed at the Al/weld interface, which is in good agreement with the previous studies [19,36,40]. However, the formation of the fine θ-Fe$_4$Al$_{13}$ and Fe$_2$Al$_5$ phases formed along α-Fe grain boundary has not been reported. The reasons for the fine IMC phases maybe partly correlated to the high Al content at the upper part of the weld zone due to the over-deep P2 in the Al alloy side [24]. In addition, the high vacancies are concentrated along the c-axis of the orthorhombic structure of the η-Fe$_2$Al$_5$ phase, and thus high content Al atoms can diffuse rapidly in this direction and even travel across the formed η-Fe$_2$Al$_5$

phase [41], which leads to the θ-Fe$_4$Al$_{13}$ phases formed in both sides of η-Fe$_2$Al$_5$ layer. When Al atoms cross η-Fe$_2$Al$_5$ phase, they continue to migrate along α-Fe grain boundaries in as much as grain boundary diffusion coefficient is much higher than bulk diffusion coefficient [42]. Al atoms diffuse along α-Fe grain boundary and react with Fe atoms existing in the α-Fe grain boundaries to form η-Fe$_2$Al$_5$ and θ-Fe$_4$Al$_{13}$ phases finally. This may explain why the fine θ-Fe$_4$Al$_{13}$ and η-Fe$_2$Al$_5$ phases are formed along α-Fe grain boundary inside the weld zones.

3.2.3. Grain Shape and Grain Size

The microstructures of the Al/weld interface and the weld for the steel/Al joints are also depicted by EBSD orientation image mapping (OIM), where the grain color specifies the orientation according to the coloring indicated in the orientation legend for the cubic symmetry. Hence, the grain shape was easily detected as the grain boundaries with angle misorientations larger than 5° were displayed, and the grain size was statistically estimated by means of EBSD software. Figure 9 displays the OIM of different locations at the Al/weld interface marked by rectangle Zones "a"–"c" in Figure 8e. It is found that the weld zone concludes some equiaxed structures and predominantly columnar grains. The Al base metal exhibits the much coarser columnar structures with an average grain size of 71.6 μm, which is noticeably coarser than those of the IMC grains formed at the Al/weld interface.

Considering the IMC grains at the Al/weld interface, the variation in locations causes the different size of the η-Fe$_2$Al$_5$ and the θ-Fe$_4$Al$_{13}$ grains. At the upper part of Zone "a", the η-Fe$_2$Al$_5$ grains are in shape of the continuous layer with an average grain size of 41.2 μm, and the needle-like θ-Fe$_4$Al$_{13}$ grains adjacent to the η-Fe$_2$Al$_5$ layers are in an average size of 6.1 μm. At the lower part of Zone "b", the η-Fe$_2$Al$_5$ and the θ-Fe$_4$Al$_{13}$ grains change into finer equiaxed morphology with a size of 3.6~24.3 μm. At the bottom part of Zone "c", the θ-Fe$_4$Al$_{13}$ and the η-Fe$_2$Al$_5$ grains are equiaxed with the smallest size of 2.14 μm and 5.6 μm. As a result, the finest θ-Fe$_4$Al$_{13}$ and η-Fe$_2$Al$_5$ grains has been obtained at the bottom part of the Al/weld interface, which undergoes higher cooling rates compared to the upper part. The noticeable decrease in grain size of the θ-Fe$_4$Al$_{13}$ and η-Fe$_2$Al$_5$ grains with increasing the distance to the steel/Al interface is probably attributable to the location-variable cooling rate experienced at the Al/weld interface. A relatively slower velocity of solidification in the upper part is due to a substantial heat for a longer period of time, whereas the bottom part experiences less heat input owing to the greater distance from the steel/Al interface [24,36]. Moreover, the fine θ-Fe$_4$Al$_{13}$ and η-Fe$_2$Al$_5$ grains formed inside the welds have an average size of 2.4~3.7 μm and 2.1~3.6 μm. Therefore, the IMC grains formed whether in the Al/weld interface or in the welds produced by dual-beam laser welding with side-by-side configuration are much finer than those of the welds and the Al base metal, which may be beneficial to improve the tensile resistance of the steel/Al joints. Now, it is a pity that the relationships between fine IMC microstructure and mechanical properties are not very clear. This observation and other aspects of the underlying relationships will be investigated further.

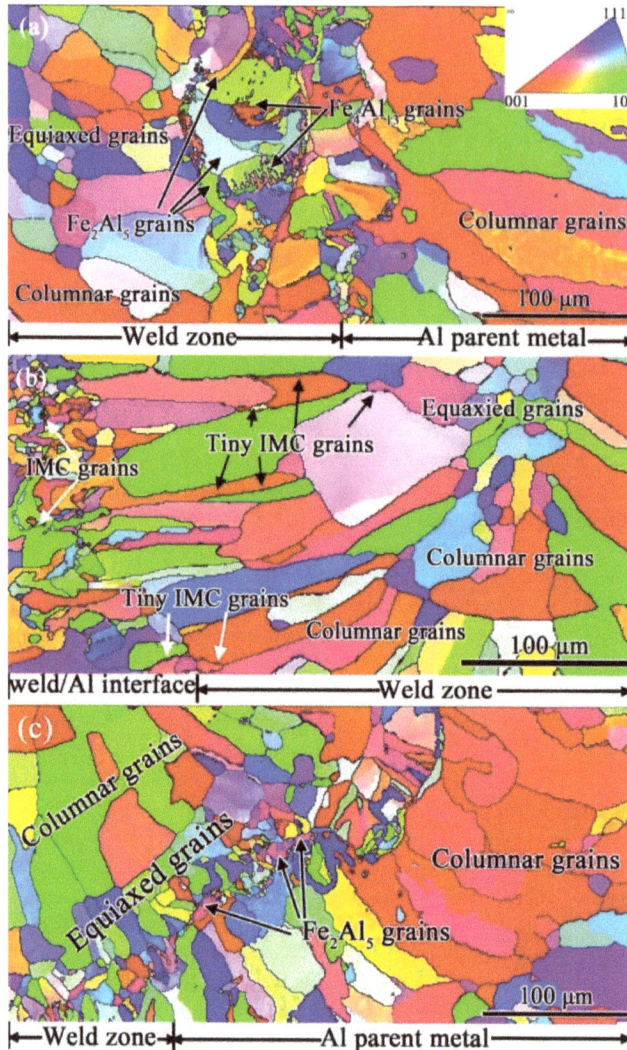

Figure 9. OIM of different zones of the steel/Al joints produced by dual-beam laser welding with side-by-side configuration: (**a**) Zone "a" marked in Figure 8e; (**b**) Zone "b" marked in Figure 8e; (**c**) Zone "c" marked in Figure 8e.

3.3. Tensile Resistance of the Steel/Al Joints

Figure 10a shows the tensile resistance of the steel/Al joints produced by dual-beam laser welding with side-by-side configuration at different Rs varied from 0.25 to 4.00. At the Rs of 0.25 and 4.00, the tensile resistance of the steel/Al joints is 48.4 N/mm, and 52.5 N/mm, which is rather low, probably due to the cracks and pores formed inside the weld zone associated with the over-deep P2 (Figure 3a,f). At the Rs of 0.50 and 1.50, the tensile resistance of the steel/Al joints is greatly increased to 84.8 N/mm and 109.2 N/mm, probably due to the suitable P2 (Figure 3b,d). Figure 10b shows the tensile resistance of the steel/Al joints with different d_1 varied from 0.5 mm to 2.00 mm. At the d_1 of 2.00 mm, tensile test of the steel/Al joints was not tested because no welded joints were obtained. It is found that increasing d_1 from 0.5 mm to 1.5 mm decreases the tensile resistance of the steel/Al joints from 100.2 N/mm to

43.3 N/mm. Thus, the maximum tensile resistance of 100.2 N/mm has been obtained at the d1 of 0.5 mm due to the optimal P2 (Figure 5a), and the lowest tensile resistance of 43.3 N/mm is obtained at the d_1 of 1.5 mm due to the insufficient P2 (Figure 5c). This indicates that the linear relationship between the P2 and the tensile resistance of the steel/Al joints. From these results, it is concluded that there is a matching relationship between the P2 and the tensile resistance of steel/Al joints, and the maximum tensile resistance of the steel/Al joints is obtained at the *Rs* of 1.50 produced by dual-beam laser welding with side-by-side configuration.

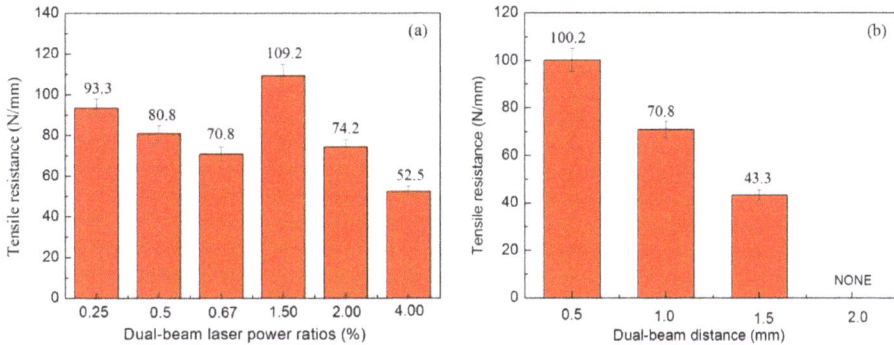

Figure 10. Effect of processing parameters on tensile resistance of the steel/Al joints produced by dual-beam laser welding with side-by-side configuration: (**a**) *Rs* varied from 0.25 to 4.00; (**b**) d_1 varied from 0.5 mm to 2.0 mm.

After the tensile test, all the steel/Al joints have been fractured at the Al/weld interface due to the brittleness of the IMC layers at the Al/weld interface. Figure 11 shows a typical strain-stress curve of the steel/Al joints obtained by three dual-beam distances with very small strain of the steel/Al joints, indicating a brittle fracture characteristic.

During the tensile test, the crack initiates from the steel/Al interface because of the higher stress concentration, and grows along the steel/Al interface to the weld border. Two different fracture propagation paths are observed depending on the P2. The first one occurred at the steel/Al joint with an over-deep P2 of 1741 μm, is shown in Figure 12. Figure 12a shows that the crack propagates along the Al/weld interface and in the weld zone above the steel/Al interface, indicating that both the Al/weld interface regions and the weld zone above the steel/Al interface are the weakening zones of the steel/Al joints. The weak weld zone above the steel/Al interface is probably due to the higher Al content resulted from the over-deep P2 [36], whereas the fractured Al/weld interface is associated with the high brittleness of the η-Fe_2Al_5 and θ-Fe_4Al_{13} phases. High magnification of the fracture path morphology of the Al/weld interface is shown in Figure 12b. It is indicated that remarkable different fracture morphologies are observed with various locations marked as rectangle "c" and "d" in Figure 12b along Al/weld interface. At Zone "c" near the steel/Al interface shown in Figure 12c, some cracks are observed across the needle-like θ-Fe_4Al_{13} phases. The fracture is occurred in the η-Fe_2Al_5 layer close to the weld, which results in lower tensile resistance of the steel/Al joints [13]. In this case, the crack path is relatively straight, which reveals the crack growth resistance in the η-Fe_2Al_5 layer is low. The η-Fe_2Al_5 layer at the Al/weld interfaces becoming a preferential cracking path are due to the cracks existed before tensile testing [24]. At Zone "d" far from the steel/Al interface, as shown in Figure 12d, only large needle-like θ-Fe_4Al_{13} phases are formed at the Al/weld interface free of cracks. At this location, a fairly jagged looking fracture path is observed and the fracture takes place along the Al/weld interface between the Al alloy and the η-Fe_2Al_5 layer, which results in higher joining strength [13]. Therefore, the fracture path varied with the morphology and the thickness of the IMC layer result in the tensile resistance variation with the location in the Al/weld interface. The factographic surface of the weld/Al interface marked by Zone "f" (Figure 12e) reveals a typical cleavage fracture mode, as shown in Figure 12f, with river pattern strips of particular orientation on the fracture surface. The η-Fe_2Al_5 layer was the most brittle region having the weakest bonding strength at the weld/Al interface of the steel/Al joints.

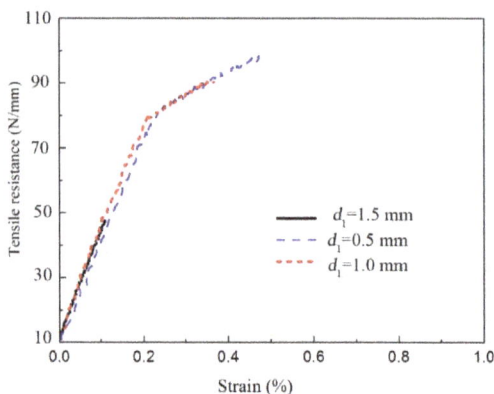

Figure 11. Stress-strain curves of the steel/Al joints obtained by dual-beam laser welding with side-by-side configuration for three different d_1.

For the second one, as shown in Figure 13a, the macrofracture profile of the steel/Al joints with an optimized P2 of 477 μm is obviously different from that of the over-deep P2. It is from the BSE images shown in Figure 13a found that the crack propagates in η-Fe_2Al_5 layer along Al/weld interface, and grows into the Al alloys at the bottom of the weld, and then connects to the fracture crack resulting in the final tearing of the steel/Al joint. This means that the weaken zones of the steel/Al joints turns into Al/weld interface regions and the Al alloy at the bottom of the weld, which induces the tensile resistance of the Al alloy at the bottom of the weld lower than the tensile strength of the weld zone in the upper part of the steel/Al joint. The fracture surfaces of Zones "b" and "c" marked by rectangle in Figure 13a are shown in Figure 13c,d, respectively. The fracture surface of the Zone "b" was full of the tearing edges and river patterns, whereas some ductile dimples were clearly observed at the fracture surface of Zone "c" indicating some plastic deformation occurred. Thus, the fracture mode of the steel/Al joint with the optimized depth of P2 was a mixed failure, which significantly enhanced its tensile resistance of the steel/Al joint. Similar fracture behavior of the steel/Al joints was reported by Chen et al. [17] during laser penetration welding an overlap steel-on-aluminum joint with/without Ni-Foil. They have demonstrated that the relatively high tensile strength was achieved when the joints exhibited a characteristic ductile fracture occurred in the parent metal or seam, whereas the tensile strength was relatively low with a brittle fracture characteristic when the joints fractured only along IMC layers in the interface. This can explain why the second fracture propagation paths occurred in the steel/Al joint with an optimized P2 exhibit a relatively high tensile resistance.

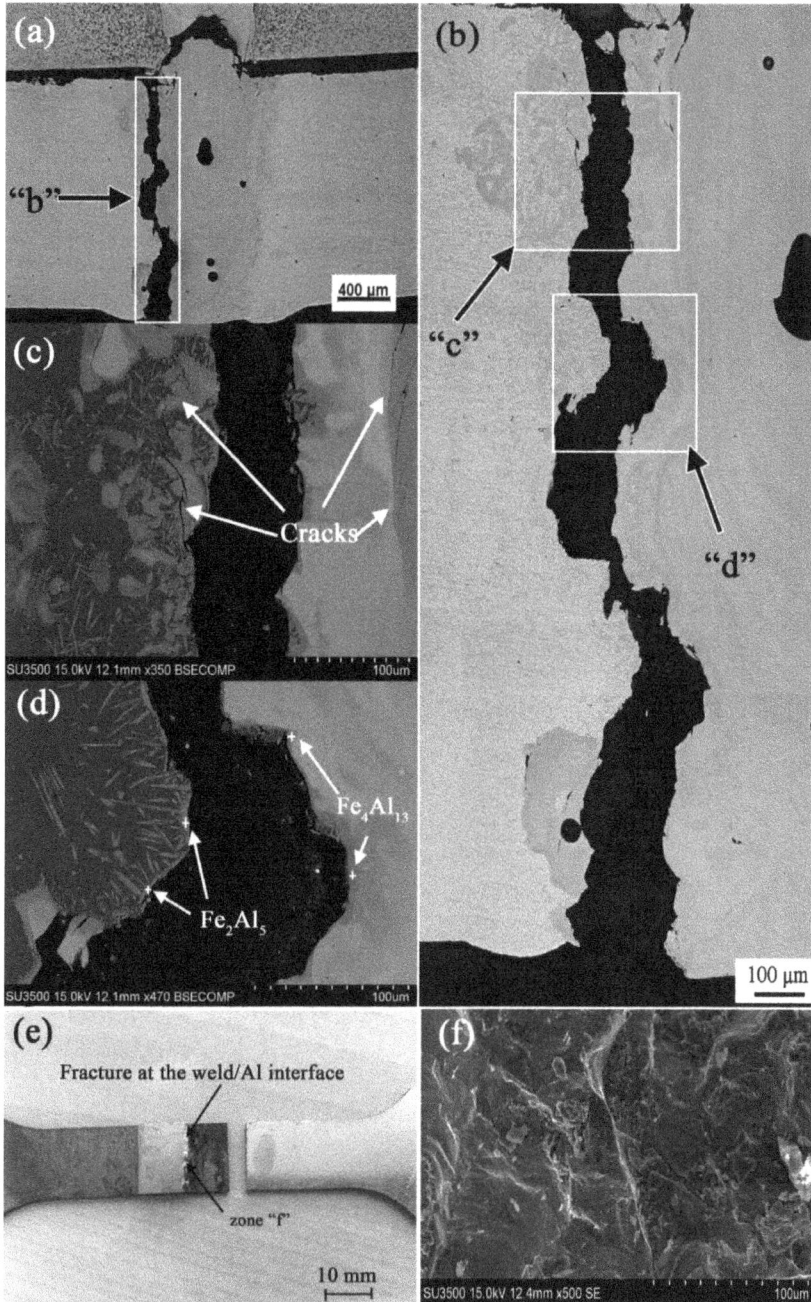

Figure 12. Fracture path morphology of steel/Al joints having an over-deep P2 of 1741 μm produced by dual-beam laser welding with side-by-side configuration arrangement: (**a**) Macroscopic fractured steel/Al joint; (**b**) High magnification of fracture path and locations of Zones "c" and "d"; (**c**) BSE image of enlarged view of Zone "c"; (**d**) BSE image of enlarged view of Zone "d"; (**e**) Macroscopic fractured steel/Al joints; (**f**) Fracture surface of Zone "f" in (**e**).

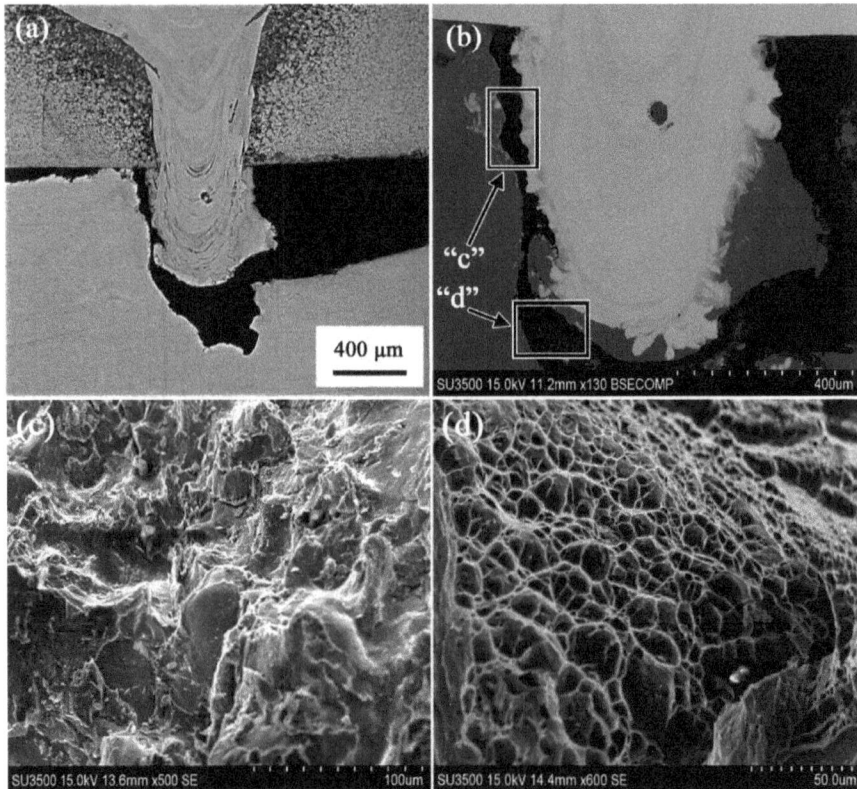

Figure 13. Fracture path morphology of steel/Al joints with an optimum P2 of 477 μm produced by dual-beam laser welding with side-by-side configuration: (**a**) Macroscopic fractured steel/Al joint; (**b**) BSE image of enlarged view of fracture path; (**c**) SEM image of fracture surface of Zone "c"; (**d**) SEM image of fracture surface of Zone "d".

4. Conclusions

A Q235 low carbon steel and 6061Al alloys of 1.5 mm thickness were welded by dual-beam fiber laser welding with side-by-side configuration. From the present results, the following major conclusions were reached:

(1) Soundly welded steel/Al lapped joints free of welding defects have been successfully achieved by using dual-beam laser welding with side-by-side configuration. The processing parameters of Rs and d_1 have a great influence on the weld appearance, the P2 and the welding defects. The good weld shape should be obtained at the relatively equal Rs of 0.50, 0.67 and 1.50 during side-by-side dual-beam laser welding of the steel/Al joints. The optimum d_1 for obtaining good weld shape is limited to 0.5 mm and 1.0 mm for side-by-side dual-beam laser welding of the steel/Al joints.

(2) The Al/weld interface microstructure in different locations consists of the island-shape structures and lath-like layer; however, the morphology and the lath-like layer thickness are different in each zone. The P2 has a significant influence on the morphology and the lath-like layer thickness, and controlling the P2 is effective to inhibit the formation of IMC layers at the Al/weld interface.

(3) EBSD phase mapping proves that the microstructures at the Al/weld interface are composed of the η-Fe_2Al_5 layers and the θ-Fe_4Al_{13} phases, and very fine θ-Fe_4Al_{13} and η-Fe_2Al_5 phases are

formed along α-Fe grain boundary inside the weld of the steel/Al joints. The η-Fe$_2$Al$_5$ layers and the needle-like θ-Fe$_4$Al$_{13}$ grains formed at the Al/weld interface are finer than those of the weld and the Al alloy.

(4) There is a matching relationship between the P2 and the tensile resistance of steel/Al joints produced by dual-beam laser welding with side-by-side configuration, and the maximum tensile resistance of the steel/Al joints is obtained at the *Rs* of 1.50 during dual-beam laser welding with side-by-side configuration.

(5) Two different fracture propagation paths are found depending on the P2. The fracture profile of the steel/Al joints with an optimized P2 exhibits a ductile fracture occurring in the parent metal or seam, resulting in a relatively high tensile resistance of the steel/Al joints.

Author Contributions: L.C. and D.H. conceived and designed the experiments; H.C. and B.C. performed the experiments; H.C. analyzed the data; L.C. wrote the paper. D.H. did the literature searching and reviewed the manuscript.

Funding: This research was funded by [the National Natural Science Foundation of China] grant number [51475006] and [the Key Program of Science and Technology Projects of Beijing Municipal Commission of Education] grant number [KZ201610005004].

Acknowledgments: Li Cui truly appreciates Li Chen (AVIC Manufacturing Technology Institute, China) for giving important insights and sharing fruitful discussion.

Conflicts of Interest: The authors declare no conflicts of interest.

References

1. Sun, Z.; Ion, J.C. Review laser welding of dissimilar metal combinations. *J. Mater. Sci.* **1995**, *30*, 4205–4214. [CrossRef]

2. Shah, L.H.; Ishak, M. Review of research progress on aluminum–steel dissimilar welding. *Mater. Manuf. Process.* **2014**, *29*, 928–933. [CrossRef]

3. Shao, L.; Shi, Y.; Huang, J.K.; Wu, S.J. Effect of joining parameters on microstructure of dissimilar metal joints between aluminum and galvanized steel. *Mater. Des.* **2015**, *66*, 453–458. [CrossRef]

4. Song, J.L.; Lin, S.B.; Yang, C.L.; Ma, G.C.; Liu, H. Spreading behavior and microstructure characteristics of dissimilar metals TIG welding–brazing of aluminum alloy to stainless steel. *Mater. Sci. Eng. A* **2009**, *509*, 31–40. [CrossRef]

5. Katayama, S. Laser welding of aluminium alloys and dissimilar metals. *Weld Int.* **2004**, *18*, 618–625. [CrossRef]

6. Meco, S.; Pardal, G.; Ganguly, S.; Williams, S.; Mcpherson, N. Application of laser in seam welding of dissimilar steel to aluminium joints for thick structural components. *Opt. Laser Eng.* **2015**, *67*, 22–30. [CrossRef]

7. Corigliano, P.; Crupi, V.; Guglielmino, E.; Sili, A.M. Full-field analysis of Al/Fe explosive welded joints for shipbuilding applications. *Mar. Struct.* **2018**, *57*, 207–218. [CrossRef]

8. Corigliano, P.; Crupi, V.; Guglielmino, E. Non linear finite element simulation of explosive welded joints of dissimilar metals for shipbuilding applications. *Ocean Eng.* **2018**, *160*, 346–353. [CrossRef]

9. Kaya, Y. Microstructural, mechanical and corrosion investigations of ship steel-aluminum bimetal composites produced by explosive welding. *Metals* **2018**, *8*, 544. [CrossRef]

10. Findik, F. Recent developments in explosive welding. *Mater. Des.* **2011**, *32*, 1081–1093. [CrossRef]

11. Xie, M.X.; Shang, X.T.; Zhang, L.J.; Bai, Q.L.; Xu, T.T. Interface characteristic of explosive-welded and hot-rolled TA1/X65 bimetallic plate. *Metals* **2018**, *8*, 159. [CrossRef]

12. Topolski, K.; Szulc, Z.; Garbacz, H. Microstructure and properties of the Ti6Al4V/Inconel 625 bimetal obtained by explosive joining. *J. Mater. Eng. Perform.* **2016**, *25*, 3231–3237. [CrossRef]

13. Borrisutthekul, R.; Yachi, T.; Miyashita, Y.; Mutoh, Y. Suppression of intermetallic reaction layer formation by controlling heat flow in dissimilar joining of steel and aluminum alloy. *Mater. Sci. Eng. A* **2007**, *467*, 108–113. [CrossRef]

14. Gao, M.; Chen, C.; Mei, S.W.; Wang, L.; Zeng, X.Y. Parameter optimization and mechanism of laser-arc hybrid welding of dissimilar Al alloy and stainless steel. *Int. J. Adv. Manuf. Technol.* **2014**, *74*, 199–208. [CrossRef]

15. Bouche, K.; Barbier, F.; Coulet, A. Intermetallic compound layer growth between solid iron and molten aluminium. *Mater. Sci. Eng. A* **1998**, *249*, 167–175. [CrossRef]

16. Wang, P.F.; Chen, X.Z.; Pan, Q.H.; Madigan, B.; Long, J.Q. Laser welding dissimilar materials of aluminum to steel: an overview. *Int. J. Adv. Manuf. Technol.* **2016**, *87*, 3081–3090. [CrossRef]

17. Chen, S.H.; Huang, J.H.; Ma, K.; Zhao, X.K.; Vivek, A. Microstructures and mechanical properties of laser penetration welding joint with/without Ni-Foil in an overlap steel-on-aluminum configuration. *Metall. Mater. Trans. A* **2014**, *45A*, 3064–3073. [CrossRef]

18. Shabadi, R.; Suery, M.; Deschamps, A. Characterization of joints between aluminum and galvanized steel sheets. *Metall. Mater. Trans. A* **2013**, *44A*, 2672–2682. [CrossRef]

19. Sun, J.H.; Yan, Q.; Gao, W. Investigation of laser welding on butt joints of Al/steel dissimilar materials. *Mater. Des.* **2015**, *83*, 120–128. [CrossRef]

20. Dharmendra, C.; Rao, K.P.; Wilden, J.; Reich, S. Study on laser welding-brazing of zinc coated steel to aluminum alloy with a zinc based filler. *Mater. Sci. Eng. A* **2011**, *528*, 1498–1503. [CrossRef]

21. Kouadri-David, A.; PSM Team. Study of metallurgic and mechanical properties of laser welded heterogeneous joints between DP600 galvanised steel and aluminium 6082. *Mater. Des.* **2014**, *54*, 184–195. [CrossRef]

22. Li, L.Q.; Chen, Y.B.; Wang, T. Research on dual-beam welding characteristics of aluminum alloy. *Chin. J. Lasers* **2008**, *35*, 1784–1788. [CrossRef]

23. Li, C.L.; Fan, D.; Wang, B. Characteristics of TIG arc-assisted laser welding-brazing joint of aluminum to galvanized steel with preset filler powder. *Rare Met.* **2015**, *34*, 650–656. [CrossRef]

24. Sierra, G.; Peyre, P.; Deschaux-Beaume, F.; Stuart, D.; Fras, G. Steel to aluminum key-hole laser welding. *Mater. Sci. Eng. A* **2007**, *447*, 197–208. [CrossRef]

25. Fabbro, R. Melt pool and keyhole behaviour analysis for deep penetration laser welding. *J. Phys. D Appl. Phys.* **2010**, *43*, 445–451. [CrossRef]

26. Milberg, J.; Trautmann, A. Defect-free joining of zinc-coated steels by bifocal hybrid laser welding. *Prod. Eng. Res. Dev.* **2009**, *3*, 9–15. [CrossRef]

27. Iwase, T.; Sakamoto, H.; Shibata, K.; Hohenberger, B.; Dausinger, F. Dual-focus technique for high-power Nd:YAG laser welding of aluminum alloys. In *SPIE High-Power Lasers in Manufacturing*; Chen, X.L., Fujioka, T.M., Matsunawa, A., Eds.; Advanced High-Power Lasers and Applications: Osaka, Japan, 1999; pp. 348–358.

28. Blackburn, J.E.; Allen, C.M.; Hilton, P.A.; Li, L. Dual focus Nd:YAG laser welding of titanium alloys. *Lasers Eng.* **2012**, *22*, 279–282.

29. Gref, W.; Russ, A.; Leimser, M.; Dausinger, F.; Huegel, H. Double-focus technique: influence of focal distance and intensity distribution on the welding process. In *First International Symposium on High-Power Laser Macroprocessing*; LAMP: Osaka, Japan, 2002. [CrossRef]

30. Ma, G.L.; Li, L.Q.; Chen, Y.B. Effects of beam configurations on wire melting and transfer behaviors in dual beam laser welding with filler wire. *Opt. Laser Technol.* **2017**, *91*, 138–148. [CrossRef]

31. Hansen, K.S.; Olsen, F.O.; Kristiansen, M.; Madsen, O. Joining of multiple sheets in a butt-joint configuration using single pass laser welding with multiple spots. *J. Laser Appl.* **2015**, *27*. [CrossRef]

32. Xie, J. Dual beam laser welding. *Weld J.* **2002**, *81*, 223–230.

33. Hsu, R.; Engler, A.; Heinemann, S. The gap bridging capability in laser tailored blank welding. *Laser Inst. Am.* **1998**, F224–F231. [CrossRef]

34. Laukant, H.; Wallmann, C.; Korte, M.; Glatzel, U. Flux-less joining technique of aluminum with zinc-coated steel sheets by a dual-spot-laser beam. *Adv. Mater. Res.* **2005**, *6–8*, 163–170. [CrossRef]

35. Shi, Y.; Zhang, H.; Takehiro, W.; Tang, J.G. CW/PW dual-beam YAG laser welding of steel/aluminum alloy sheets. *Opt. Lasers Eng.* **2010**, *48*, 732–736. [CrossRef]

36. Chen, S.H.; Zhai, Z.L.; Huang, J.H.; Zhao, X.K.; Xiong, J.G. Interface microstructure and fracture behavior of single/dual-beam laser welded steel-Al dissimilar joint produced with copper interlayer. *Int. J. Adv. Manuf. Technol.* **2016**, *82*, 631–643. [CrossRef]

37. Cui, L.; Chen, B.X.; Chen, L.; He, D.Y. Dual beam laser keyhole welding of steel/aluminum lapped joints. *J. Mater. Process. Technol.* **2018**, *256*, 87–97. [CrossRef]

38. Mohammadpour, M.; Yazdian, N.; Yang, G.; Wang, H.P.; Carlson, B.; Kovacevic, R. Effect of dual laser beam on dissimilar welding-brazing of aluminum to galvanized steel. *Opt. Laser Technol.* **2018**, *98*, 214–228. [CrossRef]

39. Xia, H.B.; Zhao, X.Y.; Tan, C.W.; Chen, B.; Song, X.G.; Li, L.Q. Effect of Si content on the interfacial reactions in laser welded-brazed Al/steel dissimilar butted joint. *J. Mater. Process. Technol.* **2018**, *258*, 9–21. [CrossRef]

40. Springer, H.; Kostka, A.; Payton, E.J.; Raabe, D.; Kaysser-Pyzalla, A.; Eggeler, G. On the formation and growth of intermetallic phases during interdiffusion between low-carbon steel and aluminum alloys. *Acta Mater.* **2011**, *59*, 1586–1600. [CrossRef]

41. Zhang, H.T.; Feng, J.C.; He, P.; Hackl, H. Interfacial microstructure and mechanical properties of aluminium-zinc-coated steel joints made by a modified metal inert gas welding-brazing process. *Mater. Charact.* **2007**, *58*, 588–592. [CrossRef]

42. Balogh, Z.; Schmitz, G. *Diffusion in Metals and Alloys*; Laughlin, D.E., Hono, K., Eds.; Physical Metallurgy: Oxford, UK, 2014; pp. 387–559.

metals

MDPI

Article

Weld Seam Geometry and Electrical Resistance of Laser-Welded, Aluminum-Copper Dissimilar Joints Produced with Spatial Beam Oscillation

Michael Jarwitz *, Florian Fetzer, Rudolf Weber and Thomas Graf

Institut fuer Strahlwerkzeuge (IFSW), Pfaffenwaldring 43, 70569 Stuttgart, Germany;
florian.fetzer@ifsw.uni-stuttgart.de (F.F.); rudolf.weber@ifsw.uni-stuttgart.de (R.W.);
thomas.graf@ifsw.uni-stuttgart.de (T.G.)
* Correspondence: michael.jarwitz@ifsw.uni-stuttgart.de; Tel.: +49-711-685-60209

Received: 8 June 2018; Accepted: 28 June 2018; Published: 3 July 2018

Abstract: Spatial beam oscillation during laser beam welding of aluminum to copper was investigated. The beam was spatially oscillated perpendicular to the direction of feed in a sinusoidal mode. The influence of the oscillation amplitude and frequency on the weld seam geometry and the implications on the electrical resistance of the joints was investigated. It was found that spatial beam oscillation allows to set the welding depth and seam width virtually independent of each other. Furthermore, low welding depths into the lower copper sheet in combination with high ratios of seam width at the interface of the two sheets to welding depth into the lower copper sheet result in low electrical resistances of the welds. Low electrical resistances were found to correlate with high mechanical strengths of the welds.

Keywords: laser beam welding; spatial beam oscillation; dissimilar metals; aluminum; copper

---

1. Introduction

Joining of aluminum (Al) to copper (Cu) is required for high-power, light-weight electrical applications, such as e-mobility or battery applications [1,2]. However, welding of the metal combination Al and Cu is challenging due to the formation of intermetallic phases in the weld seam [3] and the strongly differing thermophysical properties of the two materials, such as the melting temperature and the heat conductivity. The intermetallic phases have a higher brittleness, lower mechanical strength, and higher electrical resistivity than the base materials [4], which deteriorates both the mechanical and electrical properties of the weld joint. Therefore, the formation of these intermetallic phases during the welding process must be minimized. Reliable joining of Al to Cu with high quality by means of an unmodulated continuous wave (cw) laser welding process has proven to be challenging. When welding with a cw laser, significant mixing of the two materials and large cracks in the weld seam are a major problem [3]. A strong mixture of the materials and the formation of intermetallic phases in the weld seam is common [5]. The use of a spatial oscillation of the laser beam has proven to be a suitable measure to stabilize the welding process, especially the welding depth. The stabilization of the welding depth at low-penetration depths for laser welding of aluminum close to the deep penetration threshold is possible [6]. Furthermore, a stabilization of the welding depth against focal shift by the use of spatial beam oscillation has been observed [7]. Spatial oscillation of the laser beam has also been applied successfully for welding of Al to Cu to improve the weld quality. Homogeneous mixing of the two materials in the weld seam was demonstrated using a single-mode fiber laser and spatial beam oscillation [8]. Smaller mixing zones and less pores and cracks in the welds (compared to rectilinear welding) can be obtained using a pulsed laser and spatial beam oscillation [9]. Large joining widths at the interface and low penetration depths into the lower Cu sheet were generated by welding with a fiber laser and spatial beam oscillation [3]. The resulting welds showed a higher tensile strength than welds produced without a spatial oscillation of the laser beam. The electrical resistance of the joints is reported to decrease with an increase in the joining area. Small thicknesses (<5 µm) of the intermetallic phase layer at the interface could be achieved by the combination of temporal power modulation and spatial beam oscillation [5]. High shear strengths of the joints were obtained due to these small thicknesses of the intermetallic phase layer. In Ref. [10], it was also shown that the shear strength was reduced drastically if the thickness of the intermetallic phase layer was smaller than 3 µm. In addition, a low thickness of the intermetallic phase layer also results in a low electrical resistance of the joint, and there is a correlation between high mechanical strength and the low electrical resistance of the joints. Most of these investigations focused on welding with small focal diameters (about 50 µm or less) and comparably low laser powers (<1 kW). In contrast to the findings summarized in this section, the investigations presented in this paper focus on reproducible joining of aluminum to copper using multi-mode lasers with high power in combination with spatial beam oscillation. The weld seam geometry is known to be influenced by the spatial beam oscillation [11]. In the present paper, the influence of the oscillation parameters on the weld seam geometry was investigated. The focus was to quantify the relation between the geometry and the resultant electrical and mechanical properties of the welds.

We state that the resultant electrical resistance of the joints is mainly influenced by three factors:

(1) the seam width at the interface between the sheets,
(2) the metallurgical composition of the weld seam,
(3) the constancy of the geometrical properties of the weld seam along the welding direction,

and that spatial oscillation of the laser beam is a suitable measure to optimize these three factors. Therefore, the influence of the oscillation parameters on the geometry of the weld seam and the mixing of the two joining partners are investigated. The dependence of the electrical resistance on the geometry of the weld seam is evaluated, as well as the correlation between the electrical resistance and the mechanical strength of the connections.

2. Materials and Methods

2.1. Experimental Setup

Remote laser welding of pure aluminum (Al99.5) to pure copper (Cu-OF) in overlap configuration was investigated. A sketch of the geometry of the specimens which were used for the welding trials and the subsequent shear tests is shown in Figure 1. The thickness of each sheet was 1 mm and Al was positioned on top. Linear weld seams with a length of 35 mm were produced. Measurements of the electrical resistance (four-point measurement, detailed description in Section 2.2) and tensile shear tests (based on DIN EN ISO 14273) of the generated weld seams were performed on these samples.

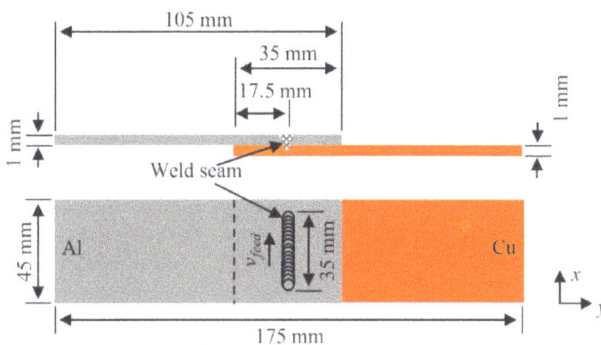

Figure 1. Sketch of the geometry of the specimens used for the welding trials and shear tests.

The used laser source was a Laserline LDF 4000.8 (Laserline GmbH, Mühlheim-Kärlich, Germany) solid-state laser with a wavelength of 1.08 µm whose beam was coupled to a transport fiber with a core diameter of 100 µm. The processing beam was focused on the top surface of the Al sheet using a Scansonic RLW-A remote welding optics (Scansonic MI GmbH, Berlin, Germany), which was inclined by an angle of 10° in direction of feed. The resulting focal diameter was 280 µm and the Rayleigh-length of the processing beam was 5 mm.

In a first step, rectilinear welds were performed at a feed rate of v_{feed} = 6 m/min and varying laser powers from 1.5 kW up to 3.25 kW. Subsequently, spatial beam oscillation perpendicular to the feed direction was applied with a trajectory corresponding to

$$\left(\begin{array}{c} x(t) \\ y(t) \end{array} \right) = \left(\begin{array}{c} v_{feed} \cdot t \\ A_y \cdot \sin(2\pi f_y \cdot t) \end{array} \right)$$ (1)

The oscillation frequency f_y was varied from 200 Hz to 1000 Hz and the oscillation amplitude A_y was varied from 0.25 mm up to 1 mm. The feed rate v_{feed} = 6 m/min was constant throughout all experiments and a constant laser power of P = 3.25 kW was used for the experiments with spatial beam oscillation. No shielding gas was used in the experiments.

2.2. Evaluation Procedure and Basic Considerations

The geometrical properties and the electrical resistances of the welds were measured to evaluate the experiments. Figure 2 shows optical microscopy (a) and SEM (b) images of typical cross-sections of rectilinear welds. The measured geometrical properties of the welds were the welding depth d into copper, the seam width w at the interface, and the fused areas A_{Al} and A_{Cu} of aluminum and copper, respectively. These quantities are indicated in the sketch of the cross-section shown in Figure 2c. From the SEM image (Figure 2b), it can be seen that there is an inhomogeneous distribution of the copper throughout the entire cross-section of the weld seam in the case of the rectilinear welding, even for low welding depths into the lower copper sheet. The averaged Cu content $X_{Cu,avg} = A_{Cu} \cdot \rho_{Cu} / (A_{Cu} \cdot \rho_{Cu} + A_{Al} \cdot \rho_{Al})$ in the weld seam was used to quantify the mixing ratio of the two metals, and it was calculated from the fused areas of aluminum and copper weighted by their densities. The fused areas were measured from the cross-sections, according to Figure 2c.

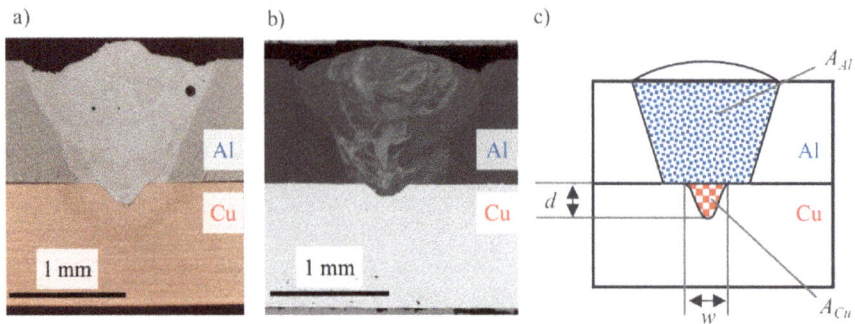

Figure 2. Images of typical cross-sections of welds produced by unmodulated rectilinear welding, optical microscopy (**a**) and SEM imaging (**b**); parameters: P = 2 kW; v_{feed} = 6 m/min; λ_{laser} = 1.08 µm; d_f = 280 µm; z_f = 0 mm, z_R = 5 mm; (**c**) sketch of a typical cross-section with indication of the analyzed geometrical quantities: welding depth d into Cu, seam width w at interface, and fused areas A_{Al} and A_{Cu} of aluminum and copper, respectively.

A four-point measurement setup was used to measure the electrical resistance of the joints. Figure 3a shows a sketch of this setup and Figure 3b shows the equivalent circuit for the measured electrical resistance $R_{measure}$.

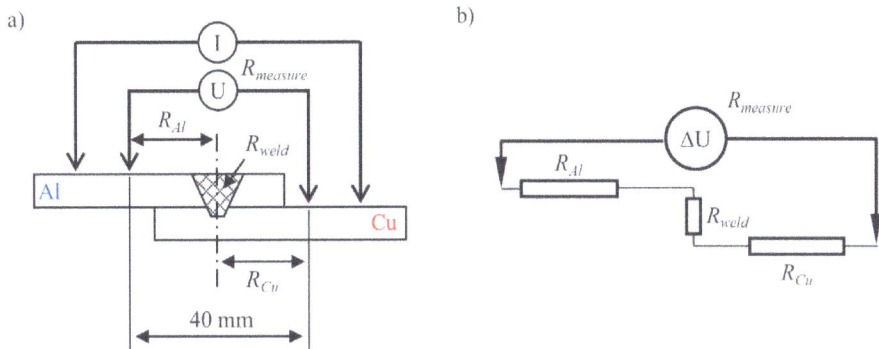

Figure 3. (**a**) Sketch of the four-point measurement setup used to determine the electrical resistance of the joints; (**b**) equivalent circuit for the measured resistance.

The measured electrical resistance $R_{measure} = R_{Al} + R_{weld} + R_{Cu}$ was considered idealized as the sum of the resistances R_{Al} and R_{Cu} of the two sheets up to the joint and the resistance of the weld seam itself R_{weld}. It is assumed that the resistances in the two base materials remain unaffected by the welding process, and therefore only the resistance of the weld seam is of interest. In a calibration measurement, the contribution of $R_{Al} + R_{Cu}$ to the measured resistance $R_{measure}$ was determined to be 21 $\mu\Omega$, considering the contribution of the base sheets between the measurement tip and the center of the weld seam to the overall electrical resistance. The offset of 21 $\mu\Omega$ was subtracted from each measured electrical resistance to obtain the electrical resistance of the weld seam R_{weld}. Current could further be conducted through the contact area between the two sheets, which can be considered as a parallel-connection to the weld. The electrical resistance of two sheets, which were not welded, but pressed together was measured to range between 10 mΩ and 50 mΩ, which is about three orders of magnitude higher as the resistance of the welded parts. The contact area between the two sheets would influence the values of $R_{measure}$ by up to 2‰ and is therefore neglected.

Figure 4 shows a sketch of a cross-section of a weld seam with indication of the path of the electrical current. The relevant cross-sectional area is considered to be the cross-sectional area at the interface between the two sheets ($w \cdot l_{weld}$) since the electrical current has to be conducted from one sheet of material to the other. With these assumptions we derive

$$R_{weld} \propto \rho_{el} \cdot \frac{s}{w \cdot l_{weld}} \qquad (2)$$

for the electrical resistance of the weld seam, where ρ_{el} is the electrical resistivity, w the seam width at the interface, l_{weld} the length of the weld seam, and s the length of the path of the electrical current through the weld seam. It should be noted that Equation (2) was derived for a rectangular-shaped weld seam, and that for a more complex geometry, the consideration of the integral across the weld seam would be better suited. Nevertheless, from Equation (2) some basic requirements can be concluded to lead to a low R_{weld}. On the one hand, the cross-sectional area at the interface has to be large. This means a large w, since the length l_{weld} of the weld seam is given by constructional boundary conditions in most cases. On the other hand, the electrical resistivity ρ_{el} of the weld seam has to be low. It is determined by the metallurgical composition of the weld seam. Since the electrical resistivity ρ_{el} of the intermetallic phases is much higher than that of the base materials [4], the amount of intermetallic phases in the weld seam should be as low as possible. Due to the strong convection in the weld pool during deep penetration laser welding, the liquid joining partners are mixed. This mixture of both materials in the liquid phase favors the formation of intermetallic phases. The averaged Cu content in the weld seam is taken as a possible measure of the amount of mixed materials, but is not sufficient to determine the amount and local distribution of the intermetallic phases. With increasing welding depth into Cu, the amount of fused Cu is expected to increase and with this the averaged Cu content. Therefore, the welding depth into Cu should be low to achieve a low ρ_{el} in the weld seam. This means that a high ratio of the seam width at the interface to the welding depth into Cu (w/d) is favorable to obtain a low R_{weld}. Moreover, the seam width and welding depth should be constant over the entire length of the weld seam.

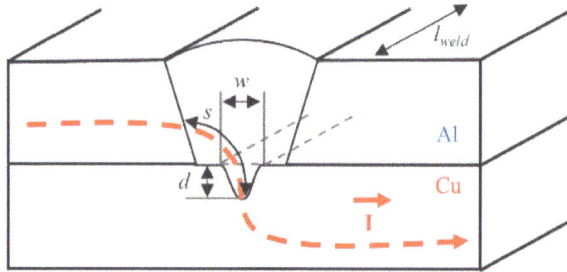

Figure 4. Sketch of a cross-section of the weld seam with indication of the path of an electrical current.

3. Results

3.1. Welding Without Spatial Oscillation of the Laser Beam

Figure 5 shows images of micrographs of cross-sections from welds produced with conventional rectilinear welding at different laser powers P.

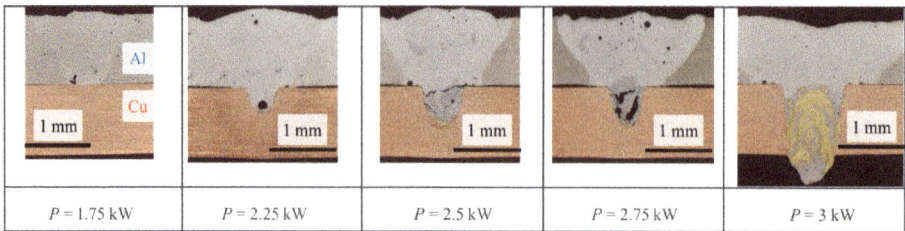

| $P = 1.75$ kW | $P = 2.25$ kW | $P = 2.5$ kW | $P = 2.75$ kW | $P = 3$ kW |

Figure 5. Micrographs of cross-sections of welds produced with conventional rectilinear welding, $v_{feed} = 6$ m/min.

As expected, the welding depth d into the Cu sheet increases with increasing laser power. Also, the fused area of Cu as well as the mixing of both materials increases (indicated by the dark grayish and golden colored areas in the cross-sections). Moreover, pores and cracks can be found in the weld seams. The cracks are located in the lower part of the weld (Cu side) and at the interface between the two sheets. Cracks are found in welds with a welding depth into the copper sheet of more than 0.5 mm and partial penetration of the Cu sheet. For small d or full penetration of the Cu sheet, no cracks were observed.

Unmodulated, rectilinear laser welding of the two metals aluminum and copper itself is a highly instable process and prone to the formation of weld defects and fluctuations of the penetration depth [12–14]. This is also true for laser welding of the combination of the two metals as can be seen from Figure 6, which shows micrographs of cross-sections of the same weld seam at two different positions. It can be seen that d and w are not constant along the seam produced by rectilinear welding, and therefore point 3 of the above stated quality factors is not fulfilled.

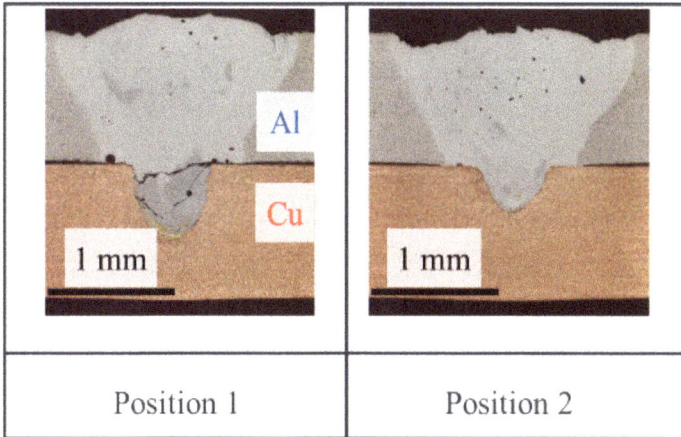

Figure 6. Micrographs of cross-sections at two positions along the same weld seam. P = 2.5 kW; v_{feed} = 6 m/min.

The dependence of the weld seam geometry on the laser power is shown in Figure 7. Both the width and the depth of the weld seams increase with increasing laser power. The welding depth shows an almost linear increase with increasing laser power until full penetration of the Cu sheet is reached at 3 kW. In contrast, the seam width shows a strong increase with laser power from 1.5 kW up to about 2.25 kW, and then is seen to converge towards an upper limit. The averaged Cu content increases in close accordance with increasing welding depth.

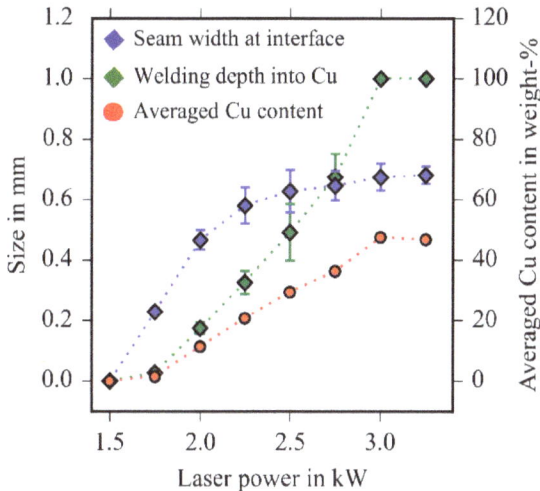

Figure 7. Seam width w at the interface (blue), welding depth d into Cu (green), and averaged Cu content (red) as a function of the laser power for welds produced with rectilinear welding at a feed rate of 6 m/min.

3.2. Welding with Spatial Oscillation of the Laser Beam

Figure 8 shows optical micrographs of cross-sections from welds produced with spatial oscillation of the laser beam for different oscillation parameters.

A_y in mm / f_y in Hz	0.25	0.5	0.75	1.0
200				
400				
600				✕
800				✕
1000			✕	✕

Figure 8. Micrographs of cross-sections from welds produced with spatial oscillation of the laser beam with 3.25 kW of laser power and a feed rate of 6 m/min.

At the smallest amplitude of A_y = 0.25 mm the weld seams resemble those obtained with conventional rectilinear welding. They exhibit a large welding depth into the lower Cu sheet and a strong mixing of the materials (indicated by the dark grayish and golden colored areas in the cross-sections). With increasing amplitude, the welding depth decreases and the seam width increases. At the same time the dark grayish and golden colored areas are significantly reduced. Cracks in the center of the weld in the Al sheet are visible in some of the welds produced with large lateral oscillation amplitudes. For large amplitudes and high frequencies, the laser power of P = 3.25 kW was not sufficient to achieve a penetration into the lower Cu sheet and a connection of the two sheets.

3.3. Geometry of Weld Seams as a Function of Oscillation Parameters

The welding depth into Cu is shown in Figure 9 as a function of the oscillation amplitude A_y and frequency f_y. The measured welding depth decreases with both, increasing amplitude and increasing frequency, whereby the impact of the amplitude is stronger. Averaged over all measurements, $\partial d / \partial A_y$ is found to be −0.2 mm/mm and $\partial d / \partial f_y$ to be −0.051 mm/100 Hz. The standard deviation is highest for the smallest oscillation amplitude, which indicates that the welding process is still instable and comparable to the rectilinear welding process without oscillation. This is in agreement with the highly fluctuating results of the rectilinear welds (compare also the cross-sections shown in Figure 8 for $A_y = 0.25$ mm and in Figure 6 for rectilinear welding). For higher oscillation amplitudes the standard deviations are very small, which indicates the stabilizing effect of the spatial oscillation of the laser beam on the geometry of the weld seam. Hence, the depth can be controlled coarsely with the oscillation amplitude and fine-tuned with the oscillation frequency.

Figure 10 shows the influence of oscillation amplitude and frequency on the seam width w at the interface of the two sheets. The measured seam width increases with increasing amplitude as long as the locally available energy suffices to guarantee melting of the lower Cu sheet. This depends on the oscillation frequency. For small amplitudes the frequency has no effect on the resultant width of the weld seam. For amplitudes exceeding 0.5 mm the width is found to decrease.

Figure 11 shows the measured widths w of the weld seams together with the respective welding depths d into the Cu sheet. For processes without beam oscillation, the measurements are found to lie on one curve (red markers). This indicates that the width and depth cannot be set independently. The values for the processes with spatial beam oscillation are spread above the values of the processes without beam oscillation. Therefore, both geometrical features can be set independently and at same depth the generation of larger seam widths is possible in welding processes with beam oscillation.

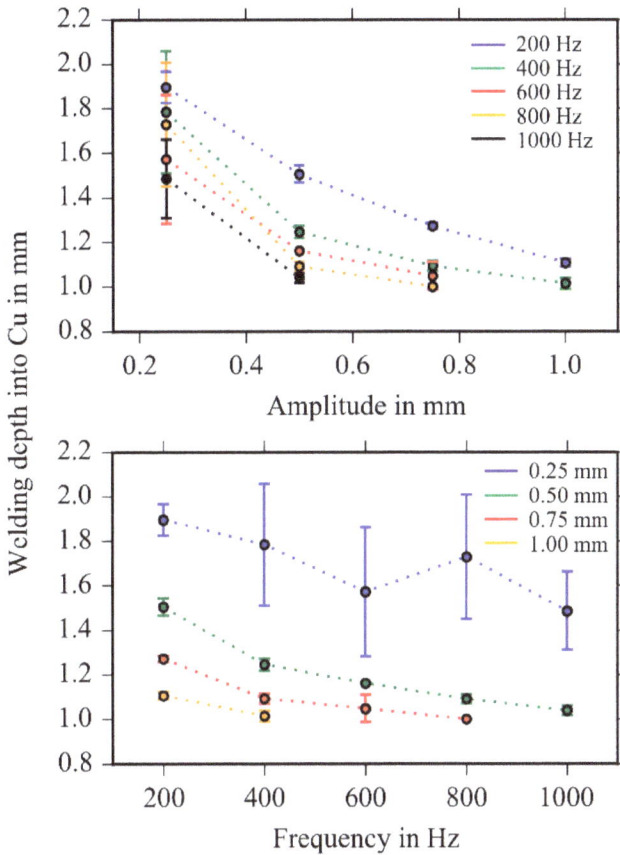

Figure 9. Welding depth into Cu in dependence on the oscillation frequency and amplitude. The error bars indicate the standard deviation.

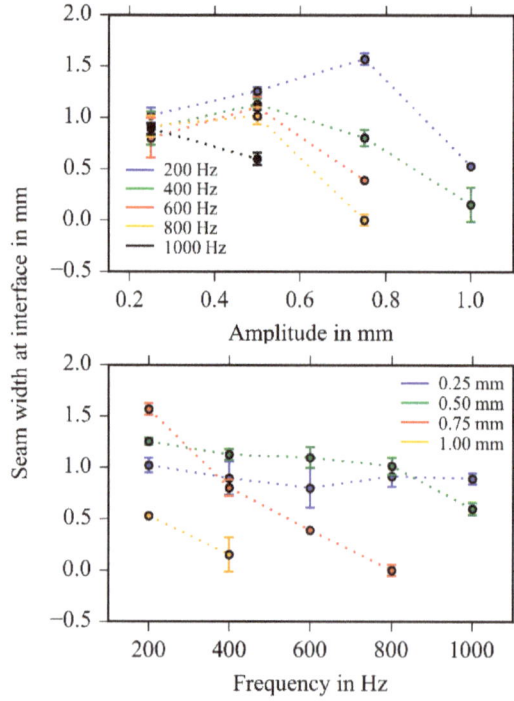

Figure 10. Seam width at the interface in dependence on the oscillation frequency and amplitude. The error bars indicate the standard deviation.

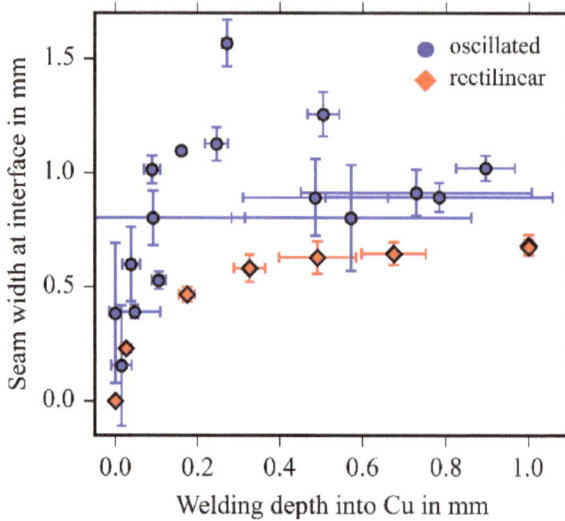

Figure 11. Seam width at interface over welding depth into Cu for welds produced with spatial beam oscillation (blue) and welding processes without beam oscillation (red). The error bars indicate the standard deviation.

3.4. Averaged Cu Content as a Function of the Geometry of the Weld Seams

In laser beam welding, the liquid joining partners are mixed due to the strong convection in the weld pool. Therefore, the averaged Cu content in the weld seam is taken as a possible measure of the amount of mixed materials. Figure 12 shows the averaged Cu content in the weld seam in dependence on the welding depth d into Cu (a) and seam width w at the interface (b). It was found that the averaged Cu content is approximately proportional to the welding depth d for both, the welding processes with and without spatial beam oscillation. A linear regression of the averaged Cu content on the welding depth into Cu (dotted lines in Figure 12a) yielded a coefficient of determination of $R^2 = 0.97$ for rectilinear welding and of $R^2 = 0.94$ in case of oscillated welding, respectively. This confirms the assumption stated before, that the amount of fused Cu, and with this the averaged Cu content in the seam, is mainly influenced by the welding depth into the lower Cu sheet. The averaged Cu content increases rapidly with the width of the seam (because also the depth increases, see Figure 11). The increase is less steep for welds produced with beam oscillation. Again, the results for welding processes without beam oscillation are on one curve, whereas the results for welding processes with beam oscillation are spread. This further indicates that welding with spatial oscillation of the laser beam allows to independently set the geometrical features as well as the averaged Cu content of the generated weld seams, which is not possible with welding processes without beam oscillation.

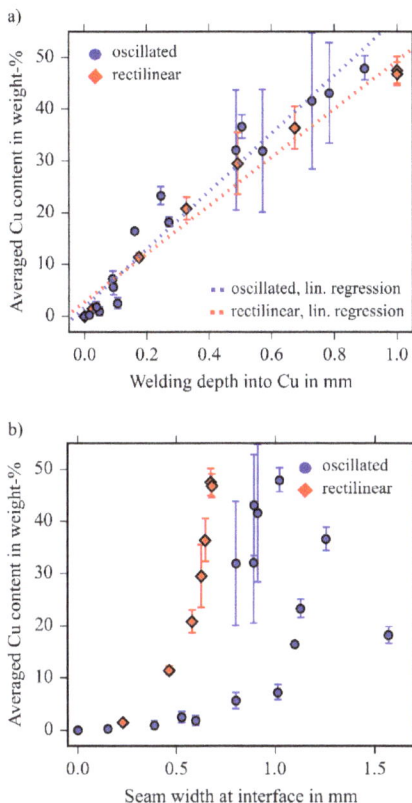

Figure 12. (a) Averaged Cu content in the weld seam in dependence on the welding depth. The markers indicate the measured values; the dotted lines represent a linear regression of the averaged Cu content on the welding depth; (b) averaged Cu content in the weld seam in dependence on the width of the seam. The error bars indicate the standard deviation.

In Figure 13, the averaged Cu content in the weld seam is plotted together with the ratio of the seam width at the interface to the welding depth into Cu (width/depth-ratio). The averaged Cu content is found to decrease with an increasing width/depth-ratio. For the same share of copper in the seam, a higher width/depth-ratio can be achieved for the welds produced with beam oscillation, and higher width/depth-ratios can be achieved for welds produced with beam oscillation in general.

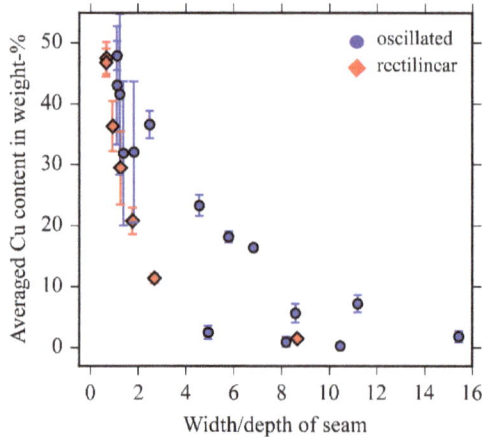

Figure 13. Dependence of the averaged Cu content in the weld seam on the ratio of seam width to depth for welds produced with spatial beam oscillation (blue) and rectilinear welds (red). The error bars indicate the standard deviation.

3.5. Electrical Resistance as a Function of the Geometry of the Weld Seams

Figure 14 shows the measured electrical resistances of the generated welds as a function of the welding depth into Cu (a) and the width/depth-ratio (b). The data measured for the welding processes with and without spatial beam oscillation are compared.

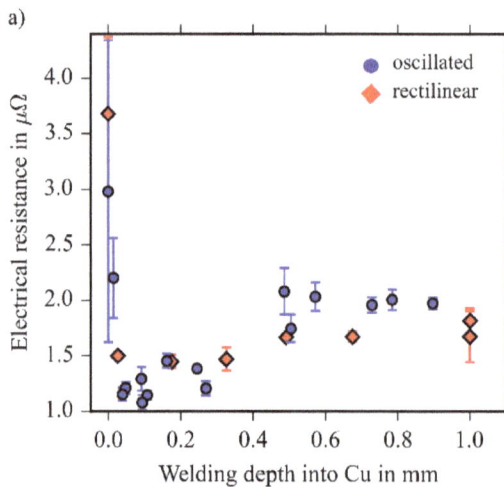

Figure 14. *Cont.*

b)

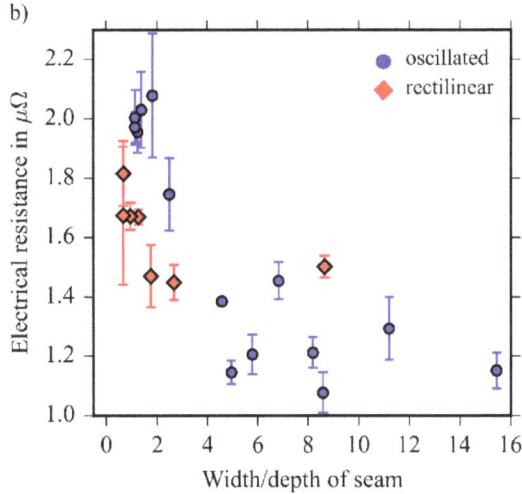

Figure 14. Electrical resistance in dependence on the welding depth into Cu (**a**) and on the ratio of weld seam width to depth (**b**).

It was found that a minimum welding depth into Cu is required to obtain reasonable resistance values. This minimum depth was found to be 0.02 mm. If the welding depth in Cu is lower than this minimum required depth, the resistance increases (>2 µΩ) due to a lack of contact. The dependence of $\partial R / \partial (w/d)$ was found to be -0.17 µΩ for the welds produced with beam oscillation and -0.025 µΩ for the rectilinear welds. For low width/depth-ratios (<3), the rectilinear welds result in lower (better) resistances. At higher width/depth-ratios (>4), the welds produced with spatial beam oscillation exhibit lower electrical resistances than the rectilinear welds. With spatial beam oscillation the electrical resistance could be reduced by about 25% compared to the welding process without spatial beam oscillation. In general, higher width/depth-ratios and correspondingly lower electrical resistances can be achieved by welding with an oscillating beam. As a consequence, it was found that a high width to depth ratio is favorable to obtain low electrical resistance. This ratio is found to represent a reasonable measure for the suitability of the generated weld seams and it is proposed to merge the quality Factors (1) and (2) stated above.

3.6. Correlation of Electrical Resistance and Mechanical Strength

Shear tests were performed on the welded samples based on DIN EN ISO 14273 and the maximum shear strength was measured to quantify the mechanical quality of the generated junctions. These values were correlated with the measured electrical resistances. In Figure 15, both the maximum shearing loads together with the electrical resistances are plotted for the welds produced with and without spatial beam oscillation.

Maximum loads of up to 3 kN were found which coincide with the lowest electrical resistance of 1.1 µΩ. The mechanical and electrical quality were found to correlate, thus for high shear strengths the lowest electrical resistances were measured and vice versa. It is therefore expected that process optimizations regarding one of these quality measures likewise optimize the other. Additionally, a measurement of the electrical resistance, which can be performed nondestructively, allows to predict the mechanical strength of the junction. These results are well consistent with correlations reported for welding of thin sheets with a single-mode laser [10].

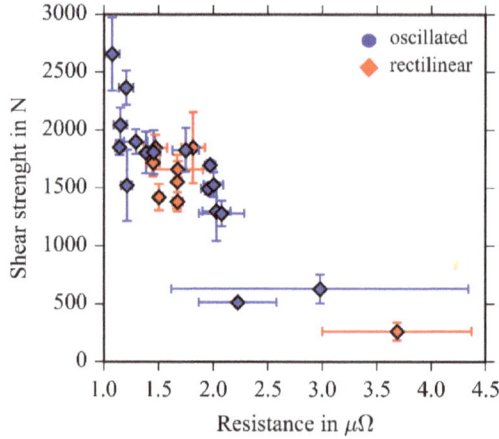

Figure 15. Correlation between electrical resistance and tensile strength for welds produced with (blue markers) and without (red markers) spatial beam oscillation.

4. Discussion

The presented investigations have shown that high width/depth-ratios are favorable together with a low and constant welding depth into the lower sheet. For the used material thickness of 1 mm, this means that weld seam widths >1 mm are required at welding depths of about 1 mm. Since the high-reflective materials (for the laser wavelength of 1 μm) aluminum and copper are used for electrical applications, deep penetration welding is preferred to achieve high thermal efficiency. We conclude that spatial beam oscillation is a well suited method to achieve this. For a conventional rectilinear welding process, a respective large beam diameter of >1 mm would be required. This in turn would lead to a higher required laser power since the threshold power for deep-penetration welding almost linearly scales with the beam diameter, as shown in Reference [15]. Moreover, at the transition from heat conduction welding to deep penetration welding, a step-like increase of the welding depth occurs, which is dependent on the size of the laser spot, as stated in Reference [16]. With increasing spot size this step also increases which makes it much more challenging to achieve low and constant welding depths in deep penetration welding for large beam diameters. In contrast, spatial oscillation of the laser beam allows to achieve lower and more stable welding depths in deep penetration welding mode compared to conventional rectilinear welding, as shown in Reference [6]. In addition, the use of two-axis scanners offers the possibility to adapt and optimize welding processes for different material and joint configurations, with the joining of battery taps for lithium-ion batteries being one possible application [1,2]. While the presented investigations quantified the influence of the processing parameters on the geometry and the resultant electrical and mechanical properties of the welds, the influence of the processing parameters on the amount and local distribution of intermetallic phases is subject to further research. Furthermore, a correlation between the electrical resistance and the tensile strength of the joints could be observed. The mechanical strength of the joints can therefore be estimated by measurements of the electrical resistance. This provides the possibility of a nondestructive method for testing the mechanical strength that is also faster and less expensive than conventional shear strength tests.

5. Conclusions

Laser beam welding of aluminum to copper with sheet thicknesses of 1 mm in overlap configuration was investigated using a high-power, multi-mode laser and spatial oscillation of the laser beam. The influence of the oscillation parameters amplitude and frequency on the geometrical

Metals **2018**, *8*, 510

features of the resultant weld seams were determined and the implications on the electrical resistance were examined. The averaged Cu content in the weld seam depended mainly on the welding depth into the lower Cu sheet. The welding depth into the lower Cu sheet decreased with both, increasing amplitude and increasing frequency, with the amplitude having a stronger influence, and the frequency allowing for a fine-tuning. Additionally, the spatial beam oscillation allowed to set the seam width at the interface and the welding depth into the lower Cu sheet virtually independent of each other, which was not possible with a rectilinear welding process. With this, high width/depth-ratios can be achieved even at low and constant welding depths into the lower Cu sheet. Low welding depths into the lower Cu sheet in combination with high width/depth-ratios (>4) are favorable to obtain low electrical resistances, but a minimum welding depth (>0.02 mm) is required for a reasonable connection. These results confirm the influencing factors for a good weld stated at the beginning of the paper.

Author Contributions: Conceptualization, M.J. and F.F.; Methodology, M.J. and F.F.; Investigation, M.J. and F.F. Data Curation, M.J. and F.F.; Writing-Original Draft Preparation, M.J. and F.F.; Writing-Review & Editing, R.W. and T.G.; Supervision, R.W.; Project Administration, R.W.; Funding Acquisition, R.W. and T.G.

Funding: The presented work was funded by the Federal Ministry of Education and Research (BMBF) and experiments were performed in the context of the "ReMiLas" project. The responsibility for this paper is taken by the authors.

Acknowledgments: The RLW-A remote welding optics was provided by Scansonic MI GmbH and the laser source Laserline LDF 4000.8 was provided by Laserline GmbH, both within the scope of the project "ReMiLas".

Conflicts of Interest: The authors declare no conflict of interest.

References

1. Schmidt, P.A.; Schweier, M.; Zaeh, M.F. Joining of lithium-ion batteries using laser beam welding: Electrical losses of welded aluminum and copper joints. In Proceedings of the International Congress on Applications of Lasers and Electro-Optics (ICALEO 2012), Anaheim, CA, USA, 23–27 September 2012.

2. Kirchhoff, M. Laser applications in battery production—From cutting foils to welding the case. In Proceedings of the 2013 3rd International Electric Drives Production Conference (EDPC), Nuremberg, Germany, 29–30 October 2013; IEEE: New York, NY, USA, 2013; pp. 1–3.

3. Smith, S.; Blackburn, J.; Gittos, M.; Bono, P.D.; Hilton, P. Welding of dissimilar metallic materials using a scanned laser beam. In Proceedings of the 32nd International Congress on Applications of Lasers & Electro-Optics (ICALEO), Miami, FL, USA, 6–10 October 2013; pp. 6–10.

4. Rabkin, D.M.; Ryabov, V.R.; Lozovskaya, A.V.; Dovzhenko, V.A. Preparation and properties of copper-aluminum intermetallic compounds. *Powder Metall. Metal Ceram.* **1970**, *8*, 695–700. [CrossRef]

5. Solchenbach, T.; Plapper, P. Mechanical characteristics of laser braze-welded aluminium-copper connections. *Opt. Laser Technol.* **2013**, *54*, 249–256. [CrossRef]

6. Sommer, M.; Weberpals, J.-P.; Müller, S.; Berger, P.; Graf, T. Advantages of laser beam oscillation for remote welding of aluminum closely above the deep-penetration welding threshold. *J. Laser Appl.* **2017**, *29*, 12001. [CrossRef]

7. Thiel, C.; Weber, R.; Johannsen, J.; Graf, T. Stabilization of a laser welding process against focal shift effects using beam manipulation. *Phys. Procedia* **2013**, *41*, 209–215. [CrossRef]

8. Kraetzsch, M.; Standfuss, J.; Klotzbach, A.; Kaspar, J.; Brenner, B.; Beyer, E. Laser beam welding with high-frequency beam oscillation. Welding of dissimilar materials with brilliant fiber lasers. *Phys. Procedia* **2011**, *12*, 142–149. [CrossRef]

9. Gedicke, J.; Olowinsky, A.; Artal, J.; Gillner, A. Influence of temporal and spatial laser power modulation on melt pool dynamics. In Proceedings of the 26th International Congress on Applications of Laser and Electro-Optics (ICALEO 2007), Orlando, FL, USA, 29 October–1 November 2007; pp. 816–822.

10. Solchenbach, T.; Plapper, P.; Wanye, C. Electrical performance of laser braze-weldedd aluminum-copper interconnects. *J. Manuf. Processes* **2014**, *16*, 183–189. [CrossRef]

11. Fetzer, F.; Jarwitz, M.; Stritt, P.; Weber, R.; Graf, T. Fine-tuned remote laser welding of aluminum to copper with local beam oscillation. *Phys. Procedia* **2016**, *83*, 455–462. [CrossRef]

12. Heider, A.; Sollinger, J.; Abt, F.; Boley, M.; Weber, R.; Graf, T. High-speed X-ray analysis of spatter formation in laser welding of copper. *Phys. Procedia* **2013**, *41*, 112–118. [CrossRef]
13. Heider, A.; Stritt, P.; Hess, A.; Weber, R.; Graf, T. Process stabilization at welding copper by laser power modulation. *Phys. Procedia* **2011**, *12*, 81–87. [CrossRef]
14. Hagenlocher, C.; Fetzer, F.; Weber, R.; Graf, T. Benefits of very high feed rates for laser beam welding of AlMgSi aluminum alloys. *J. Laser Appl.* **2018**, *30*, 12015. [CrossRef]
15. Graf, T.; Berger, P.; Weber, R.; Hügel, H.; Heider, A.; Stritt, P. Analytical expressions for the threshold of deep-penetration laser welding. *Laser Phys. Lett.* **2015**, *5*, 56002. [CrossRef]
16. Hügel, H.; Graf, T. *Laser in der Fertigung*; Vieweg + Teubner, GWV Fachverlage GmbH: Wiesbaden, Germany, 2009.

![metals logo] *metals*

MDPI

Article

Effect of Revolutionary Pitch on Interface Microstructure and Mechanical Behavior of Friction Stir Lap Welds of AA6082-T6 to Galvanized DP800

Shuhan Li [1,2], Yuhua Chen [2,*], Jidong Kang [1,*], Babak Shalchi Amirkhiz [1] and Francois Nadeau [3]

[1]   CanmetMATERIALS, Natural Resources Canada, 183 Longwood Road South,
     Hamilton, ON L8P 0A5, Canada; shuhanli@outlook.com (S.L.); babak.shalchi_amirkhiz@canada.ca (B.S.A.)
[2]   School of Aerospace Manufacturing Engineering, Nanchang Hangkong University, 696 Fenghe Road South,
     Nanchang 330063, China
[3]   National Research Council of Canada (NRC), Saguenay, QC G7H 8C3, Canada;
     Francois.Nadeau@cnrc-nrc.gc.ca
*   Correspondence: ch.yu.hu@163.com (Y.C.); jidong.kang@canada.ca (J.K.); Tel.: +86-133-3006-7995 (Y.C.);
     +1-(905)-645-0820 (J.K.)

Received: 24 October 2018; Accepted: 5 November 2018; Published: 9 November 2018

Abstract: Friction stir lap welding of 1.5-mm thick 6082-T6 aluminum alloy to 2-mm thick galvanized DP800 steel (Zn-coated) was carried out. Optimal welding conditions were obtained aiming to defect-free joints with good mechanical properties. The interfacial intermetallic compounds (IMCs) at the stir zone and hook zone were characterized under different revolutionary pitches. With a revolutionary pitch of 1.0 mm/rev, maximum joint strength reached 71% of that of the aluminum alloy. In the meantime, the average thickness of IMC layer is less than 1 μm; $Al_{3.2}Fe$ in the Al-rich side and Al_5Fe_2 in the Fe-rich side at the interfaces of stir zone while Al_6Fe and nanocrystalline close to $Al_{3.2}Fe$ at the interface of the hook zone. At a relatively lower revolutionary pitch (0.5 mm/rev), Zn was found with the aggregation of Si and Mn at the hook-zone interface, leading to the generation of Al-Fe-Si phase thus decreasing the thickness of the IMC layer. In the stir zone, the revolutionary pitch has a significant influence on the interfacial microstructures. The interfacial IMC layer at 1.0 mm/rev is simple and flat, but the one at 0.5 mm/rev becomes thicker and more complex. Stir zone aluminum under different revolutionary pitches is similar in microhardness and tensile behavior. The mechanical response of joints was modeled based on linear mixture law with an iso-strain assumption and neglection of the IMC layer. The modeling results are in good agreement with the experimental ones indicating the resultant interfaces act as good as the good boundaries between dissimilar Al/steel joints.

Keywords: Al/steel dissimilar materials; friction stir welding; interface; intermetallic compounds

---

1. Introduction

With the increasing use of aluminum alloy for lightweighting (replacing steel) in the automotive industry, dissimilar aluminum to steel joining has received substantial attention [1]. However, the differences in thermal properties (i.e., melting point, heat capacity, thermal expansion, and thermal conductivity) and formation of brittle Al/Fe intermetallics (i.e., $AlFe_3$, $AlFe$, Al_2Fe, Al_3Fe, Al_5Fe_2, and Al_6Fe), welding distortion, cavities, and cracks cause fusion welding of aluminum to steel to be challenging [2].

As a solution, the use of friction stir welding (FSW) [3] helps in reducing intermetallic compounds (IMCs) in Al/steel joints. For example, the subframe assembly of the Honda Accord 2013 lap joined aluminum to steel using the FSW technique [4]. Bozzi et al. [5] made friction stir spot welds of

AA6016/IF-steel. Al_3Fe, Al_5Fe_2, and Al_2Fe were identified within an optimal IMC layer with a thickness of 8 μm. Uzun et al. [6] friction stir welded AA6013-T4 to 304 stainless steel successfully and the joints could reach approximately 70% base aluminum alloy with an interfacial layer less than 1 μm thick.

However, there are sparse data in the open literature on the FSW of aluminum alloy to advanced high strength steels (AHSS [7]), which are more desirable for automotive applications [8]. Liu et al. [8] friction stir welded AA6061-T6511 (1.5 mm) to TRIP 780/800 sheets of steel (1.4 mm) in butt joint configuration. It was reported that the maximum ultimate tensile strength (UTS) could reach 85% of the base aluminum alloy and the interfacial IMCs layer of AlFe and $AlFe_3$ has a thickness less than 1 μm. Zhao et al. [9] investigated the tool geometry effects on AA6061/TRIP 780/800 (1.5 mm thick) FSW joints. They demonstrated the tool geometry determines the thickness of the IMC layer. With an appropriate tool geometry, the UTS of the joint could be higher than 80% of the base Al alloy. Further, an electrically-assisted FSW system was developed for butt joining AA6061/TRIP 780 by Liu et al. [10].

It should be highlighted that researchers mostly focused on the investigation on butt configuration, while insufficient work has done with friction stir lap welding of Al/AHSS [8–10]. Compared to the butt configuration, the advantage of a lap joint is that it provides no lateral gap. As no filler wire is used in friction stir welding, approximately 20% of the thickness in terms of maximum gap allowed is typically required [11]. It also depends on the underfill value after welding to pass the requirements. Also, the positioning in lap joints is more robust than in butt joints where lap joining does not need to follow the joint line. It can be shifted off slightly without quality compromise and simplify part positioning in production.

In the present study, AA6082-T6 and Zn-coated DP800 were friction stir lap welded for automotive application. Various welding parameters were tested aiming for a process window with optimal welding conditions. The resultant interfacial microstructures were then characterized using transmission electron microscopy (TEM) in addition to optical and scanning electron microscopy (OM and SEM) to reveal the effect of welding parameters on those results. Further, the role of interface microstructure on mechanical properties of the FSWs was demonstrated through experiments and modeling of a series of modified shear tests in addition to lap shear tests.

2. Experimental Procedures

The materials used in this work were 1.5-mm thick 6082-T6 aluminum alloy and 2-mm thick galvanized DP800 dual-phase steel. These grades come from Neuman Aluminium through Europe. The surface of DP800 was coated with a 19-μm thick zinc layer. The nominal chemical compositions of the base metals are listed in Table 1. Welding process was carried out using a gantry friction stir welding machine where the pin-tip of the tool slightly penetrated into the steel sheet [12,13] (see welding setup illustrated in Figure 1). As the sub-steel is a high strength steel, it may cause significant wear to the tool pin-tip. To avoid this, the FSW tool material was selected as WC-20%Co with a cylindrical geometry and rounded pin end. Travel speeds were 300, 750, and 1250 mm/min, and rotation speeds were 500, 750, 1000, 1250, and 1500 RPM. Process parameters are proprietary to the NRC research consortium ALTec. All welded joints were prepared using an MTS I-Stir PDS gantry-type FSW machine and specimens were extracted in the steady-state region of the welds, away from plunge and exit zones. The welding parameters were varied in a wide range aiming for a process window with optimal welding conditions. In addition, the longitudinal force during welding was recorded for the potential application of robotic welding.

Table 1. Nominal chemical compositions of base materials (wt. %).

AA6082	Mg	Si	Mn	Cu	Al	Fe	Others
	1.10	0.96	0.41	0.02	96.99	0.43	Bal.
DP800	C	Si	Mn	Ni	Cr	Fe	Others
	0.15	0.22	1.79	0.03	0.41	97.10	Bal.

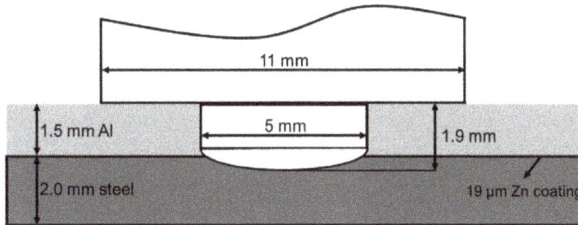

Figure 1. Schematic illustration of the welding setup and details of the tool and materials.

After welding, some of the FSWs were cross-sectioned following the standard metallographic procedure. Microstructural observations were performed via an Olympus OM and a FEI Nova Nano 650 SEM equipped with energy dispersive X-ray spectrometry (EDS). Furthermore, to characterize the interfacial IMC, focused ion beam (FIB) milling was used to prepare the interface specimens using the FEI Nova Nano 650 SEM. Then the specimens were characterized via a FEI Tecnai Osiris TEM for investigation. Selected area diffraction (SAD) was used for phase identification. Scanning transmission electron microscopy (STEM) mode using the bright field (BF) and high angle annular dark field (HAADF) detectors were performed in combination with EDS. Vickers hardness along the cross-centerline of Al sheet was tested every 0.5 mm using a load of 100 g with a dwell time of 10 s. Lap shear specimens were cut in a nominal width of 25 mm and tested using an MTS test frame.

For the mechanical behavior evaluation of joints' stir zone, a modified shear test specimen developed by Kang et al. [14] was adopted to obtain the mechanical response of different stack-ups. The modified shear test specimen with the shear zone central at the weld was designed as shown in Figure 2 (volume fraction of steel is 15%). A commercially available optical strain mapping system based on digital image correlation (DIC), Aramis, was used to follow the strain development during the shear test. Details of this method are given in past papers [14,15]. After the testing, to measure the actual thickness aluminum and steel in the shear zone thus calculate the actual volume fraction of each material, the shear test specimens were cut cross the shear zone followed a standard metallographic procedure for optical microscopy. In addition, mini-tensile specimens of stir-zone Al/steel were designed as shown in Figure 3. All mechanical tests were carried out at a crosshead speed of 3 mm/min at room temperature. Note that the number of the replicate mechanical test specimens is three. Wire electrical discharge machining (WEDM) and sink EDM were used to machine these samples.

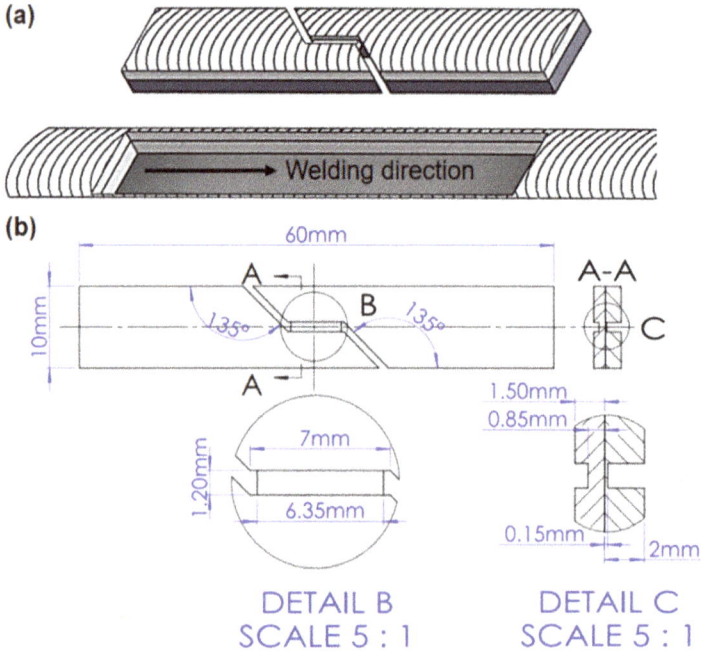

(a)

Welding direction

(b)

60mm

10mm

A

B

135o

135o

A-A

C

A

1.50mm

0.85mm

7mm

1.20mm

6.35mm

0.15mm

2mm

DETAIL B
SCALE 5 : 1

DETAIL C
SCALE 5 : 1

Figure 2. A shear test specimen with a volume fraction of 15% steel and 85% Al (**a**) schematic illustration of sampling location and (**b**) specimen geometry.

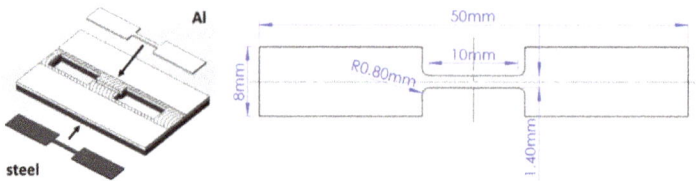

Al

steel

8mm

50mm

10mm

R0.80mm

40mm

Figure 3. Mini-tensile specimens for measuring stir zone Al mechanical responses.

3. Results and Discussion

3.1. Welding Process Optimization

In general, process optimization mainly targets three objectives: producing defect-free welds, maximizing joining efficiency, and minimizing welding time [16]. In the present study, for a potential robotic welding application, the longitudinal tool force under various welding conditions was also taken into account.

Figure 4 shows the combined effect of rotation speed and travel speed on the joint formation. The points in black rhombuses indicate the good welds without visible macro defects by visual inspection and optical microscopy. The points in open circle indicate the joints with volumetric defects and the solid circle in gray color indicates the joints needing high longitudinal force for welding. The results shown in Figure 4 indicated that the joint formation is quite dependent on the rotation speed but independent on the travel speed. The defects mainly occurred when the rotation speed is less than 800 RPM. Sound joints were obtained with a wide range of welding parameters: rotation speeds ranging from 1000 to 1500 RPM and travel speeds ranging from 300 to 1250 mm/min.

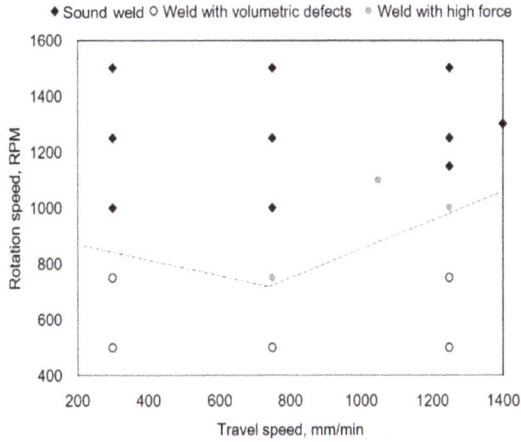

Figure 4. Process window of joint formation.

In the following discussion, we eliminate those data points with defects (points in open circles in Figure 4). For the robotic solution of friction stir welding, the effect of the revolutionary pitch (RP) on the longitudinal force is shown in Figure 5. The major issue with robotic FSW process is the high loads induced by the process itself. In our study, the targeted robotic FSW equipment needs a payload that is estimated to be less than 300 kg (3 kN). Thus, an approximate admissible load of 2.25 kN, i.e., 75% of the payload limit, is chosen as the guideline. The joints with RP no more than 0.3 mm/rev were produced under a low force rate, suggesting aluminum was highly plasticized with high heat input during welding. RP ranging from 0.4 to 1.1 mm/rev resulted in a steady rate of the longitudinal force, which almost reached the threshold of the longitudinal force in robotic FSW. It is noteworthy that all three data points in the solid gray circle in Figure 5 were at a lower level of rotation speeds (750–1100 RPM) compared with others joints at RP values approximately 1 mm/rev. This also means that, with the same RP, decreasing rotation speed may significantly increase the longitudinal force during welding processing as the rotational speed has a strong influence on the specific energy which is related to the temperature distribution within a workpiece generated during FSW [17]. As process temperature decreases, the longitudinal forces increase accordingly.

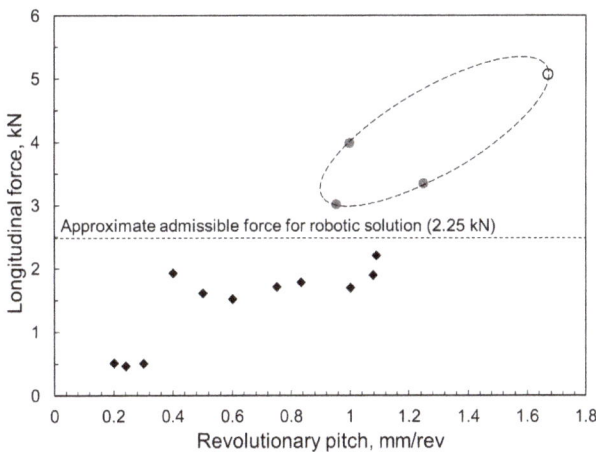

Figure 5. Longitudinal force versus revolutionary pitch.

Figure 6 shows the relationship between RP and joints' lap shear performance. The maximum lap shear strength was obtained at RP of 1.0 mm/rev, which is 71% of the respective weaker base material (AA6082-T6). Joints with good lap shear performances (more than 360 N/mm) were obtained at revolutionary pitches from 0.6 to 1.0 mm/rev except using rotation speeds below 1200 RPM. As seen in Figure 5, the longitudinal force at 0.5 mm/rev is at a similar rate compared with those at 0.6 to 1.0 mm/rev.

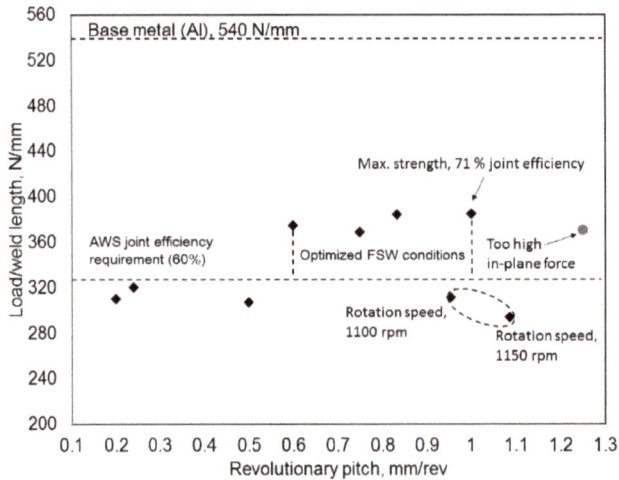

Figure 6. Lap shear test results in a function of revolutionary pitch (RP) value.

Considering joint appearance, tool payload, and maximum joint load capacity, 0.6 to 1.0 mm/rev were selected as the optimal revolutionary pitch range in this work. Thus, to understand the welding condition effect on interfacial microstructures and mechanical properties of the FSW Al/steel joints, further investigation was conducted focusing on the optimized joint with the highest RP (1.0 mm/rev). For comparison, the one with less maximum joint load capacity (0.5 mm/rev) was also selected. The detailed welding parameters are shown in Table 2.

Table 2. Detailed welding parameters for further investigation.

Sample	Rotation Speed, RPM	Travel Speed, mm/min	Revolutionary Pitch, mm/rev
A	1250	1250	1.0
B	1500	750	0.5

3.2. Interfacial Microstructure

Figure 7 illustrates the macrostructures of cross weld section of the two FSWs. When looking at the steel hook on the retreating side, the widths of the thermomechanically affected zone (TMAZ) and SZ of both specimens are different, indicating the potential different interfacial characteristics with different welding heat input.

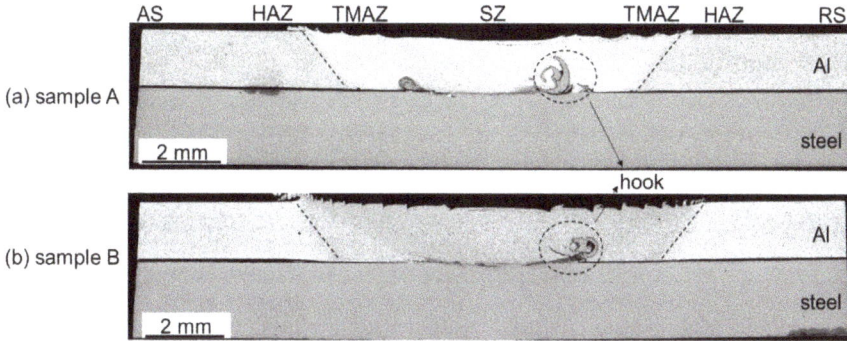

Figure 7. Cross-sectional macrostructures of (**a**) specimen A and (**b**) specimen B. (AS: advancing side, RS: retreating side).

Another feature seen in Figure 7 is the black crack-like layer in the hook zone. It was reported that the hook geometry significantly influenced the joint strength [18]. In addition, the interfacial brittle intermetallic compound plays a key role in the joints' interface strength. Borrisutthekul R et al. [19] proposed that, in order to get satisfactory joints, the thickness of the reaction layer should be limited to several micrometers. Therefore, a detailed SEM investigation was performed to obtain microstructures as well as the thickness of the interfacial IMCs layers in different locations (SZ and hook-zone interfaces). The results are shown in Figure 8. As seen in Figure 8a,b, at an RP of 1.0 mm/rev, the mean thickness of IMC layer is less than 1 μm, and the layer at the hook zone of the retreating side is slightly thicker than the one in the stir zone. Figure 8c presents the center zone interface of 0.5 mm/rev. In comparison with Figure 8a, this result indicates that higher revolutionary pitch results in a thinner IMCs layer which is in agreement with previous findings. However, the mean thickness of the interface at the hook zone of 0.5 mm/rev (see Figure 8d) is 0.97 μm which is thinner than the center zone interface at the same welding condition. More interesting is that it is approximately equal to the interface of the hook zone at the higher revolutionary pitch (1.0 mm/rev).

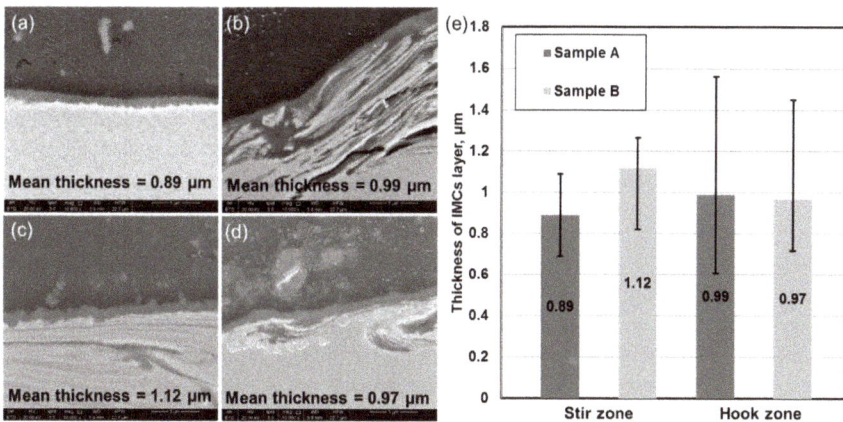

Figure 8. SEM images for evaluating the thickness of intermetallic compounds: (**a**) stir zone of sample A, (**b**) hook zone at retreating side of sample A, (**c**) stir zone of sample B, (**d**) hook zone at retreating side of sample B, and (**e**) thicknesses of interfacial layers.

The variation of the interfacial IMCs thickness at hook zone is unreasonable in general understanding. It was reported that zinc has a significant effect on the performance of Al/steel joints, but this effect may be positive or negative [20–23], suggesting Zn plays a possible influence on interfacial IMCs. In this study, during the welding process, the pin tip was slightly inserted into the steel sheet with a zinc coating layer. In the stir zone with drastic deformation, the zinc coating was likely to be stirred into the aluminum matrix. That means the center interface hardly contain zinc layer. To confirm this, EDS element distribution maps were carried out to trace the zinc element (Figure 9). The results indicate there is no aggregation of zinc in the interfacial zone. The presence of zinc was further checked by TEM investigation. EDS element distribution maps of the hook zone of sample A are present in Figure 10. As seen from the results, the black crack-like layer in Figure 7 is distinguished as the zinc coating. That is to say, at the hook zone, the zinc coating was lifting up and away from the steel surface. The nearly intact morphology of the zinc layer indicates there was no significant dissolution of zinc at the interfacial zone prior to the lift-up of the zinc layer.

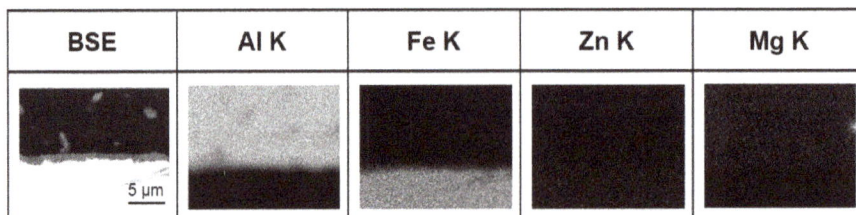

Figure 9. EDS element mapping of the center interface of sample A.

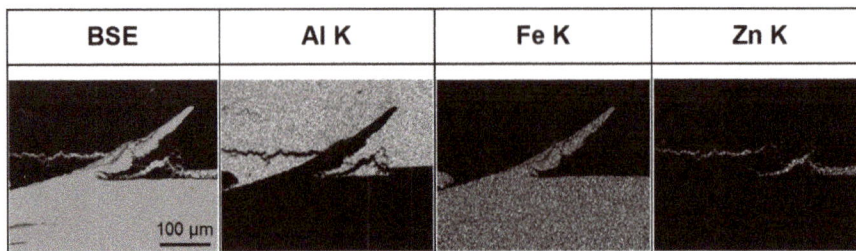

Figure 10. EDS element mapping of the hook zone at the retreating side of sample A.

However, the situation is different for the case with the lower revolutionary pitch (0.5 mm/rev). It should be noted that the abnormal thickness of IMCs layer in Figure 8d was observed in the region I in Figure 11a. To clarify whether this is a real reflection of the specimen, the hook zone structures and EDS element distribution maps in different regions were also investigated. Similar to Figure 10, zinc layer rising was also observed in Figure 11a. Figure 11b,c presents the EDS element distribution maps of regions I and II. As seen from Figure 11b, the zinc coating is highly raised up in region I and there is no zinc left. However, as shown in Figure 11c, the rising zinc coating in region II shows signs of dissolution in the aluminum matrix. During the welding process, in comparison with region II, region I is relatively closer to the rotating pin that is easier to be affected by the mechanical effect of the pin. Therefore, in the region I, the rapid rise of the zinc layer causes no dissolution at the interface because the mechanical effect is stronger than the heat effect. In contrast, in region II, due to the limited lifting of zinc, the dissolution of the zinc layer occurred at the interfacial zone. Since no zinc at the interfacial in the region I as seen in the SEM analysis, the abnormal IMCs layer in this region is yet to explain. Thus, region I was further analyzed using TEM and compared with other locations.

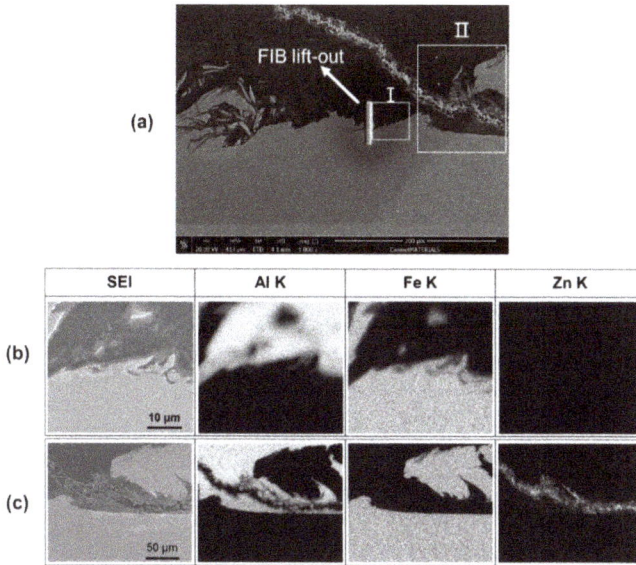

Figure 11. Hook zone at the retreating side of sample B: (**a**) SEM image of hook zone and energy dispersive X-ray spectrometry (EDS) element mapping of (**b**) region I and (**c**) region II.

Figure 12 shows the TEM images of the interface in SZ of specimen A. The interfacial IMCs layer reveals a simple flat microstructure (see Figure 12a), which thanks to the high travel speed. The HAADF image of the interface is given in Figure 12b and the relevant EDS combination map is presented in Figure 12c. While the elemental distribution shows no zinc at the interface the thickness of the IMCs layer is 0.5 ± 0.1 µm, which is slightly thinner than the SEM results. In addition, two kinds of IMCs were detected by SAD patterns: $Al_{3.2}Fe$ in the Al-rich side (position d in Figure 12b) and Al_5Fe_2 (position e in Figure 12b) in the Fe-rich side. The distribution of interfacial IMCs is consistent with the reported results of Al/steel friction stir welded joints [12,24].

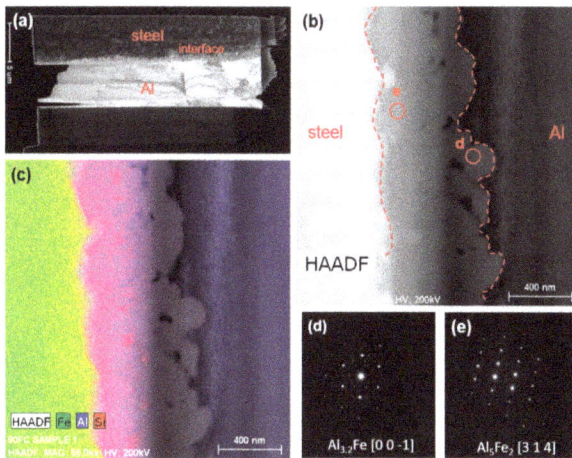

Figure 12. TEM micrographs of the interfacial microstructure at SZ of sample A: (**a**) whole view, (**b**) HAADF interfacial image, (**c**) EDS combination elemental map and SAD patterns of (**d**) Position d and (**e**) Position e in (**b**).

The TEM images of the SZ interfacial layer in specimen B are given in Figure 13. Similar to specimen A (Figure 12), $Al_{3.2}Fe$ and Al_5Fe_2 were also identified at the interface at the same order. In comparison with Figure 12a, a significant difference as seen in Figure 13a is the needle-shaped objects scattered in the aluminum matrix (higher magnification view in Figure 13b,c), which were identified as $Al_{3.2}Fe$. The interfacial reaction layer is thicker than that in specimen A which also agrees with the SEM analysis (Figure 8). Note that no zinc was found in both elemental distribution maps as shown in Figures 12c and 13c. That is to say, in the stir zone, with the pin inserted into the steel, the zinc coating was apart from the original location and no zinc remained at the interface.

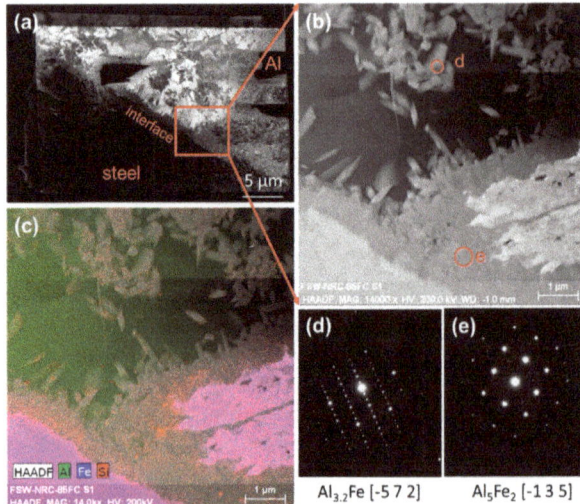

Figure 13. TEM micrographs of the interfacial microstructure at the stir zone of sample B: (**a**) low magnification STEM images, (**b**) high angle annular dark field (HAADF) interfacial image, (**c**) EDS combination elemental map and SAD patterns of (**d**) $Al_{3.2}Fe$ and (**e**) Al_5Fe_2.

Adjacent to the steel hook at the retreating side, the interfacial zone was also examined using TEM analysis (Figures 14 and 15). As seen in Figure 14a for specimen A, the sampling position is very close to the Zn ascent. However, from the EDS elemental map (Figure 14c), no Zn was found in the interfacial zone. Several interfacial positions (i.e., Al matrix, IMCs layer, and steel matrix) for EDS analysis were also investigated. The maximum Zn content was found to be 1.06% at the IMCs layer where is near the Al-rich side. This means, with a high revolutionary pitch of 1.0 mm/rev, the low welding heat input does not cause significant dissolution of Zn in Al. Also, as seen in Figure 14b,c, the interface shows a flat microstructure; the EDS combination elemental map found the thickness of the IMCs layer to be approximately 600–700 nm. The general interfacial microstructure and thickness are similar to the interface at the stir zone (see Figure 12), indicating the similarity of the heat history at the two positions. Two kinds of IMCs were identified by SAD: Al_6Fe at the Al-rich side and nanocrystalline, which is close to $Al_{3.2}Fe$, at the Fe-rich side.

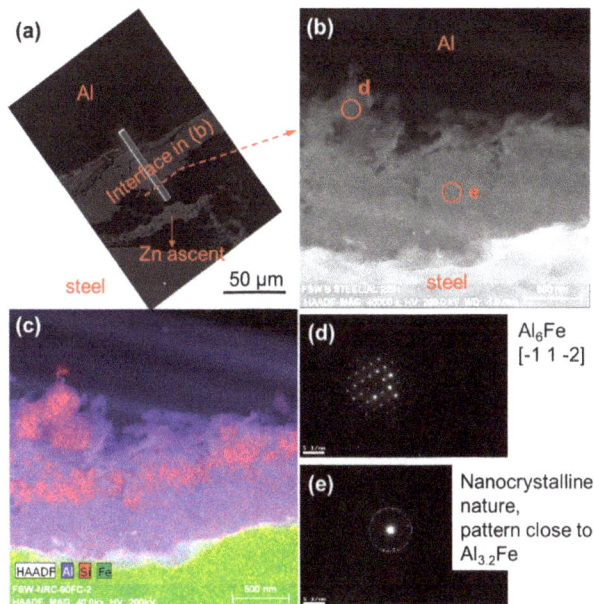

Figure 14. TEM micrographs of the interfacial microstructure at the hook zone of sample A: (**a**) focused ion beam (FIB) sampling position, (**b**) HAADF image, (**c**) EDS combination elemental map and SAD patterns of (**d**) Al6Fe and (**e**) nanocrystalline close to Al$_{3.2}$Fe.

The hook zone TEM images of specimen B are shown in Figure 15. From the sampling position as shown in Figure 15a, the Zn layer ascended under the impact of stirring pin at the steel hook zone. Figure 15b shows the TEM overview of the FIB specimen, the red rectangle indicates the EDS analysis location. EDS analyses of different interfacial regions in Figure 15c are presented in Table 3. Only a very small amount of Zn (0.2%) was found at location 1, i.e., for RP of 0.5 mm/rev, the dissolution of Zn is trivial or not in the Al matrix. Zn appeared at the interface from the EDS combination elemental map as shown in Figure 15d. Within the IMC layer, location 4 contains 38.3% Zn, which is much higher than the Al matrix. These results indicate that interfacial Zn is not caused by the dissolution of Zn. It could be due to the residue of Zn-coating when it was ascending with the pin rotating. This phenomenon does not occur in specimen A which is potentially due to the lower welding heat input of sample A. Note that the mean thickness of the IMCs layer is about 500 nm, which is thinner than the interface at the stir zone (see Figure 13) of sample B. It was reported that Zn promoted solid solubility of Fe in Al, which probably contributed to decreasing the intermetallic compounds in the layer structures [23]. In this work, the solubility of Fe in the Al matrix does not significantly increase, as seen in the EDS results (position 1). However, it is interesting to note that Si (~6%) and Mn (~1%) are found to diffuse from the Fe-side to the Al-side which causes aggregation in the interface (Figure 15e,g). The SAD pattern of the interface is shown in Figure 15h, but the interfacial IMC could not be identified based on the current database. From EDS analyses (location 2 and 3), it could be speculated that the IMC is Al-Fe-Si-based with ~6% Si. It was reported that, in Al/steel brazing joints, Si could suppress the diffusion of Fe from the steel-side into the weld, thus reducing the thickness of the IMC layer [25,26] and even generating Al-Fe-Si-based IMC [26]. This means the abnormal thin IMC layer could be related to the aggregation of Si and Mn and the generation of Al-Fe-Si-based IMC in the interface.

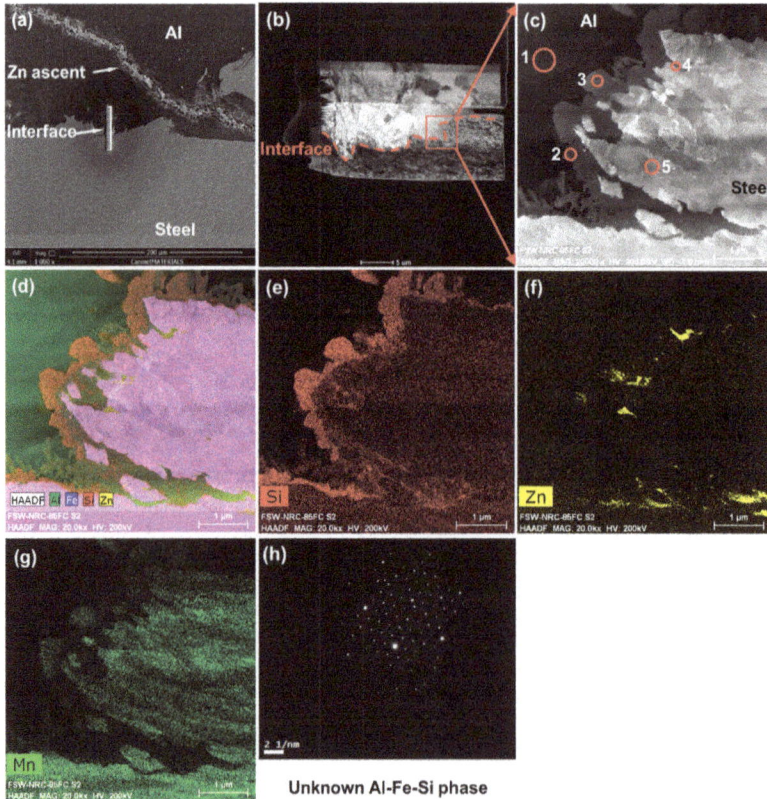

Figure 15. Interfacial microstructure at the hook zone of sample B: (**a**) FIB sampling position, (**b**) TEM overview of the interfacial zone, (**c**) HAADF image, (**d**) EDS combination elemental map, individual elemental maps of (**e**) Si, (**f**) Zn, and (**g**) Mn, and (**h**) SAD pattern.

Table 3. EDS spot analyses of the interfacial zone (see Figure 15c).

Position	Element (at. %)							
	Al	Fe	Si	Zn	Mn	Cr	Cu	Mg
1	98.2	0.1	0.2	0.2	-	-	0.9	0.4
2	74.0	17.4	5.9	0.1	1.1	0.1	1.5	-
3	75.4	15.8	5.6	0.1	1.5	0.1	1.4	-
4	18.0	21.8	0.6	38.3	0.6	-	3.5	17.3
5	0.6	93.5	0.7	-	1.5	0.2	3.5	-

3.3. Mechanical Properties

Owing to the fact that the steel is thicker and of much higher strength than aluminum, sound lap joints are unlikely to fracture at the steel sheet. We concentrated on the microhardness distribution of the aluminum sections and results are presented in Figure 16. In general, when friction stir welding age-hardening aluminum alloys, due to the welding heat input, the precipitates are dissolved or coarsened in the weld nugget and HAZ, these regions would be softer than the base metal. As seen from the hardness profiles, similar results have been observed. The softest region in the aluminum sheet is the junction between HAZ and TMAZ. The above results are in good agreement with the reported study of AA6082 FSWs [27].

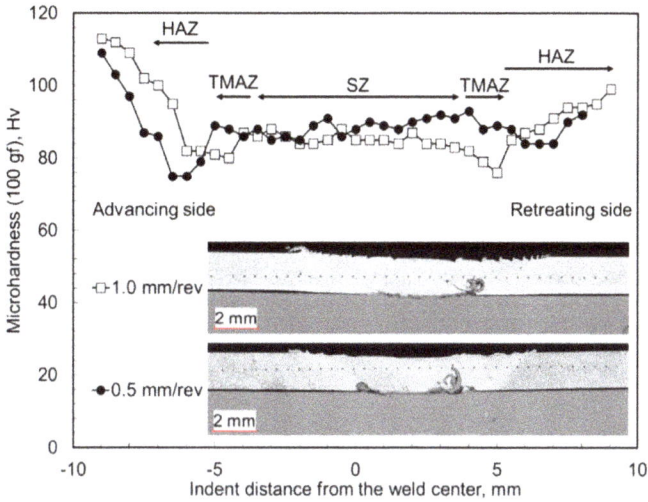

Figure 16. Microhardness profiles of the aluminum sections with different RPs.

However, it is interesting to note the similar microhardness values of the stir zone at 1.0 and 0.5 mm/rev. To validate these results, a tensile test was carried out and the results are presented in Figure 17. Although the hardness of the stir zone is significantly lower than that of the base metal, the stir zone shows good mechanical properties. The ultimate tensile strength (UTS) of the stir zone did not decrease significantly and the elongations were better than that of the base metal. It should be noted that the strength efficiency of the stir zone is higher than the reported result of AA6082-T6 friction stir welded butt joints [27]. In that report, the stir zone at RP of 0.25 mm/rev performed a strength efficiency of 77%. As a comparison, in this work, RPs of 0.5 mm/rev and 1.0 mm/rev resulted in strength efficiencies of 87% and 89%, respectively.

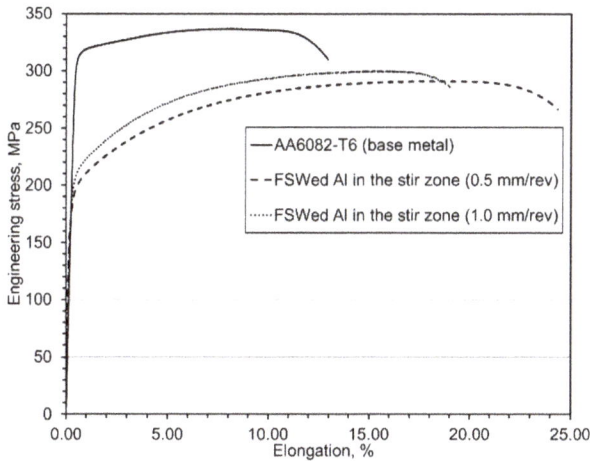

Figure 17. Engineering stress–strain curves of the aluminum in the stir zone compared with the base material.

For joints at 1.0 and 0.5 mm/rev, given the strong mechanical properties of SZ and the similar hardness profile of the welded Al sheet, the two joints are expected to exhibit similarly good lap shear performance considering the interfacial IMCs layers are thinner than several micrometers and possibly enhanced the joints. However, as mentioned in Figure 6, the experimental result is different from the above guess. Figure 18 presents the lap shear curves of the two welding conditions. As can be seen, the maximum lap shear loads of three joints at 1.0 mm/rev are at a stable level. In contrast, the results of the three joints at 0.5 mm/rev are fluctuant. IMC plays a key role on the mechanical properties of the joints [28]. The thickness of the brittle IMC usually should be limited to several micrometers or even thinner to obtain acceptable joint strength. As seen from Figures 12 and 14, the revolutionary pitch determines the IMC thickness at the interfaces. In contrast, we also found that the welding parameter changes have little effect on the mechanical performance of the aluminum in the stir zone (from Figures 16 and 17). It should be noted that no interfacial fracture occurred in the specimens of 1.0 mm/rev, but interfacial fracture occurred in the specimens of 0.5 mm/rev. Therefore, we could conclude that revolutionary pitch determines the joint fracture location and lap shear specimen performance by determining the interfacial microstructure.

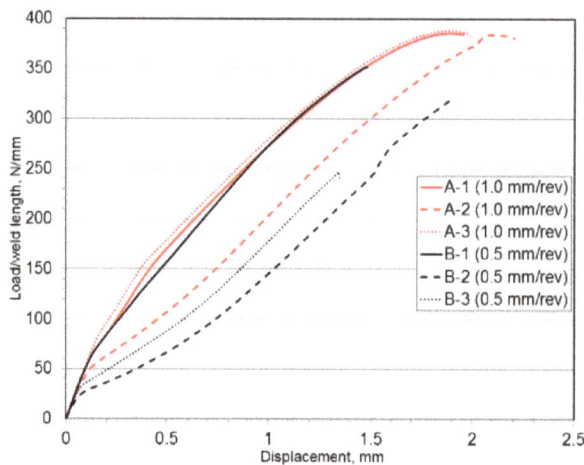

Figure 18. Lap shear curves at the different revolutionary pitch, joint strength expressed as the maximum failure load per millimeter of the weld.

3.4. Estimation of Stir Zone Mechanical Behavior

With the development of FSW and the goal of understanding the joints, researchers developed several models to predict the joint's mechanical properties such as hardness [29], UTS [30–32], elongation [32], and fatigue lifetime [33]. So far, the prediction of mechanical behavior is rarely reported. In this part, based on the experimental data of stir zone aluminum and steel, we tried to model the stir zone mechanical behavior in different aluminum/steel thickness combinations. According to the modeling results, the role of interface microstructure on mechanical properties of the FSWs was demonstrated.

The most commonly used mixture law, the iso-strain assumption, is used for modeling [34]. In this assumption, the strain of each constituent is taken equally. Correspondingly, in the stir zone, the constituents could be divided into three kinds: welded Al, welded steel, and interfacial IMCs. For easily running the modeling, we assume that no phase transition during the deformation and ignore the slight content of Zn in the interface. Thus, the stress equation could be given by

$$\sigma(\varepsilon) = V_{Fe} \cdot \sigma_{Fe}(\varepsilon) + V_{Al} \cdot \sigma_{Al}(\varepsilon) + (1 - V_{Fe} - V_{Al}) \cdot \sigma_{IMC}(\varepsilon) \tag{1}$$

where $\sigma_{Fe}(\varepsilon)$, $\sigma_{Al}(\varepsilon)$, and $\sigma_{IMC}(\varepsilon)$ are the flow stresses of steel, aluminum, and IMC, respectively, V_{Fe} and V_{Al} are the volume fractions of steel and aluminum, respectively.

It should be highlighted that different kinds of IMCs have different mechanical responses so it is difficult to quantify those in practice. In order to continue to simplify the modeling, we could treat the microscale IMCs layer as the strong boundary between Al constituent and steel constituent. Thus, the equation could be given as

$$\sigma(\varepsilon) = V_{\text{Fe}} \cdot \sigma_{\text{Fe}}(\varepsilon) + (1 - V_{\text{Fe}}) \cdot \sigma_{\text{Al}}(\varepsilon) \tag{2}$$

By conventional tensile and modified shear testing, the stress–strain curves of basic constituents were obtained and presented in Figure 19. It should be noted that the true strains of tensile data are smaller than those of the shear test. For comparison, by calculating the reduction of area at fracture, the stress–strain curves of tensile data were thus lengthened. Results clearly show that the aluminum stresses of tensile data are higher than those of shear-test data. The mechanical behavior of welded steel is quite dependent on the weld parameters. The experimental data of aluminum in the stir zone was used to model the mechanical response of friction stir welded aluminum. The conventional tensile experimental data of steel in the stir zone was used to model the mechanical response of friction stir welded steel.

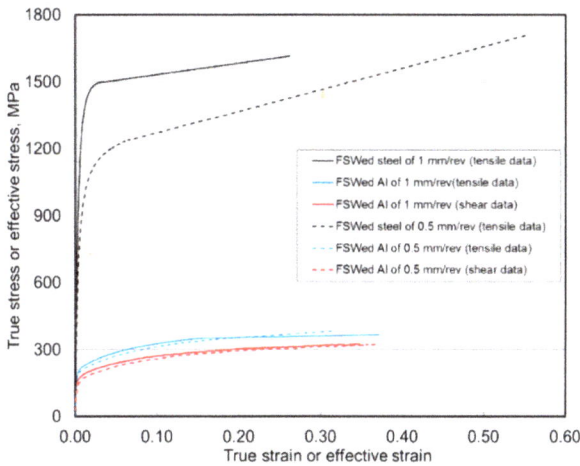

Figure 19. True and effective stress–strain curves of constituents in the stir zone obtained from tensile and shear test.

For tests on specimens with different RPs, the modeling results are presented in Figure 20 along with the experimental ones. It should be noted that no delamination occurred during shear testing owing to the mentioned microscale interfacial IMCs layer. As we know, FSW process is a solid-state welding process with intense mixing and stirring. Thus, the rotating pin has the potential to introduce some small steel fragments from the steel matrix into the aluminum matrix. Note that the volume fractions of the constituents are measured by OM as previously described. The local volume fractions of two constituents may vary within the joint. In addition, we ignored the mechanical responses of Zn and IMCs. In consideration of these factors, the modeling results are in good agreement with the experimental ones indicating the resultant interfaces act as good as grain boundaries. This part work prospects an easy way to estimate the joint mechanical response using a modified shear test.

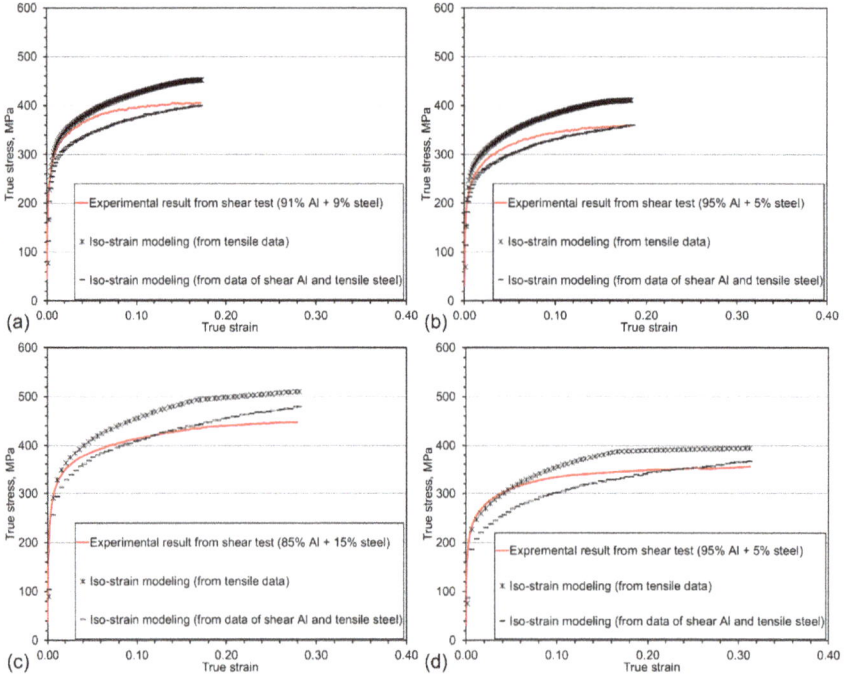

Figure 20. Linear modeling compared with the experimental results at different RPs. (**a,b**) at RP of 1.0 mm/rev with different volume fractions and (**c,d**) at RP of 0.5 mm/rev with different volume fractions.

4. Conclusions

The dissimilar lap joining of 1.5-mm thick AA6082-T6 sheet to 2.0-mm thick galvanized Zn-coated DP800 sheet was successfully conducted using friction stir welding process. The interfacial IMCs and their effect on the mechanical properties of FSWs were evaluated. The conclusions can be drawn as follows.

- It is feasible to lap join the 6082-T6 aluminum alloy to DP800 dual-phase steel sheets. Under the welding conditions in the present study, maximum joint strength reaches 71% of that of the base Al material with a RP value of 1 mm/rev (1250 RPM and 1250 mm/min).
- In the stir zone, $Al_{3.2}Fe$ in the Al-rich side and Al_5Fe_2 in the Fe-rich side were detected for two welding conditions studied. In the hook zone, however, Al_6Fe was detected in the Al-rich side and nanocrystalline pattern close to $Al_{3.2}Fe$ at RP of 1.0 mm/rev. Under a relatively low RP (0.5 mm/rev) in the hook-zone, zinc was found at the interface with the aggregation of Si and Mn elements at the Al-rich side of the interface which leads to the generation of Al-Fe-Si phase thus decreases the thickness of IMCs layer.
- In the stir zone, RP has a significant influence on the interfacial microstructures. The interfacial IMCs layer at an RP of 1.0 mm/rev is simple and flat but the one at RP of 0.5 mm/rev becomes thicker and more complex where IMCs are scattered in the Al matrix. However, the stir zone aluminum, under different RP values, is similar in microhardness value and tensile behavior.
- The iso-strain-based linear mixture law was used to model the stir zone mechanical response. The modeling results are in good agreement with the experimental ones, indicating the microscale IMCs act good as strong boundaries of dissimilar materials.

Author Contributions: Funding acquisition, J.K., F.N., and Y.C.; Investigation, J.K., S.L., B.S.A., and F.N.; Methodology, J.K., S.L., and F.N.; Supervision, J.K. and Y.C.; Writing—original draft, S.L.; Writing—review & editing, J.K., Y.C., and F.N.

Funding: The authors acknowledged the financial support from Canadian Federal Government Interdepartmental Program on Energy R&D (PERD) and CanmetMATERIALS, Natural Resources Canada. Yuhua Chen and Shuhan Li also express thanks for the financial support from the National Natural Science Foundation of China, grant number 51865035; Fund for Jiangxi Distinguished Young Scholars (2018ACB21016), the Project for Jiangxi Advantageous Scientific and Technological Innovation Team (20171BCB24007 & 20181BCB19002); and Aviation Science Funds of China, grant number 2017ZE56010.

Conflicts of Interest: The authors declare no conflicts of interest.

References

1. Shah, L.H.; Ishak, M. Review of Research Progress on Aluminum—Steel Dissimilar Welding. *Mater. Manuf. Process.* **2014**, *29*, 928–933. [CrossRef]
2. Atabaki, M.M.; Nikodinovski, M.; Chenier, P.; Ma, J.; Harooni, M.; Kovacevic, R. Welding of aluminum alloys to steels: An overview. *J. Manuf. Sci. Prod.* **2014**, *14*, 59–78. [CrossRef]
3. Thomas, W.; Nicholas, E.; Needham, J.C.; Murch, M.; Templesmith, P.; Dawes, C. *Friction Stir Welding, International Patent Application No. PCT/GB92102203 and Great Britain Patent Application*; TWI Ltd.: Cambridge, UK, 1991.
4. Ohhama, S.; Hata, T.; Yahaba, T.; Kobayashi, T.; Miyahara, T.; Sayama, M. Application of an FSW Continuous Welding Technology for Steel and Aluminum to an Automotive Subframe. In Proceedings of the SAE 2013 World Congress & Exhibition, Detroit, MI, USA, 16–18 April 2013.
5. Bozzi, S.; Helbert-Etter, A.; Baudin, T.; Criqui, B.; Kerbiguet, J. Intermetallic compounds in Al 6016/IF-steel friction stir spot welds. *Mater. Sci. Eng. A* **2010**, *527*, 4505–4509. [CrossRef]
6. Uzun, H.; Donne, C.D.; Argagnotto, A.; Ghidini, T.; Gambaro, C. Friction stir welding of dissimilar Al 6013-T4 To X5CrNi18-10 stainless steel. *Mater. Des.* **2005**, *26*, 41–46. [CrossRef]
7. Matlock, D.K.; Speer, J.G. Third generation of AHSS: Microstructure design concepts. In *Microstructure and Texture in Steels*; Springer: Berlin, Germany, 2009; pp. 185–205.
8. Liu, X.; Lan, S.; Ni, J. Analysis of process parameters effects on friction stir welding of dissimilar aluminum alloy to advanced high strength steel. *Mater. Des.* **2014**, *59*, 50–62. [CrossRef]
9. Zhao, S.; Ni, J.; Wang, G.; Wang, Y.; Bi, Q.; Zhao, Y.; Liu, X. Effects of tool geometry on friction stir welding of AA6061 to TRIP steel. *J. Mater. Process. Technol.* **2018**, *261*, 39–49. [CrossRef]
10. Liu, X.; Lan, S.; Ni, J. Electrically assisted friction stir welding for joining Al 6061 to TRIP 780 steel. *J. Mater. Process. Technol.* **2015**, *219*, 112–123. [CrossRef]
11. Threadgill, P.L.; Leonard, A.J.; Shercliff, H.R.; Withers, P.J. Friction stir welding of aluminium alloys. *Int. Mater. Rev.* **2009**, *54*, 49–93. [CrossRef]
12. Elrefaey, A.; Gouda, M.; Takahashi, M.; Ikeuchi, K. Characterization of aluminum/steel lap joint by friction stir welding. *J. Mater. Eng. Perform.* **2005**, *14*, 10–17. [CrossRef]
13. Lee, C.Y.; Choi, D.H.; Yeon, Y.M.; Jung, S.B. Dissimilar friction stir spot welding of low carbon steel and Al–Mg alloy by formation of IMCs. *Sci. Technol. Weld. Join.* **2009**, *14*, 216–220. [CrossRef]
14. Kang, J.; McDermid, J.R.; Bruhis, M. Determination of the constitutive behaviour of AA6022-T4 aluminium alloy spot welds at large strains. *Mater. Sci. Eng. A* **2013**, *567*, 95–100. [CrossRef]
15. Kang, J.; Wilkinson, D.S.; Jain, M.; Embury, J.D.; Beaudoin, A.J.; Kim, S.; Mishira, R.; Sachdev, A.K. On the sequence of inhomogeneous deformation processes occurring during tensile deformation of strip cast AA5754. *Acta Mater.* **2006**, *54*, 209–218. [CrossRef]
16. Mishra, R.S.; De, P.S.; Kumar, N. *Friction Stir Welding and Processing: Science and Engineering*; Springer: Berlin, Germany, 2014.
17. Colligan, K.J.; Mishra, R.S. A conceptual model for the process variables related to heat generation in friction stir welding of aluminum. *Scr. Mater.* **2008**, *58*, 327–331. [CrossRef]
18. Badarinarayan, H.; Shi, Y.; Li, X.; Okamoto, K. Effect of tool geometry on hook formation and static strength of friction stir spot welded aluminum 5754-O sheets. *Int. J. Mach. Tools Manuf.* **2009**, *49*, 814–823. [CrossRef]

19. Borrisutthekul, R.; Yachi, T.; Miyashita, Y.; Mutoh, Y. Suppression of intermetallic reaction layer formation by controlling heat flow in dissimilar joining of steel and aluminum alloy. *Mater. Sci. Eng. A* **2007**, *467*, 108–113. [CrossRef]

20. Chen, Y.; Nakata, K. Effect of the surface state of steel on the microstructure and mechanical properties of dissimilar metal lap joints of aluminum and steel by friction stir welding. *Metall. Mater. Trans. A* **2008**, *39*, 1985. [CrossRef]

21. Ratanathavorn, W.; Melander, A. Influence of zinc on intermetallic compounds formed in friction stir welding of AA5754 aluminium alloy to galvanised ultra-high strength steel. *Sci. Technol. Weld. Join.* **2017**, *22*, 673–680. [CrossRef]

22. Suhuddin, U.; Fischer, V.; Kostka, A.; Santos, J.D. Microstructure evolution in refill friction stir spot weld of a dissimilar Al–Mg alloy to Zn-coated steel. *Sci. Technol. Weld. Join.* **2017**, *22*, 658–665. [CrossRef]

23. Elrefaey, A.; Takahashi, M.; Ikeuchi, K. Friction-stir-welded lap joint of aluminum to zinc-coated steel. *Q. J. Jpn. Weld. Soc.* **2005**, *23*, 186–193. [CrossRef]

24. Chen, C.; Kovacevic, R. Joining of Al 6061 alloy to AISI 1018 steel by combined effects of fusion and solid state welding. *Int. J. Mach. Tools Manuf.* **2004**, *44*, 1205–1214. [CrossRef]

25. Dong, H.; Hu, W.; Duan, Y.; Wang, X.; Dong, C. Dissimilar metal joining of aluminum alloy to galvanized steel with Al–Si, Al–Cu, Al–Si–Cu and Zn–Al filler wires. *J. Mater. Process. Technol.* **2012**, *212*, 458–464. [CrossRef]

26. Xia, H.; Zhao, X.; Tan, C.; Chen, B.; Song, X.; Li, L. Effect of Si content on the interfacial reactions in laser welded-brazed Al/steel dissimilar butted joint. *J. Mater. Process. Technol.* **2018**, *258*, 9–21. [CrossRef]

27. Scialpi, A.; de Filippis, L.; Cavaliere, P. Influence of shoulder geometry on microstructure and mechanical properties of friction stir welded 6082 aluminium alloy. *Mater. Des.* **2007**, *28*, 1124–1129. [CrossRef]

28. Li, S.; Chen, Y.; Zhou, X.; Kang, J.; Huang, Y.; Deng, H. High-strength titanium alloy/steel butt joint produced via friction stir welding. *Mater. Lett.* **2019**, *234*, 155–158. [CrossRef]

29. Zhao, Y.H.; Lin, S.B.; He, Z.Q.; Wu, L. Microhardness prediction in friction stir welding of 2014 aluminium alloy. *Sci. Technol. Weld. Join.* **2006**, *11*, 178–182. [CrossRef]

30. Zhang, Z.H.; Li, W.Y.; Li, J.L.; Chao, Y.J. Effective predictions of ultimate tensile strength, peak temperature and grain size of friction stir welded AA2024 alloy joints. *Int. J. Adv. Manuf. Technol.* **2014**, *73*, 1213–1218. [CrossRef]

31. Dewan, M.W.; Huggett, D.J.; Liao, T.W.; Wahab, M.A.; Okeil, A.M. Prediction of tensile strength of friction stir weld joints with adaptive neuro-fuzzy inference system (ANFIS) and neural network. *Mater. Des.* **2016**, *92*, 288–299. [CrossRef]

32. Sundaram, N.S.; Murugan, N. Tensile behavior of dissimilar friction stir welded joints of aluminium alloys. *Mater. Des.* **2010**, *31*, 4184–4193. [CrossRef]

33. Sun, G.; Chen, Y.; Chen, S.; Shang, D. Fatigue modeling and life prediction for friction stir welded joint based on microstructure and mechanical characterization. *Int. J. Fatigue* **2017**, *98*, 131–141. [CrossRef]

34. Bouaziz, O.; Buessler, P. Mechanical behaviour of multiphase materials: An intermediate mixture law without fitting parameter. *Rev. Métall. Int. J. Metall.* **2002**, *99*, 71–77. [CrossRef]

![metals logo] *metals*

MDPI

Article

Friction Stir Spot Welding-Brazing of Al and Hot-Dip Aluminized Ti Alloy with Zn Interlayer

Xingwen Zhou [1,2], Yuhua Chen [1,*], Shuhan Li [1,*], Yongde Huang [1], Kun Hao [1] and Peng Peng [2]

[1] School of Aerospace Manufacturing Engineering, Nanchang Hangkong University, Nanchang 330063, China; xingwenzhou@buaa.edu.cn (X.Z.); huangydhm@nchu.edu.cn (Y.H.); 13362624883@163.com (K.H.)

[2] School of Mechanical Engineering and Automation, Beihang University, Beijing 100191, China; ppeng@buaa.edu.cn

* Correspondence: ch.yu.hu@163.com (Y.C.); shuhanli@outlook.com (S.L.); Tel.: +86-133-3006-7995 (Y.C.); +86-156-7912-8826 (S.L.)

Received: 23 October 2018; Accepted: 5 November 2018; Published: 8 November 2018

Abstract: Friction stir spot welding (FSSW) of Al to Ti alloys has broad applications in the aerospace and automobile industries, while its narrow joining area limits the improvement of mechanical properties of the joint. In the current study, an Al-coating was prepared on Ti6Al4V alloy by hot-dipping prior to joining, then a Zn interlayer was used during friction stir joining of as-coated Ti alloy to the 2014-Al alloy in a lap configuration to introduce a brazing zone out of the stir zone to increase the joining area. The microstructure of the joint was investigated, and the joint strength was compared with the traditional FSSW joint to confirm the advantages of this new process. Because of the increase of the joining area, the maximum fracture load of such joint is 110% higher than that of the traditional FSSW joint under the same welding parameters. The fracture load of these joints depends on the joining width, including the width of solid-state bonding region in stir zone and brazing region out of stir zone.

Keywords: friction stir spot welding; friction stir spot brazing; joining area; fracture load

1. Introduction

Lightweight hybrid structures have attracted increasing attention in the aerospace and automobile industries to enhance performance efficiency as well as to reduce environmental impact [1–4]. In this context, the dissimilar joining of aluminum (Al) and titanium (Ti) alloys has been widely used in the manufacturing of lightweight hybrid structures [4–6]. However, joining these two alloys is a considerable challenge because of their different physical properties, limited mutual solubility, and formation of brittle intermetallic phases in the Al–Ti alloying system [7–10].

Several solid-state joining processes, such as friction stir welding (FSW) [11]; diffusion bonding [12,13]; and fusion welding with filler metals, such as laser welding [10,14,15] and arc welding [8,16], were thus developed for dissimilar joining to overcome the difficulties mentioned above. Among these, an innovative solid-state spot welding process, friction stir spot welding (FSSW), is a promising method that can reduce the formation of intermetallic compounds (IMCs) because of its low processing temperature, and has various advantages, such as a simple process, as well as being environment-friendly and energy efficient [5,6,17]. In a typical FSSW, a rotating cylindrical pin tool is plunged into the workpiece to be welded. Frictional heat is generated in the plunging and stirring stages, and thus the materials adjacent to the tool are heated, softened, and mixed in the stirring stage where a solid-state joint will be formed [18]. Up to now, many reports on FSSW of similar or dissimilar metals mainly focus on the process, microstructural characteristics, and numerical simulations to optimize the mechanical properties of the joint [17–21]. Unavoidably, the traditional FSSW would leave a keyhole that reduces its bonding area [22]. Although some modified processes for FSSW, such as refill FSSW [23], flat FSSW [24], and short travel FSSW [25], have attempted to eliminate this visible keyhole, they are complicated and time-consuming owing to the rigorous welding conditions and/or complex tool design. Recently, friction stir brazing (FSSB) has been developed by using a pinless tool and adding brazing filler metal to induce metallurgical reaction at the interface with the aid of friction heat and forging pressure, instead of plastic flow [22]. This technique has successfully introduced a brazing zone in dissimilar joints of steel/Al [26], Cu/Al [27], and Al /Mg [28], while FSSB of Al and Ti alloys is yet to be reported.

This work aims to find an effective approach to improve the joining area of the Al/Ti joint by combining FSSW and FSSB. Zinc (Zn) is an optional filler for FSSB of Al alloys as a result of its low melting point (420 °C), high solubility in Al, as well as absence of formation of IMCs in the Al–Zn alloying system [29–32]. During our preliminary study, a Zn interlayer was placed between the interface of 2014 Al and Ti6Al4V alloys for the FSSB process to determine their weldability. It was found that the high affinity of Ti towards oxygen led to the formation of non-protective oxide scales on the Ti6Al4V [33], resulting in poor wettability of Zn to Ti alloys and, therefore, failure to join Al and Ti alloys. Recently, using an electroplating process before the FSB process would produce a copper layer on the surface of graphite, which converted the graphite–copper lap joint to the similar joining of copper to copper [34]. This gives us an idea that pretreating the base metal will be a possible benefit for FSSB of Al and Ti.

In the present paper, we utilized hot dip to aluminize T6Al4V alloys and developed a friction stir spot joining process, called friction stir spot welding-brazing (FSSW-B), for joining of 2014-T4 Al and aluminized T6Al4V alloys with a Zn interlayer. The effect of tool rotation speed on the interfacial structure and mechanical properties of joints was investigated in detail and the relation between microstructure, mechanical behavior, and process parameters was established.

2. Experimental

The substrate materials under study were aluminum alloy 2014-T4 and annealed titanium alloy Ti6Al4V sheets, both 3-mm thick and with dimensions of approximately 80 mm × 35 mm. Table 1 lists the nominal chemical composition of the two alloys.

Table 1. Chemical compositions of the 2014-T4 aluminum alloy and Ti6Al4V titanium alloy (wt.%).

Alloys	Cu	Si	Mn	Mg	Fe	Zn	Ti	Ni	Al	V	C	N	H	O
2014-T4	4.3	1.0	0.73	0.55	0.3	0.08	0.02	0.02	Bal.	-	-	-	-	-
Ti6Al4V	-	-	-	-	0.026	-	Bal.	-	6.0	4.0	0.015	0.008	0.007	0.06

A molten bath consisting of pure Al (99.9 wt.%) was used for the hot-dip aluminizing of the Ti6Al4V alloy. Prior to the hot-dip aluminizing, the specimens were treated with an acid pickling solution containing 10 vol.% HF, 10 vol.% HNO$_3$, and 80 vol.% H$_2$O for 10–15 s followed by washing in water and acetone, and then drying. The molten Al bath was kept at 780 °C, and the treated Ti6Al4V specimens were immersed in the bath for 25 min. The aluminized Ti6Al4V samples were subsequently removed from the molten bath and then dried in air. Figure 1a shows the cross-section micrograph of the hot-dipped Al coating on the Ti6Al4V substrate. The coating has a thickness of about 200 μm and dispersed particles that have an average diameter of 5.2 μm. These particles have an atomic ratio of Ti to Al of about 1:3 (composition of 22.4 Ti-77.6 Al (in at. %)), which is consistent with the composition of TiAl3. Furthermore, the interface between the alloy and the coating is not flat but undulating, and a continuous TiAl3 layer with a mean thickness of 7.3 μm formed at the interface (see Figure 1b). The content of TiAl3 in this Al coating could be controlled by varying the hot-dip aluminizing process [35,36].

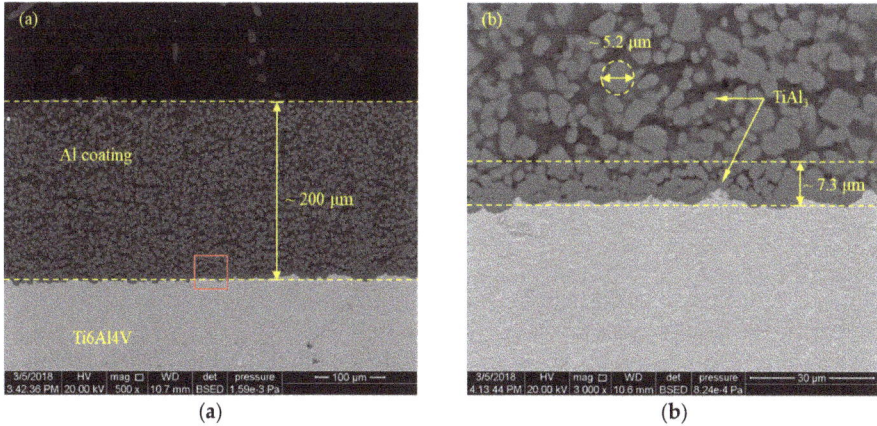

(a)   (b)

Figure 1. Cross-sectional morphology of the hot-dip aluminized Ti6Al4V alloy; (**a**) low magnification and (**b**) high magnification (red square marked in (**a**)).

Figure 2 shows the schematic illustration of the FSSW-B setup. All the FSSW-B operations were conducted in lap joint configuration so that the Al plate was always on top of the Ti sheet with a 35 mm × 35 overlap between two sheets. The welding machine used in this study was modified from an X53K milling machine (Tonmac, Nantong, China). The stir tool was made of a GH4169 superalloy, with a tool shoulder of 16 mm and a circle threaded pin 5 mm in diameter. The pin length was 4 mm, exceeding the thickness of the Al top plate, to induce sufficient mechanical stirring in these two base metals. Joining was performed with a tool rotation speed of 900 rpm to 1500 rpm (900 rpm, 1200 rpm, and 1500 rpm), a constant tool penetration depth of 0.3 mm, and a constant dwell time of 15 s.

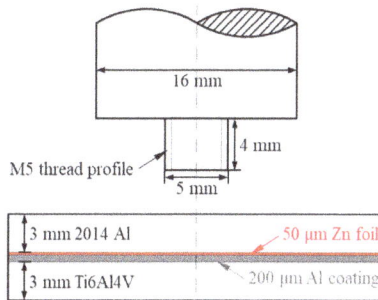

Figure 2. Schematic of the set-up used in the friction stir spot welding-brazing (FSSW-B) process.

To study the characteristics of the joint, the samples were sectioned transversely and polished for metallographic examinations. The microstructure and chemical composition of the cross-section of joints were examined using an FEI Inspect S50 scanning electron microscopy (SEM, ThermoFisher, Hillsboro, OR, USA) equipped with an Oxford Inca X-Act energy dispersive spectrometry (EDS, ThermoFisher, Hillsboro, OR, USA).

The tensile shear properties of the FSSW-B joint and FSSW joint at same welding conditions were tested to compare the mechanical properties. The dimension of the specimens was 125 mm in length and 35 mm in width. A supporting plate was placed at each end of the tensile specimen to maintain the joint weld region parallel to the tensile loading direction. The lap-shear tensile tests were conducted at room temperature on a WDW-50 tensile test machine with a constant crosshead speed of 1 mm/min. The fracture surfaces after tensile testing were examined by SEM and EDS.

3. Results and Discussion

3.1. Macrostructure

The cross-section of the FSSW-B joint produced at a tool rotation speed of 1200 rpm is shown in Figure 3. Macroscopically, the Al coating is still visible at the interface, while the Zn interlayer is hard to observe. Retraction of the tool caused a visible keyhole at the center of the joint and a shallow indentation at the top of upper Al alloy. A partially bonded region (commonly referred to as the hook [37,38]) formed by upward bending of the interface is clearly observed as well. In a traditional FSSW joint, the distance from the tip of the hook to keyhole interface (including stir zone and hook region) is addressed as the bond width of weld due to the existence of an unbonded region outside of the hook [39]. In this study, another joining zone under the shoulder is detected outside of the hook as shown in Figure 3, which is named as the brazing zone and will be analyzed in detail later.

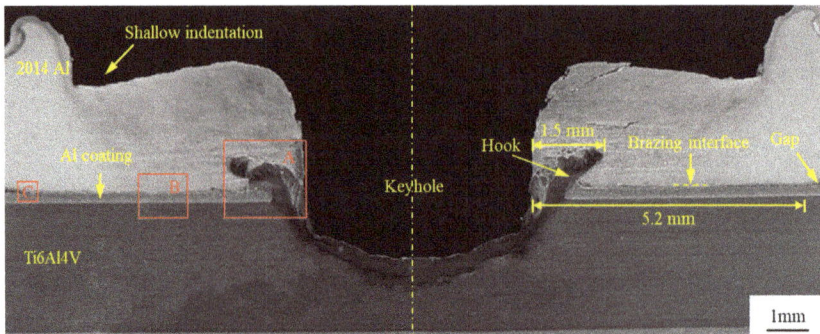

Figure 3. Typical cross-sectional of the FSSW-B joint made at the rotation speed of 1200 rpm. The labeled dimensions were measured via the optical microscope measurement.

3.2. Microstructure

Figure 4a shows the microstructure of the stir zone (region A marked in Figure 3) with several areas of interest (S1, S2, and S3) as highlighted by rectangular boxes. EDS element distribution maps of these selected regions in the stir zone are present in Figure 5. It indicates that the darker region consists of the Al-rich phase and the lighter region consists of the Ti-rich phase.

One of the important factors that can influence the strength of a dissimilar weld is the formation and distribution of IMCs [40–42]. The formation of IMCs during FSW of Al and Ti alloys has been discussed in detail elsewhere [7,43]. Generally, $TiAl_3$ IMC is preferable in FSWed Al/Ti joints because the $TiAl_3$ phase is the only transient phase when the reaction temperature is lower than the melting point of Al [44]. Thermodynamically, the free energy of formation of $TiAl_3$ is also the minimum among Ti_3Al, $TiAl$, and $TiAl_3$ compounds [8]. In the current study, $TiAl_3$ IMC has also been observed in the stir zone, which exists in two features, layer-like and particle-like. The layer-like $TiAl_3$ as marked in Figure 4b is about 10 μm in thickness, showing a mechanical interlocking with the base metals. It is formed because of the mixing of Al and Ti at a high temperature during friction stir processing [45]. Apparently, the particle-like $TiAl_3$ in the stir zone comes from original Al coating, which has been stirred into the Al alloy by the rotating pin. However, they aggregate at the interface outside the hook (see Figure 4c) and the tip region of the hook (see Figure 4d) caused by vigorous mixing and deformation. Interestingly, no Zn element was detected in these regions, indicating the added Zn interlayer has been removed from the stir zone (see Figure 5). Therefore, the Zn interlayer and Al coating have no significant influence in the stir zone of the FSSW-B joint.

Figure 6 displays the microstructure of a reaction layer with an average thickness of about 22 μm between the upper Al alloy and Al coating (region B marked in Figure 3). There are numerous TiAl$_3$ particles present in this reaction layer, and these particles maintain their original shape. EDS line scanning results reveal that the matrix of this reaction layer mainly consists of Al and Zn elements (see Figure 6b,c). Therefore, most of the TiAl3 particles are likely to remain intact and no obvious IMC is generated in the reaction layer during the FSSW-B process. According to the presence of Zn in this layer, it is reasonable to conclude that the Al alloy and Al coating were brazed with the Zn interlayer. Moreover, the thickness of the Al coating, about 200 μm (as shown in Figure 6a) in this brazing zone, is close to that before the joining process, suggesting that this FSSW-B process could not cause significant changes in the Al coating underneath the shoulder. At the end of the brazing zone (region C marked in Figure 3), as shown in Figure 6d, a large amount of lamellar and island Al–Zn eutectic structures, including the dark Al-rich phase and light Zn-rich phase, are observed. This confirms the melting of the Zn foil during the FSSW-B process [29].

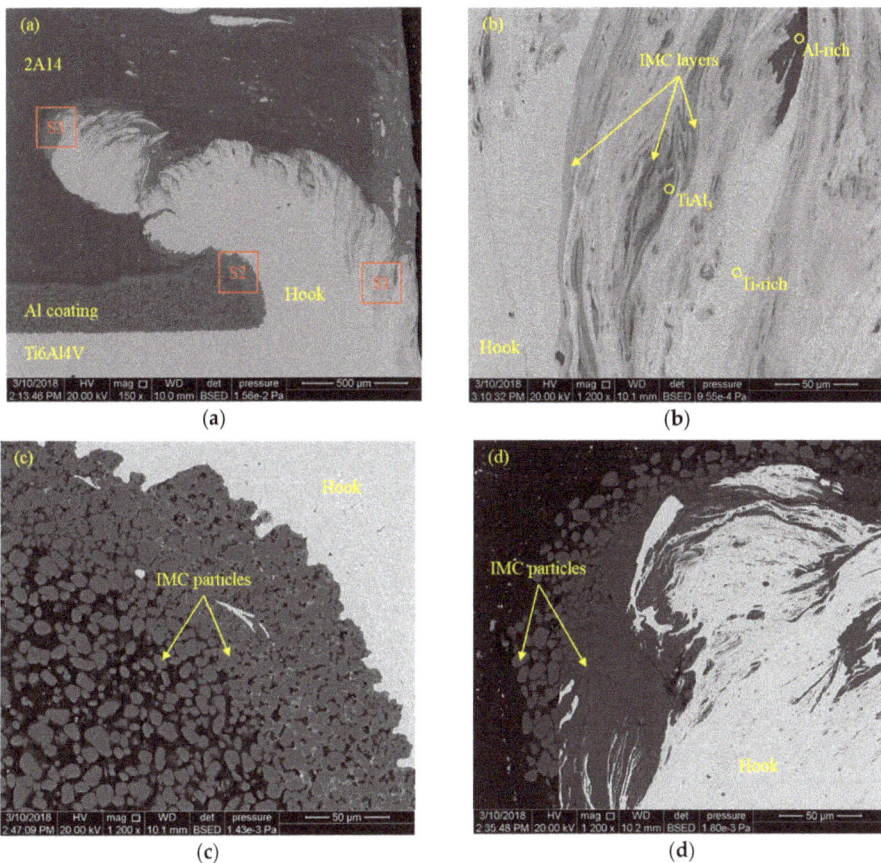

Figure 4. Scanning electron microscopy (SEM) image showing different regions in the stir zone of the FSSW-B joint; (**a**) high magnification view of the hook zone (region A marked in Figure 3a); (**b**) region S1 indicating the interlocking; (**c**) region S2 indicating the tip of the hook; and (**d**) magnified view of region S3. IMC—intermetallic compound.

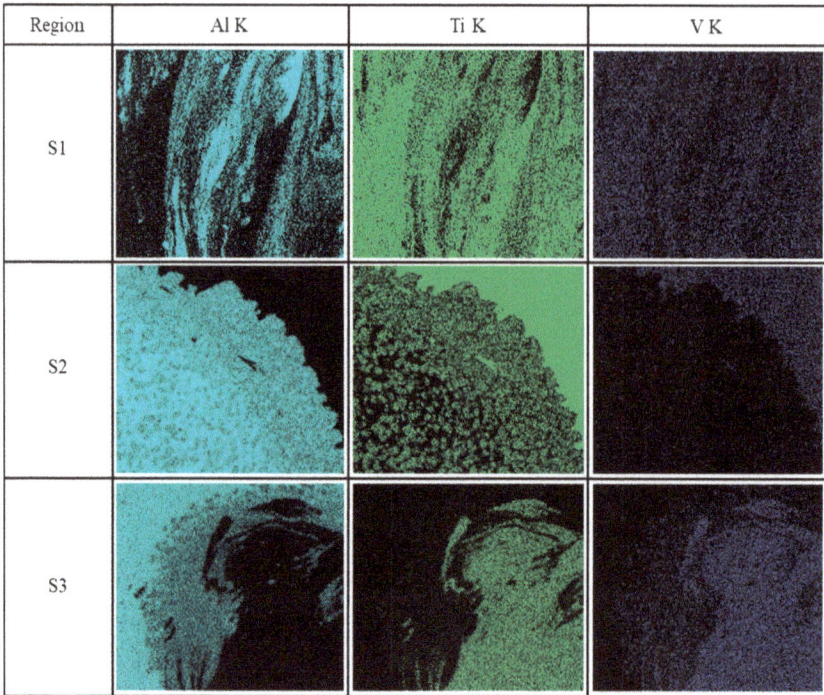

Figure 5. Energy dispersive spectrometry (EDS) element distribution maps of selected regions in the stir zone (red square marked in Figure 4a).

Herein, the formation process of the FSSW-B joint can be elucidated as follows: When the pin penetrates into the upper Al sheet, the temperature at the interface underneath the stirring pin rises rapidly with the gradual penetration as a result of the combined action of frictional heat and plastic deformation. The Zn interlayer would melt if the local temperature reaches its melting point. Meanwhile, the liquid Zn diffuses into the upper Al sheet and lower Al coating resulting, in the formation of the Al–Zn eutectic. However, when the stir pin comes into contact with the top surface of the lower Ti6Al4V sheet, most of the interfacial structure underneath the stirring pin starts to be destroyed as the tool penetrates further. At the same time, the heated and softened material adjacent to the pin deforms plastically, resulting in mixing and solid-state bonding between the upper Al and lower Ti alloys. When the shoulder comes into contact with and penetrates into the upper Al alloy, Zn foil under the shoulder melts and reacts with the upper Al sheet and lower Al coating, also resulting in the formation of the Al–Zn eutectic. Notably, the Al–Zn eutectics formed above are created along the interface between the upper Al sheet and Al coating. They can be squeezed into the gap out of the shoulder area as a result of extrusion force, which would help to remove the oxides from metal surfaces [22,46]. Finally, similar to the FSSB process [22,29,47], a brazing zone is formed underneath the shoulder. This brazing layer has a curved interface (see Figure 6a) because of the uneven forging force created by the shoulder [48].

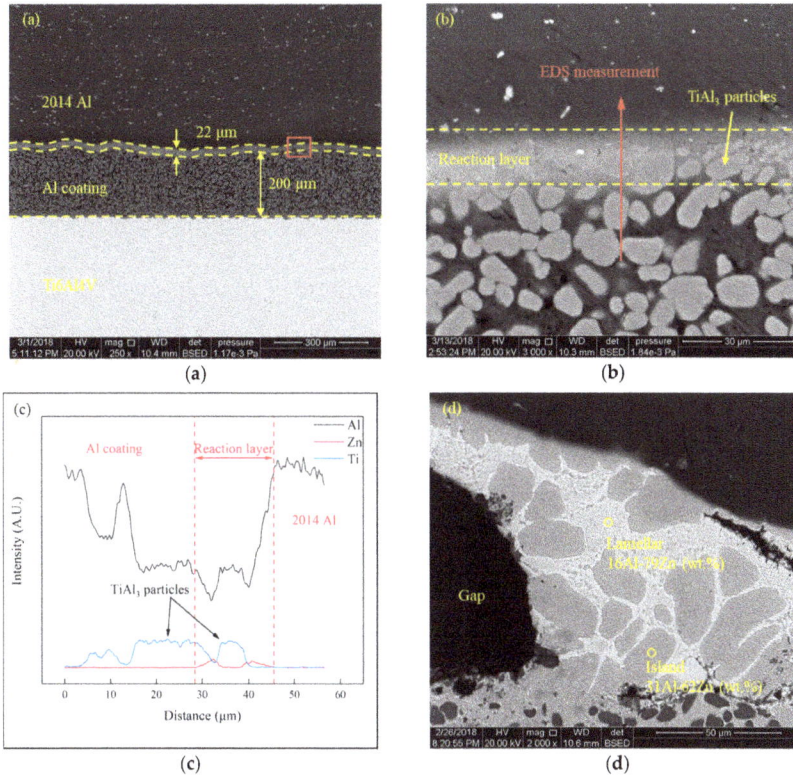

Figure 6. Microstructure of the brazing zone of the joint with (**a**) low magnification (region B marked in Figure 3a) and (**b**) high magnification (red square marked in (**a**)); (**c**) EDS results across the reaction layer (red line marked in (**b**)); (**d**) microstructure of the region at the end of the brazing zone (region C marked in Figure 3a), showing the Al–Zn eutectic structure. The thickness of the reaction layer was based on the measured distribution of Zn element.

3.3. Mechanical Property

Thanks to the formation of such a brazing zone, an enlarged bonding area, the lap shear fracture load of this FSSW-B joint is significantly increased compared with the FSSW joint. Figure 7a plots typical load–displacement curves of the joint produced by FSSW-B and FSSW at the same processing parameters (1200 rpm tool rotation speed). It can be seen that they both have no obvious yield platform, while the fracture load and fracture strain of FSSW-B joint are higher than those of the FSSW joint.

Figure 7b depicts the fracture surface (Ti6Al4V side) of the FSSW-B joint after the tensile shear test. It clearly reveals that the primary crack initiates from the brazing zone, then propagates through the stir zone, and finally through the keyhole to the other side. Meanwhile, as marked in Figure 7b, two identifiable regions are observed in the stir zone (region I and II) and brazing zone (region III and IV), respectively.

Figure 8 displays the details of all four regions of the fracture surface. It can be seen that all these four regions exhibit brittle fracture morphology, confirming the brittle fracture in the loading curve. The EDS analysis in region I (see Figure 8a) shows 70 wt.% Ti, indicating the separation of this region was mainly through the hook. TiAl3 particles can be observed in region II (see Figure 8b) and III (see Figure 8c), suggesting the separation was along the hook and through the reaction layer, respectively. The fracture surface of region IV is composed of about 36 at.% Al and 62 at.% Ti and appears to be flatter microscopically than other regions (see Figure 8d). It suggests that the fracture was at the continuous TiAl$_3$ layer between Al coating and Ti6Al4V alloy. It is worth noting that the aluminizing could help with increasing the joining area, while TiAl$_3$ in this Al coating would affect the mechanical properties of the joint.

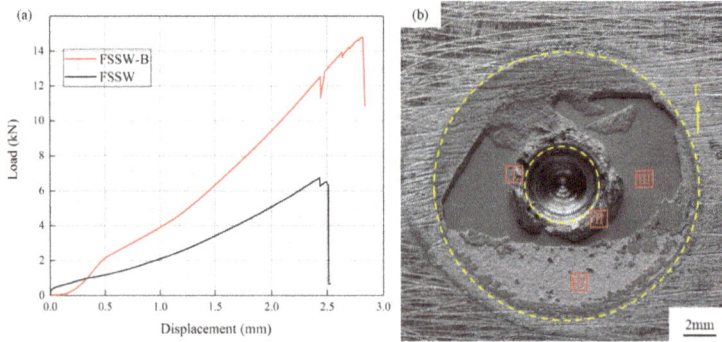

Figure 7. (**a**) Comparison of the lap shear fracture load of FSSW-B and FSSW joint at the same process parameter (1200 rpm); (**b**) macrograph of the fracture surface of FSSW-B joint (Ti6Al4V side). The yellow dotted lines in (**b**) mark the effective joining zone.

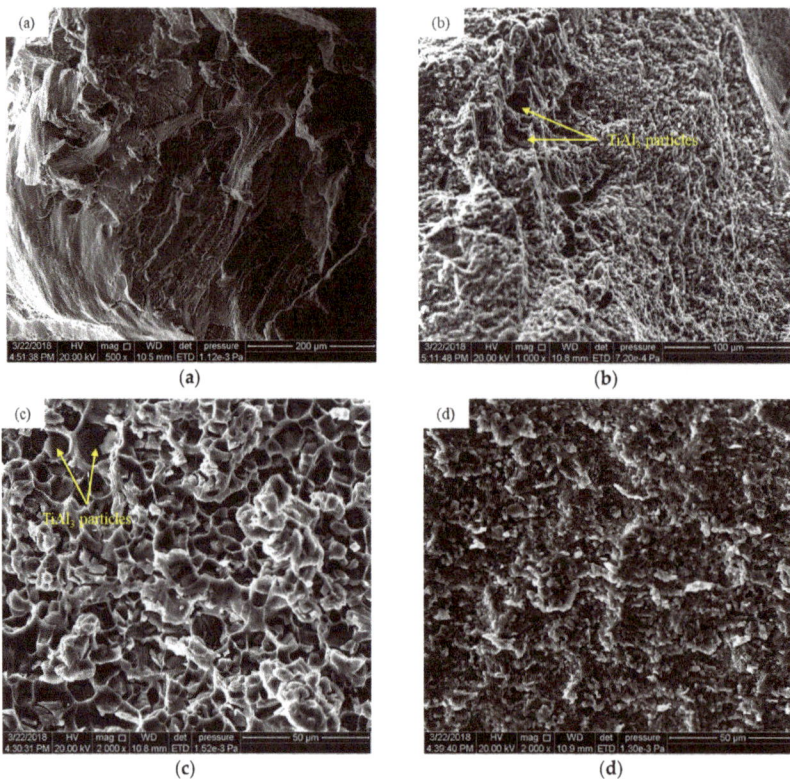

Figure 8. Highlighting the details of the fracture surface of each region marked with the red square in Figure 7b: (**a**) region I; (**b**) region II; (**c**) region III; and (**d**) region IV.

3.4. Parametric Study

Undoubtedly, the mechanical property of joints significantly depends on the process parameters. Generally, the heat input can be increased by increasing penetration force, angular velocity, and sampling time [49]. As the increase of penetration depth and rotation speed could increase the formation of IMCs [39,50], only the influence of rotation speed on the fracture load of joints was considered in the current study.

Figure 9 compares the fracture load of joints produced by FSSW-B and FSSW, showing that the change of failure load of FSSW-B joints is significantly greater than that of FSSW. At a low rotation speed, 900 rpm, their average fracture loads are quite close. The fracture load of the FSSW-B joint is much higher than that of the FSSW joint when the rotation speeds are 1200 rpm and 1500 rpm, respectively.

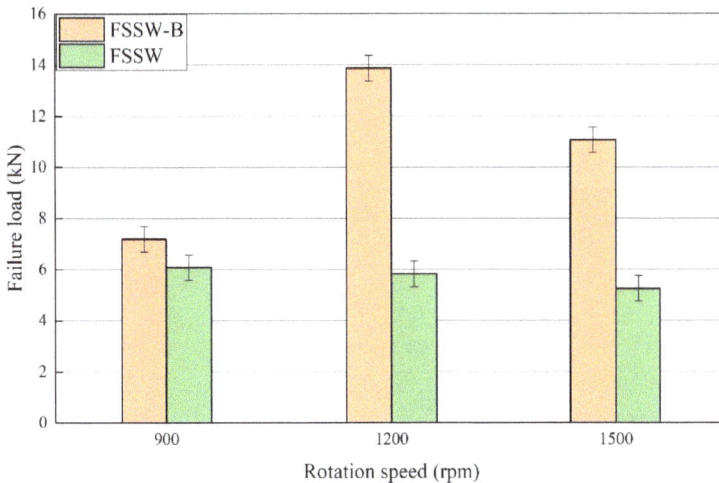

Figure 9. Comparison of lap shear fracture load of FSSW-B and FSSW joint at the same tool rotation speeds.

To study the variation in the fracture load, Figure 10 compares the microstructure of the brazing zone formed at the rotation speed of 900 rpm and 1500 rpm, respectively. As shown in Figure 10a, an unbonded region exists in the reaction layer because of insufficient heat input at a low rotation speed for the brazing process in this area, resulting in limited contribution to the fracture load increment. A continuous brazing zone is successfully fabricated as the rotation speed is increased to 1200 rpm and 1500 rpm. The thickness of reaction layer reduces from 22 μm (see Figure 6a) to 17 μm (see Figure 10b) because more liquid has been squeezed out of the brazing zone, caused by increased heat input, which is agreement with that of the Al/Cu joint made by pinless FSW using a Zn filler [51]. Therefore, sufficient energy input is necessary to ensure the successful brazing of Al alloy and Al coating, which play a key role in determining the joining width of the joint.

Similar to the traditional FSSW [37,39,52], the geometrical features (including indentation depth, hook height, and orientation) vary with rotation speed, as shown in Figure 11. When the tool rotation speed is 1500 rpm, compared with rotation speeds of 1200 rpm and 900 rpm, the hook terminates much closer to the keyhole. Although the tool penetration depth remains constant in the current study, more material is drawn out with the tool as the heat input increases [53,54]. This could cause the decreasing of the effective top sheet thickness. These varied geometrical features reduce the solid-state bonded width of the joint and thereby reduce the failure load of the FSSW joint from 6.6 kN to 4.3 kN (see Figure 9). Obviously, it would also affect the failure load of the FSSW-B joint.

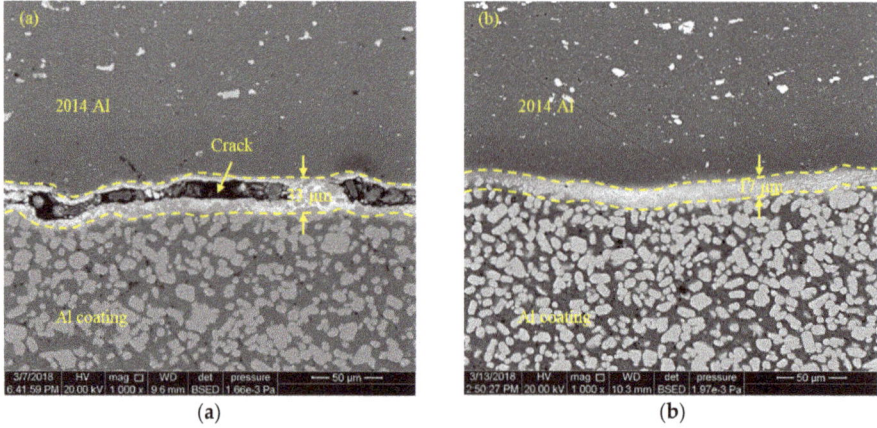

Figure 10. Microstructure of the brazing zone of the FSSW-B joint made at the tool rotation speed of (**a**) 900 rpm and (**b**) 1500 rpm.

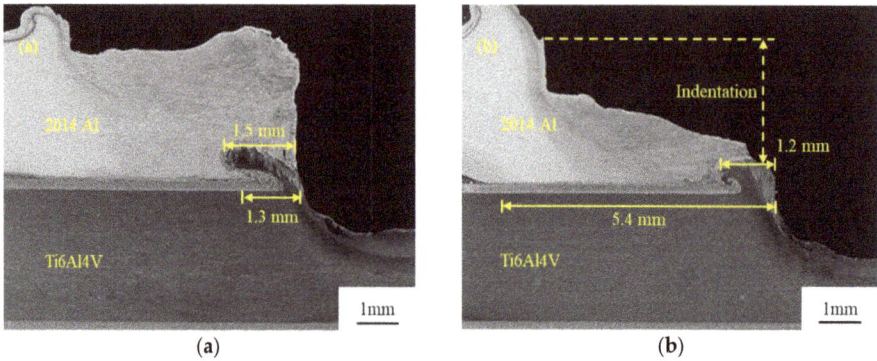

Figure 11. Microstructure of the stir zone of the FSSW-B joint made at the tool rotation speed of (**a**) 900 rpm and (**b**) 1500 rpm. The labeled dimensions were measured via the optical microscope measurement.

In the current study, the joining zone is a ring-like area that extends radially from the keyhole interface to the end of the brazing zone. Here, the diameter of this joining zone is described as the joining width. When the rotation speed is 900 rpm, the joining width is approximately 1.3 mm as measured in Figure 11a, which is smaller than the solid-state bonding width, 1.5 mm. This suggests that the brazing zone is almost negligible and does not contribute to the fracture load of the FSSW-B joint in comparison with the FSSW joint. Upon increasing the rotation speed to 1200 rpm, the joining width increases to approximately 5.2 mm (see Figure 3). Here, the sum of this joining width (5.2 mm) and the radius of keyhole (about 2.5 mm) is close to the radius of the used shoulder (8 mm) when the rotation is 1200 rpm. Therefore, this size might have reached the maximum value owing to the limited thermo-mechanically affected zone determined by the size of the shoulder. Therefore, as the rotation speed further increases to 1500 rpm, the joining width slightly increases to 5.4 mm (see Figure 11b). However, the reduction of solid-state bonding width, from 1.5 mm to 1.2 mm, results in the decreasing of joining width and thereby reduces the fracture load. This fact explains why the fracture load of the FSSW-B joint is first increased and then decreased with the increase of rotational speed.

4. Conclusions

This study provides an improved friction stir spot joining process for the joining of Al and Ti alloys. The following conclusions are derived:

- In addition to the solid-state joining of 2014-T4 Al and Ti6Al4V alloys via the FSSW technique, brazing between the Al alloy and Al coating on Ti6Al4V alloy was successfully introduced by the addition of a Zn interlayer.
- The Zn interlayer and Al coating have no significant influence in the stir zone of the FSSW-B joint. Because of the extrusion force during the joining process, the introduced TiAl$_3$ particles from Al coating are squeezed into the brazing zone, while the formed Zn–Al eutectic is squeezed out of the brazing zone.
- The formation of the brazing zone significantly increases the joining area, causing the highest fracture load of the FSSW-B joint to be improved by 110% compared with that of the traditional FSSW joint.
- Fracture load of the joint was first increased and then decreased with increasing of the rotational speed, which was rationalized to the varied effective joining areas.

Author Contributions: Investigation, X.Z., S.L., and K.H.; Resources, Y.H.; Supervision, Y.C.; Writing—original draft, X.Z.; Writing—review & editing, S.L. and P.P.

Funding: The research was financially supported by the National Natural Science Foundation of China (No. 51865035), the Fund for Jiangxi Distinguished Young Scholars (2018ACB21016), the Project for Jiangxi Advantageous Scientific and Technological Innovation Team (20171BCB24007 & 20181BCB19002), and Aviation Science Funds of China (2017ZE56010).

Conflicts of Interest: The authors declare no conflict of interest.

References

1. Micari, F.; Buffa, G.; Pellegrino, S.; Fratini, L. Friction Stir Welding as an Effective Alternative Technique for Light Structural Alloys Mixed Joints. *Proced. Eng.* **2014**, *81*, 74–83. [CrossRef]
2. Mohammadi, J.; Behnamian, Y.; Mostafaei, A.; Izadi, H.; Saeid, T.; Kokabi, A.H.; Gerlich, A.P. Friction stir welding joint of dissimilar materials between AZ31B magnesium and 6061 aluminum alloys: Microstructure studies and mechanical characterizations. *Mater. Charact.* **2015**, *101*, 189–207. [CrossRef]
3. Mallick, P.K. *Materials, Design and Manufacturing for Lightweight Vehicles*; Woodhead Publishing Limited: New Delhi, India, 2010; pp. 275–330. ISBN 978-1-84569-463-0.
4. Vacchi, G.S.; Plaine, A.H.; Silva, R.; Sordi, V.L.; Suhuddin, U.F.H.; Alcântara, N.G.; Kuri, S.E.; Rovere, C.A.D. Effect of friction spot welding (FSpW) on the surface corrosion behavior of overlapping AA6181-T4/Ti-6Al-4V joints. *Mater. Des.* **2017**, *131*, 127–134. [CrossRef]
5. ÇApar, N.; Kumru, U.; BaŞEr, T.A.; TekİN, G.; Saray, O. Friction Stir Spot Welding for Automotive Applications. *Int. J. Adv. Auto. Technol.* **2017**, *1*, 114–118. [CrossRef]
6. Mironov, S.; Sato, Y.S.; Kokawa, H. Friction-stir welding and processing of Ti-6Al-4V Titanium alloy: A review. *J. Mater. Sci. Technol.* **2018**, *34*, 58–72. [CrossRef]
7. Plaine, A.H.; Suhuddin, U.F.H.; Afonso, C.R.M.; Alcântara, N.G.; dos Santos, J.F. Interface formation and properties of friction spot welded joints of AA5754 and Ti6Al4V alloys. *Mater. Des.* **2016**, *93*, 224–231. [CrossRef]
8. Sun, Q.J.; Li, J.Z.; Liu, Y.B.; Li, B.P.; Xu, P.W.; Feng, J.C. Microstructural characterization and mechanical properties of Al/Ti joint welded by CMT method—Assisted hybrid magnetic field. *Mater. Des.* **2017**, *116*, 316–324. [CrossRef]
9. Chen, Y.C.; Nakata, K. Microstructural characterization and mechanical properties in friction stir welding of aluminum and titanium dissimilar alloys. *Mater. Des.* **2009**, *30*, 469–474. [CrossRef]
10. Baqer, Y.M.; Ramesh, S.; Yusof, F.; Manladan, S.M. Challenges and advances in laser welding of dissimilar light alloys: Al/Mg, Al/Ti, and Mg/Ti alloys. *Int. J. Adv. Manuf. Technol.* **2018**, 1–17. [CrossRef]
11. Mishra, R.S.; Ma, Z.Y. Friction stir welding and processing. *Mater. Sci. Eng. R: Rep.* **2005**, *50*, 1–78. [CrossRef]

12. Ren, J.; Li, Y.; Tao, F. Microstructure characteristics in the interface zone of Ti/Al diffusion bonding. *Mater. Lett.* **2002**, *56*, 647–652.

13. Li, Y.J.; Gerasimov, S.A.; Wang, J.; Ma, H.J.; Ren, J.W. A study of vacuum diffusion bonding and interface structure of Ti/Al dissimilar materials. *Mater. Sci. Technol.* **2007**, *15*, 206–210.

14. Chen, S.; Li, L.; Chen, Y.; Huang, J. Joining mechanism of Ti/Al dissimilar alloys during laser welding-brazing process. *J. Alloy Compd.* **2011**, *509*, 891–898. [CrossRef]

15. Leo, P.; D'Ostuni, S.; Casalino, G. Low temperature heat treatments of AA5754-Ti6Al4V dissimilar laser welds: Microstructure evolution and mechanical properties. *Opt. Laser Technol.* **2018**, *100*, 109–118. [CrossRef]

16. Lv, S.; Cui, Q.; Huang, Y.; Jing, X. Influence of Zr addition on TIG welding–brazing of Ti–6Al–4V toAl5A06. *Mater. Sci. Eng. A* **2013**, *568*, 150–154. [CrossRef]

17. Mehta, K.P.; Badheka, V.J. A Review on Dissimilar Friction Stir Welding of Copper to Aluminum: Process, Properties, and Variants. *Mater. Manuf. Process.* **2015**, *31*, 233–254. [CrossRef]

18. Yang, X.W.; Fu, T.; Li, W.Y. Friction Stir Spot Welding: A Review on Joint Macro- and Microstructure, Property, and Process Modelling. *Adv. Mater. Sci. Eng.* **2014**, *2014*, 1–11. [CrossRef]

19. Padhy, G.K.; Wu, C.S.; Gao, S. Friction stir based welding and processing technologies—processes, parameters, microstructures and applications: A review. *J. Mater. Sci. Tech.* **2018**, *34*, 1–38. [CrossRef]

20. Kim, D.; Badarinarayan, H.; Ryu, I.; Kim, J.H.; Kim, C.; Okamoto, K.; Wagoner, R.H.; Chung, K. Numerical simulation of friction stir spot welding process for aluminum alloys. *Metals Mater. Int.* **2010**, *16*, 323–332. [CrossRef]

21. Kim, J.-R.; Ahn, E.-Y.; Das, H.; Jeong, Y.-H.; Hong, S.-T.; Miles, M.; Lee, K.-J. Effect of tool geometry and process parameters on mechanical properties of friction stir spot welded dissimilar aluminum alloys. *Int. J. Precis. Eng. Manuf.* **2017**, *18*, 445–452. [CrossRef]

22. Zhang, G.; Zhang, L.; Kang, C.; Zhang, J. Development of friction stir spot brazing (FSSB). *Mater. Des.* **2016**, *94*, 502–514. [CrossRef]

23. Zhou, L.; Luo, L.Y.; Zhang, T.P.; He, W.X.; Huang, Y.X.; Feng, J.C. Effect of rotation speed on microstructure and mechanical properties of refill friction stir spot welded 6061-T6 aluminum alloy. *Int. J. Adv. Manuf. Technol.* **2017**, *92*, 3425–3433. [CrossRef]

24. Sun, Y.F.; Fujii, H.; Takaki, N.; Okitsu, Y. Microstructure and mechanical properties of dissimilar Al alloy/steel joints prepared by a flat spot friction stir welding technique. *Mater. Des.* **2013**, *47*, 350–357. [CrossRef]

25. Swamy, M.M.; Muthukumaran, S.; Kiran, K. A Study on Friction Stir Multi Spot Welding Techniques to Join Commercial Pure Aluminum and Mild Steel Sheets. *Trans. Indian Inst. Metals* **2016**, *70*, 1221–1232. [CrossRef]

26. Yu, W.; Shao, J.; Chen, D.; Liu, S. Effect of transitional layer on property and microstructure of FSB joints of aluminum and steel. *Trans. China Weld.* **2015**, *36*, 85–89.

27. Huang, G.; Feng, X.; Shen, Y.; Zheng, Q.; Zhao, P. Friction stir brazing of 6061 aluminum alloy and H62 brass: Evaluation of microstructure, mechanical and fracture behavior. *Mater. Des.* **2016**, *99*, 403–411. [CrossRef]

28. Gan, R.; Jin, Y. Friction stir-induced brazing of Al/Mg lap joints with and without Zn interlayer. *Sci. Technol. Weld. Join.* **2017**, *23*, 164–171. [CrossRef]

29. Zhang, G.-F.; Zhang, K.; Guo, Y.; Zhang, J.-X. A Comparative Study of Friction Stir Brazing and Furnace Brazing of Dissimilar Metal Al and Cu Plates. *Metall. Microstuct. Anal.* **2014**, *3*, 272–280. [CrossRef]

30. Xu, R.Z.; Ni, D.R.; Yang, Q.; Liu, C.Z.; Ma, Z.Y. Influence of Zn coating on friction stir spot welded Magnesium-Aluminium joint. *Sci. Technol. Weld. Join.* **2016**, *22*, 512–519. [CrossRef]

31. Boucherit, A.; Avettand-Fènoël, M.N.; Taillard, R. Effect of a Zn interlayer on dissimilar FSSW of Al and Cu. *Mater. Des.* **2017**, *124*, 87–99. [CrossRef]

32. Zhou, W.B.; Teng, G.B.; Liu, C.Y.; Qi, H.Q.; Huang, H.F.; Chen, Y.; Jiang, H.J. Microstructures and Mechanical Properties of Binary Al-Zn Alloys Fabricated by Casting and Heat Treatment. *J. Mater. Eng. Perform.* **2017**, *26*, 3977–3982. [CrossRef]

33. Zhang, Z.G.; Peng, Y.P.; Mao, Y.L.; Pang, C.J.; Lu, L.Y. Effect of hot-dip aluminizing on the oxidation resistance of Ti–6Al–4V alloy at high temperatures. *Corros. Sci.* **2012**, *55*, 187–193. [CrossRef]

34. Ebrahimian, A.; Kokabi, A.H. Friction stir soldering: A novel route to produce graphite-copper dissimilar joints. *Mater. Des.* **2017**, *116*, 599–608. [CrossRef]

35. Sadeq, F.O.; Sharifitabar, M.; Afarani, M.S. Synthesis of Ti–Si–Al coatings on the surface of Ti–6Al–4V alloy via hot dip siliconizing route. *Surf. Coat. Technol.* **2018**, *337*, 349–356. [CrossRef]

36. Zhang, Z.G.; Teng, X.; Xiang, H.F.; Sheng, Y.G.; Zhang, X.J. Preparation of TiAl3 Coating on γ-TiAl through Hot-dip Aluminizing and Subsequent Interdiffusion Treatment: High Temperature Materials and Processes. *High Temp. Mater. Process.* **2009**, *28*, 115–119. [CrossRef]

37. Badarinarayan, H.; Shi, Y.; Li, X.; Okamoto, K. Effect of tool geometry on hook formation and static strength of friction stir spot welded aluminum 5754-O. sheets. *Int. J. Mach. Tool. Manuf.* **2009**, *49*, 814–823. [CrossRef]

38. Yin, Y.H.; Sun, N.; North, T.H.; Hu, S.S. Hook formation and mechanical properties in AZ31 friction stir spot welds. *J. Mater. Process. Technol.* **2010**, *210*, 2062–2070. [CrossRef]

39. Rao, H.M.; Yuan, W.; Badarinarayan, H. Effect of process parameters on mechanical properties of friction stir spot welded Magnesium to aluminum alloys. *Mater. Des.* **2015**, *66*, 235–245. [CrossRef]

40. Zhou, X.; Huang, Y.; Chen, Y.; Peng, P. Laser joining of Mo and Ta sheets with Ti6Al4V or Ni filler. *Opt. Laser Technol.* **2018**, *106*, 487–494. [CrossRef]

41. Zhou, X.; Chen, Y.; Huang, Y.; Mao, Y.; Yu, Y. Effects of niobium addition on the microstructure and mechanical properties of laser-welded joints of NiTiNb and Ti6Al4V alloys. *J. Alloy. Compd.* **2018**, *735*, 2616–2624. [CrossRef]

42. Pouquet, J.; Miranda, R.M.; Williams, S. Dissimilar laser welding of NiTi to stainless steel. *Int. J. Adv. Manuf. Technol.* **2012**, *61*, 205–212. [CrossRef]

43. Huang, Y.; Lv, Z.; Wan, L.; Shen, J.; dos Santos, J.F. A new method of hybrid friction stir welding assisted by friction surfacing for joining dissimilar Ti/Al alloy. *Mater. Lett.* **2017**, *207*, 172–175. [CrossRef]

44. Song, Z.; Nakata, K.; Wu, A.; Liao, J.; Zhou, L. Influence of probe offset distance on interfacial microstructure and mechanical properties of friction stir butt welded joint of Ti6Al4V and A6061 dissimilar alloys. *Mater. Des.* **2014**, *57*, 269–278. [CrossRef]

45. Su, P.; Gerlich, A.; North, T.H.; Bendzsak, G.J. Intermixing in Dissimilar Friction Stir Spot Welds. *Metall. Mater. Trans. A* **2007**, *38*, 584–595. [CrossRef]

46. Gale, W.F.; Butts, D.A. Transient liquid phase bonding. *Sci. Technol. Weld. Join.* **2013**, *9*, 283–300. [CrossRef]

47. Zhang, G.F.; Zhang, K.; Zhang, L.J.; Zhang, J.X. Approach to disrupting thick intermetallic compound interfacial layer in friction stir brazing (FSB) of Al/Cu plates. *Sci. Technol. Weld. Join.* **2014**, *19*, 554–559. [CrossRef]

48. Zhang, G.F.; Su, W.; Zhang, J.; Zhang, J.X. Visual observation of effect of tilting tool on forging action during FSW of aluminium sheet. *Sci. Technol. Weld. Join.* **2013**, *16*, 87–91. [CrossRef]

49. Su, P.; Gerlich, A.; North, T.H.; Bendzsak, G.J. *Energy Generation and Stir Zone Dimensions in Friction Stir Spot Welds*; SAE Technical Paper Series; SAE International: Warrendale, PA, USA, 2006.

50. Li, S.; Chen, Y.; Zhou, X.; Kang, J.; Huang, Y.; Deng, H. High-strength titanium alloy/steel butt joint produced via friction stir welding. *Mater. Lett.* **2019**, *234*, 155–158. [CrossRef]

51. Kuang, B.; Shen, Y.; Chen, W.; Yao, X.; Xu, H.; Gao, J.; Zhang, J. The dissimilar friction stir lap welding of 1A99 Al to pure Cu using Zn as filler metal with "pinless" tool configuration. *Mater. Des.* **2015**, *68*, 54–62. [CrossRef]

52. Yin, Y.H.; Sun, N.; North, T.H.; Hu, S.S. Influence of tool design on mechanical properties of AZ31 friction stir spot welds. *Sci. Technol. Weld. Join.* **2010**, *15*, 81–86. [CrossRef]

53. Gerlich, A.; Su, P.; North, T.H. Tool penetration during friction stir spot welding of Al and Mg alloys. *J. Mater. Sci.* **2005**, *40*, 6473–6481. [CrossRef]

54. Choi, D.H.; Lee, C.Y.; Ahn, B.W.; Choi, J.H.; Yeon, Y.M.; Song, K.; Park, H.S.; Kim, Y.J.; Yoo, C.D.; Jung, S.B. Frictional wear evaluation of WC–Co alloy tool in friction stir spot welding of low carbon steel plates. *Int. J. Refract. Metals Hard Mater.* **2009**, *27*, 931–936. [CrossRef]

metals

MDPI

Article

Dissimilar Friction Stir Butt Welding of Aluminum and Copper with Cross-Section Adjustment for Current-Carrying Components

Nima Eslami [1], Alexander Harms [1], Johann Deringer [1], Andreas Fricke [1] and Stefan Böhm [2,*]

[1]   Volkswagen AG, Corporate Research, Berliner Ring 2, 38440 Wolfsburg, Germany;
     nima.eslami@volkswagen.de (N.E.); alexander.harms@volkswagen.de (A.H.); j.deringer@gmx.de (J.D.);
     andreas.Fricke2@volkswagen.de (A.F.)
[2]   Department for Cutting and Joining Manufacturing Processes, University of Kassel, Kurt Wolters Str. 3,
     34125 Kassel, Germany
*   Correspondence: s.boehm@uni-kassel.de; Tel.: +49-561-804-3141

Received: 9 August 2018; Accepted: 23 August 2018; Published: 24 August 2018

Abstract: Manufacturing dissimilar joints of aluminum and copper is a challenging task. However, friction stir welding (FSW) was found to be a suitable technique to produce aluminum–copper joints. Due to different electrical conductivities between aluminum and copper, an adjustment of the cross-section is required to realize electrical conductors free of resistive losses. Taking this into account, this paper presents initial results on the mechanical and electrical properties of friction stir butt welded aluminum and copper blanks having thicknesses of 4.7 mm and 3 mm, respectively. Three different approaches were investigated with the aim to produce sound welds with properties similar to those of the used base materials. Friction stir welding has been conducted at a welding speed of 450 mm/min. Subsequently, the welded specimens were subjected to metallographic analysis, tensile testing, and measurements of the electrical conductivity. The ultimate tensile force of the best joints was about 10 kN, which corresponds to joint efficiencies of approximately 72% of the aluminum base material. The analysis of electrical joint properties led to very promising results, so that the potential of FSW of Al–Cu butt joints with sheets having different thicknesses could be confirmed by the investigations carried out.

Keywords: friction stir welding; aluminum; copper; cross-section adjustment; mechanical properties; electrical properties

---

1. Introduction

Copper-based materials are increasingly used for electrical components due to the high ductility, the excellent electrical and thermal conductivities, as well as the good creep and corrosion resistance of copper. However, the use of copper is disadvantageous due to its high cost and weight. Since aluminum features the most efficient ratio of electrical conductivity to density among all conducting materials and is also significantly cheaper than copper, aluminum–copper joints offer a great potential for reduced-cost and reduced-weight current-carrying components [1–3]. In order to produce electrical contacts, it is well-known that firmly bonded joining techniques lead to better electrical properties than force-locking and interlocking methods [4]. The joining of aluminum to copper by means of conventional fusion welding is a challenging task due to the different melting temperatures, the high thermal conductivities and the tendency to intermetallic phase formation [5]. As an alternative, various authors report unanimously about the suitability of friction stir welding (FSW) to manufacture dissimilar aluminum–copper joints [6–10]. FSW is a solid-state welding method that was developed and patented in 1991 by Wayne Thomas [11]. The principle of this joining method is based on plasticization and local plastic deformation of the workpieces [12,13]. To achieve firm bonding, a rotating cylindrical tool consisting of a shoulder and a pin is pressed between the butting edges of the joining partners and then moved along the joint edge in order to plasticize and stir them [14].

In the field of friction stir butt welding of dissimilar aluminum–copper joints, most of the studies focus on the influence of the welding parameters on the resultant microstructure and mechanical properties. For example, Xue et al. [7] investigated the effect of positioning the workpieces on the advancing side (AS) and retreating side (RS), respectively, when joining 5 mm thick 1060 aluminum and commercially pure copper. The authors recommend placing the copper plate on the AS to ensure the formation of sound welds with ultimate tensile strengths of up to 110 MPa. Furthermore, it was shown in the published results that the weld quality can be improved by offsetting the FSW tool laterally in the direction to the softer aluminum material.

Due to the differences in electrical conductivity between aluminum and copper, adjustments of the sheet thicknesses must be considered for aluminum–copper joints with homogeneous current-carrying behavior. In order to manufacture friction stir welds with blanks of different thicknesses, several approaches are available in the relevant literature. The easiest way is to choose a lap joint configuration. However, the extra material to generate the overlap and the resultant gaps that lead to a higher risk of corrosion are found to be disadvantageous [15]. Friction stir butt welding of aluminum sheets having different thicknesses was investigated by Fratini et al. [16]. In addition to the usual tilt angle in friction stir welding, they used a second angle towards the thicker workpiece. This leads to joint efficiencies up to 80% for joining partners with thickness ratios between 1 and 2. Sahu et al. [17] mention the possibility to adapt a second nuting angle by tilting the machine bed. Moreover, different patents and patent applications deal with friction stir butt welding of materials having different thicknesses. The patent application EP1825946A1 [18] describes a two-step process, in which firstly material is transferred from the thicker to the thinner workpiece by using a columnar tool which is rotatingly moved along the butting edges. As a result, a convex substitute surface is formed. In the second step, conventional FSW is conducted on the convex surface. Patent application DE102014001050A1 [19] outlines an apparatus and a method for friction stir butt welding sheets of different thicknesses using a tool which comprises a special welding shoe. Further methods for FSW of workpieces having different thicknesses with conventional friction stir welding tools are presented in a patent of Werz et al. [20]. It is proposed to compensate the difference regarding sheet thickness between the joining partners by using a combined lap and butt joint configuration or by folding, wrapping or edge bending the thinner workpiece. Based on this patent application, in this study three different approaches were investigated to manufacture dissimilar aluminum–copper friction stir welds with joining partners having different thicknesses.

2. Materials and Methods

In this investigation, the applied materials were EN CW008A-240 and EN AW-1050A-H14. The chemical compositions of both materials are listed in Table 1.

Table 1. Chemical compositions of the applied copper and aluminum blanks [21,22].

Materials	Al	Fe	Si	Mn	Mg	Zn	Ti	Pb	Bi	Cu
EN AW-1050A	≥99.50	≤0.40	≤0.25	≤0.05	≤0.05	≤0.07	≤0.05	-	-	≤0.05
CW008A	-	-	-	-	-	-	-	≤0.005	≤0.0005	≥99.95

Dimensions of the aluminum blanks were 1300 mm, 100 mm, and 4.7 mm (length, width, thickness). The copper blanks had an original thickness of 3 mm. To achieve a similar electrical conductivity between the two materials it is necessary to realize a cross-section adjustment with an approximate ratio of 1.56. Therefore, aluminum plates with a thickness of 4.7 mm were required. Because of good availability 5 mm thick aluminum blanks were purchased. In the next step, the blanks were milled down from 5 mm to 4.7 mm to ensure the needed cross-section adjustment. In order to achieve the same cross-sections at the joint edge, the copper blanks were edge bent on one side.

The friction stir welds were manufactured with a PTG Powerstir portal system from the Department for Cutting and Joining Manufacturing Processes at the University of Kassel (Kassel, Germany).

Two different shoulder shapes were used to examine the three different approaches. One tool had a flat and the other one had a concave shoulder. Both shoulder diameters were 14 mm and threaded pins with diameters of 5 mm were used for all investigations. The tools were made of heat treated steel (X40CrMoV5-1). Both types of tools were used in preliminary studies. Furthermore, a stainless steel plate (X5CrNi18-10) was placed under the edge bent copper blank to compensate the difference in thickness between the two workpieces.

A configuration setup for the first approach is shown in Figure 1a. This first technical examination included to place the copper blank on the AS and to choose an offset towards the aluminum blank. The objective was to create sound welds following similar studies that deal with conventional friction stir butt welding of aluminum and copper. This setup contained the welding of seams 1–76. In the second approach the effect of changed plate positions was analyzed. The aluminum blanks were placed on the AS and the edge bent copper workpieces were positioned on the RS. For this approach the offset was also set in the direction of the aluminum blank. The intention of this setup was to investigate a significant material flow difference between the first and the second experimental design. Welds 77–99 were produced with this configuration setup. A schematic setup for this described approach is illustrated in Figure 1b.

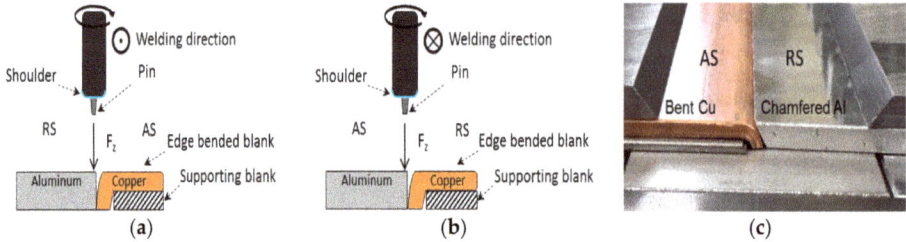

Figure 1. (**a**) Friction stir welding (FSW) setup with Cu on advancing side (AS); (**b**) Setup with Al on AS; and (**c**) setup with chamfer on Al edge.

The third approach of this investigation examines the effect of a chamfer on the edge of the aluminum blank. The edge bent copper was placed on the AS. Within the setup, which is shown in Figure 1c, the abutting end of the aluminum led to a slight overlap with the edge bent copper. The idea was to reduce the material deficit in the joining zone. The chamfers were manufactured manually by a belt grinder and nearly achieved 45 degrees at the edge. Furthermore, the copper and aluminum blanks were cut into shorter pieces (300 mm). This circumstance was advantageous to simplify chamfering the aluminum blanks. Friction stir welds 100–110 were produced using this approach.

The second approach did not result in any useful samples. All parametric tests, including the change of shoulder shapes, led to an insufficient compression of the weld seam and voids. The plasticized material was pressed out of the weld zone during the friction stir welding process. The weld seams of approach 2 were therefore not taken into consideration for further analyses. The used welding parameters for approaches 1 and 3 are listed in Table 2.

Table 2. Used welding parameters for approaches 1 and 3.

Approach	Rotational Speed (rpm)	Feed Rate (mm/min)	Plunge Depth (mm)	Dwell Time (s)	Tool Tilt Angle(°)	Tool Pin Length (mm)	Tool Shoulder Shape	Offset (mm)
A1	1100	450	4.25	2	4	3.7	flat	2.0
A3	1100	450	4.20	2	3	3.8	flat	1.8

According to DIN EN ISO 25239-5 [23] transversal sections of the friction stir welds were detached by water jet cutting to evaluate the mechanical properties. The distance to the plunging position of the tool was 20 mm for first tensile specimen. This value deviates from the mandatory 50 mm described in DIN EN ISO 25239-5. The shape of the samples accorded with the DIN EN ISO 4136 [24] whereby the length was 190 mm. All specimens were slightly grinded on their sides with an abrasive paper to avoid the influence of notches on the detached edges during tensile testing. Other than that, no further processing of the samples was carried out. Tensile testing was conducted by a tensile test machine Zwick Z100 (Zwick GmbH & Co. KG, Ulm, Germany) at an operating speed of 10 mm/min. Furthermore, the results of tensile testing for the different approaches were compared to tensile test results of the corresponding base materials.

In addition to tensile testing metallographic analyses of the weld seams were carried out to assess the quality of the welds. Sections directly next to the extracted tensile specimens were used to prepare the metallographic samples. Metallographic samples were hot mounted with an epoxy resin. The next step contained the grinding of the metallographic specimens with SiC abrasive papers up to grain sizes of 4000 μm. In particular, the grinding was done by hand to avoid a shifting of copper particles into the aluminum side. The last step of preparing the samples included an ion-milling process.

In order to examine the electrical properties of the dissimilar joints, a four point resistance measurement method was applied (Figure 2). The specimens were also detached by water jet cutting. The width of the rectangular samples was 37 mm and the length was 190 mm. During measurements the edges of the dissimilar FSW joints were contacted with a copper sheet to provide the test current of 180 A. The four measuring tips had a distance of 40 mm to each other. Via the first two tips the resistance of the copper base material was measured while tips 4 and 5 were used to determine the resistance of the aluminum base material. The friction stir welded seam was positioned between measuring tips 2 and 4. This setup allowed measuring the electrical resistances of the weld seam and of the two base materials simultaneously. The testing consisted of ten measurements for each specimen with the determined values being averaged subsequently.

| (a) | (b) |

Figure 2. (**a**) Test specimen installed on measuring system. (**b**) Test setup with measuring tips.

The last part of this section describes the labelling of the samples. The label describes the approach type, the number of the weld, the executed testing method and the removal area. More precisely, A1.64/T/a stands for a specimen which was welded with the setup from approach 1. The investigation is related to weld seam 64 whereby a tensile test has been carried out. The last part of the sample label refers to the area where the specimens were detached, starting from position a at the beginning of the weld seam with 20 mm distance to the FSW tool plunging spot. The following detached specimens were named in an alphabetic order. For further specification Table 3 includes all the different variations.

Table 3. Tabular list of specimen labeling.

Approach	FSW No.	Method	Order
A1	1–76	T = Tensile testing	a–i
A3	100–110	M = Metallographic analysis	a–c
-	-	R = Electrical resistance	-

3. Results and Discussion

3.1. Appearance of the Al–Cu Friction Stir Welds

Figure 3 shows a friction stir weld produced by the use of approach 3. It was evident that only a small burr occurred on the advancing as well as on the RS. The weld seam did not show any defects at the surface.

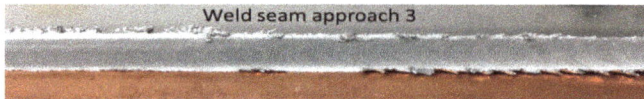

Figure 3. Surface morphology of Al–Cu friction stir weld seam using approach 3.

For further investigations macrographs of approach 1 and 3 were examined. Therefore, areas of each approach were selected. Within approach 1 the specimen A1.67/M/a was analyzed. For approach 3 the samples A3.106/M/a,c were chosen. According to DIN EN ISO 25239-5 [23] a set of acceptance criteria for friction stir butt welds has to be fulfilled. The first criterion is the underfill maximum. The permissible value for blanks with a thickness of 4.7 mm is 0.67 mm. Furthermore, the permissible size of cavities was determined for each sample by the thickness of the weld seam multiplied with 0.2. The third criterion describes a complete joint penetration depth. With a lack of penetration the seam quality is invalid. Figure 4 shows the whole weld seam area of samples A1.67/M/a, A3.106/M/a, and A3.106/M/c. Figure 5 shows an enlarged area of the stir zone of specimens A1.67/M/a and A3.106/M/c.

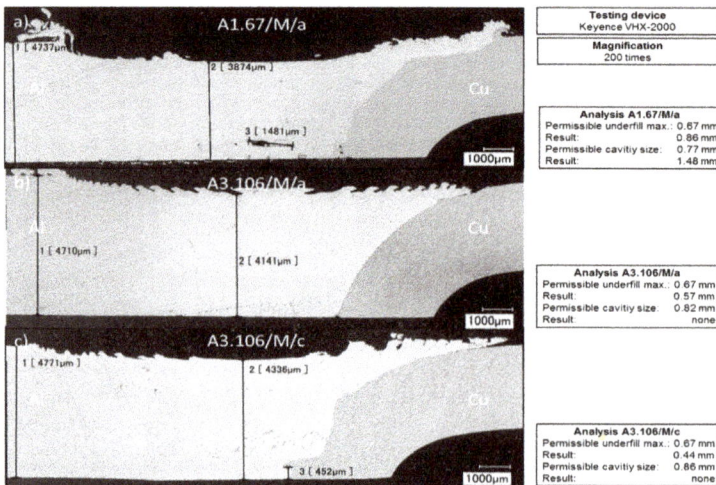

Figure 4. Cross-sectional macrostructures of Al–Cu friction stir welds: (**a**) specimen A1.67/M/a; (**b**) specimen A3.106/M/a; and (**c**) specimen A3.106/M/c.

Specimen A1.67/M/a did not fulfill the criteria of underfill maximum. A cavity was located in the middle of the stir zone. The size of this defect exceeded the acceptance criteria by almost twice the value. Because of the offset in the aluminum direction the tool pin only scratched a tiny part of the copper blank. This area can clearly be detected through the deformation at the edge bent copper. Furthermore in Figure 5 the scratched copper particles that were transported into the aluminum matrix can be seen. The diameter of these particles is between 0.15 mm and 0.24 mm. The penetration depth of this seam lacked of 0.5 mm. The macrostructure for specimen A3.106/M/a in Figure 4b shows an unsuccessful joining process. The edge bent copper blank does not show any deformation. Consequently, the pin did not scratch the copper. This is probably due to a failure in the positioning of the workpieces. The results for specimen A3.106/M/c show an underfill of 0.44 mm. This value is below the maximum permissible limit. In both cross-sectional images for sample A3.106/M/c no significant cavities are present. The observation of the joint penetration depth led to the same findings as for approach 1. The pin stirs a greater amount of large copper particles into the stir zone which was due to the lower offset. This can be seen by the severe material abrasion on the edge bent copper blank. By extension, the offset and positioning of the FSW tool created a more homogeneous stir zone compared to A1.67/M/a. Another observation of the better stirring effect can be seen in the position of the copper particles. These are spread throughout the whole stir zone. Nevertheless the FSW seam was inadequate because of the lack of penetration by 0.45 mm.

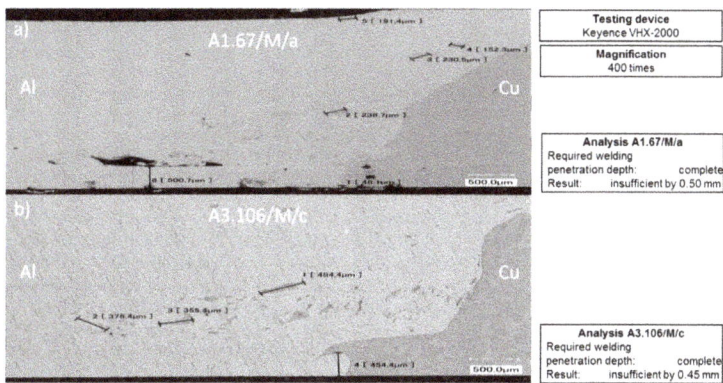

Figure 5. Magnified view on the macrostructures of Al–Cu friction stir welds: (**a**) specimen A1.67/M/a and (**b**) specimen A3.106/M/c.

Both approaches did not fulfill the criteria from DIN EN ISO 25239-5 [23]. Approach 1 failed with respect to all three relevant criteria. The macrostructural analysis of the samples that were welded according to approach 3 showed two completely different findings. While the results for A3.106/M/c were promising, the FSW process for specimen A3.106/M/a was only performed in the aluminum base material. An irregularity in the positioning and clamping of the blanks evidently led to insufficient mixing of the materials in this area of the weld seam. In contrast, the evaluation of the macrostructure of sample A3.106/M/c led to acceptable results in case of underfill and cavities. Regardless to this fact, the lack of penetration as a criterion for exclusion led to an invalid friction stir weld quality for this approach. However, the lack of penetration can be easily addressed in further experiments by using a longer pin.

3.2. Results of Tensile Testing

The results of the tensile tests for each weld seam are shown in Figure 6. For all welds examined in this study, failure occurred in the stir zone. Within the evaluation of the tensile tests the results were grouped into approach 1 and approach 3. In addition, Figure 6 shows mean values for each weld seam whenever more than one tensile specimen from a weld was tested. The highest average ultimate tensile force (UTF) of all friction stir welds was achieved by specimens A3.104/T/a-c with 10.45 kN. However, the lowest value was also measured for a weld that was manufactured according to approach 3. Weld seam A3.103/T/a-c obtained an average force of 6.73 kN. In general, the results for approach 1 were lower than those of approach 3. The highest overall measured tensile force was 11.09 kN. This can be seen in the range of specimen A3.106/T/a-c while sample c achieved the highest ultimate tensile force. The mean values for elongations at the maximum tensile forces are shown as black squares in the diagram. The highest values are achieved with approach 3 by specimens A3.104/T/a-c (2.12 mm) while the lowest value was as well identified for approach 3 by specimens A3.103/T/a-c (0.30 mm). The deviation is also higher for approach 3 (0.72 mm) compared to approach 1 (0.27 mm).

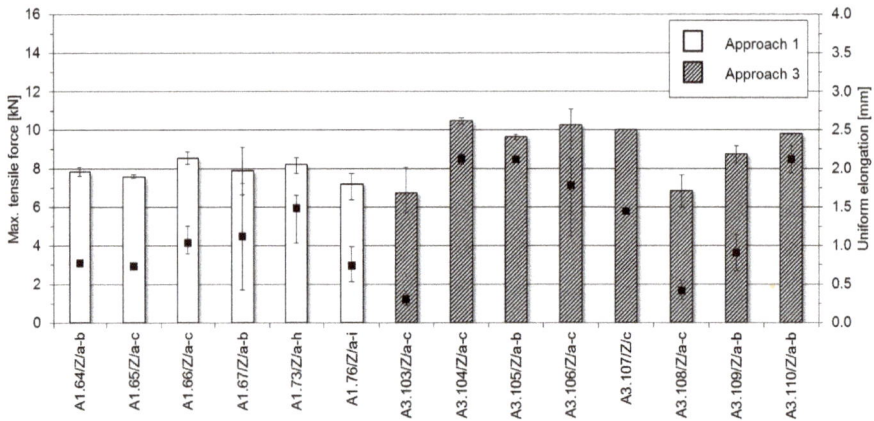

Figure 6. Ultimate tensile forces and elongations for specimens that were welded according to approach 1 and approach 3.

After the individual welds had been compared with each other, the tensile test results for the respective approaches 1 and 3 were summarized and compared with results for the used base materials. (Figure 7) Approach 1 reached 7.78 kN in UTF, while approach 3 led to a better result with an average UTF of 8.89 kN. Consequently, there is a difference of 1.17 kN in terms of UTF between approaches 1 and 3. However, the results of approach 1 were more consistent. The standard deviation was 0.43 kN while the outcome of approach 3 indicated a standard deviation of 1.39 kN. The aluminum base material had a mean UTF of 13.73 kN which was examined by five tensile specimens, while the UTF of the copper base material was at 18.45 kN. The average UTF of approach 1 reached approximately 57% of the aluminum base material. In contrast, approach 3 reached a joint efficiency of 65%. Within the analysis of approach 3 the highest measured UTF was at 81% of the aluminum base material for sample A3.106/T/c. The tensile test results for the two different approaches represent encouraging findings. However, especially the evaluation for approach 3 leads to the conclusion that a highly precise preparation of the chamfer, material, and tool positioning is required. These circumstances correspond to the macrostructural analysis of the welds. The highest individual UTF of all samples was achieved by specimen A3.106/T/c while the value for the first part of this weld seam (A3.106/T/a) was 24% lower due to uneven blank preparation and positioning. Excluding the two lowest values for approach 3, the UTF reached 72% compared to the aluminum base material. By using a longer tool pin resulting into a complete penetration depth, even better results can be expected in this regard.

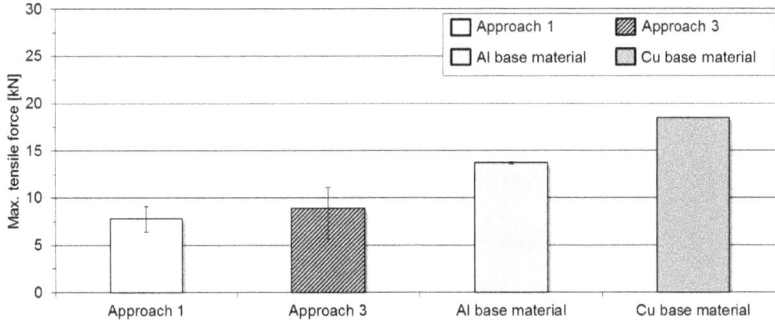

Figure 7. Averaged ultimate tensile forces for approach 1, approach 3, aluminum, and copper base material.

3.3. Electrical Properties of the Friction Stir Welds

The results of the electrical resistance measurements for each sample are shown in Figure 8. The values represent the electrical resistances of the weld seams that were determined between tips 2 and 4. It is recognizable that the electrical resistances of almost all samples that were welded according to approach 1 are slightly higher than the resistances of approach 3. Exceptions are specimens A1.65/R/c, A1.66/R/a and A1.67/R/a. The range of the electrical resistances for approach 1 is approximately between 0.0071 mΩ and 0.0078 mΩ while the average value for this approach is 0.0075 mΩ. For approach 3 the values are in the range of 0.0067 mΩ and 0.0072 mΩ. The average value for all nine measurements of this approach is 0.0070 mΩ. In comparison to the determined resistances of the weld seams, the average electrical resistance of the copper material that was measured between tips 1 and 2 is 0.0067 mΩ. For the aluminum base material, an average electrical resistance of 0.0074 mΩ was measured. The aim of this study was to manufacture Al–Cu friction stir welds with a homogenous current-carrying behavior on the basis of 3 mm thick copper sheets. There is a difference in electrical resistance of 0.00071 mΩ between the respective base materials. Therefore it is evident, that the applied cross-section adjustment was not sufficient. The required cross-section adjustment was calculated following the electrical conductivities of elemental copper (58×10^6 S/m) and aluminum (37×10^6 S/m). This leads to the conclusion that the small amount of alloying additions in the high purity materials used influences the electrical properties in a way that the realized cross-section adjustment was not precise enough. In order to obtain similar electrical conductivities between the respective base materials an adjustment factor of 1.73 is required for the present material combination. However, especially the results for approach 3 can be considered as very positive. These results deviate only slightly from the results of the copper base material. This deviation is suspected not to be only due to the insufficient cross-section adjustment, but also because of the underfill which is a result of the shoulder penetration. The underfill reduces the current-carrying cross-section which leads to an increase of the electrical resistance. Taking this into account, for a homogenous current-carrying behavior aluminum sheets with thicknesses of at least 5.2 mm are needed when the task is to manufacture cross-section-adjusted Al–Cu butt joints on the basis of 3 mm thick copper sheets.

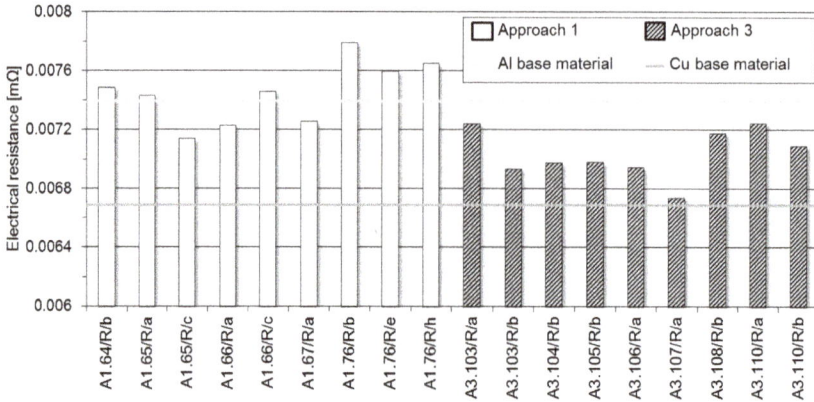

Figure 8. Electrical resistance measurement for approach 1 and approach 3.

Besides analyzing the measured electrical resistances, the electrical properties of the friction stir welds can be assessed by calculating the resistance factor k_u for each specimen. The resistance factor permits a material- and process-independent evaluation of the electrical joint properties. The dimensionless resistance factor k_u is calculated by doubling the electrical resistance of the weld seam and dividing the summed resistances of the respective base materials [25]:

$$k_u = \frac{2 \cdot R_{AlCu}}{(R_{Al} + R_{Cu})}. \tag{1}$$

The resistance factors in Figure 9 were calculated using the electrical resistances of the weld seams and the respective base materials from Figure 8. Ideally, the resistance factor of butt joints can reach the minimum value 1. If the resistance factor k_u is lower than 1.5, the joint accomplishes the criteria of long-term stability [26]. This criterion is met for all examined specimens. Therefore, a long-term stable electrical behavior can be expected for both approaches 1 and 3. The highest calculated resistance factor was 1.11 for specimen A1.76/R/b while the lowest value was calculated for specimen A3.107/R/a. The resistance factor for this sample (0.97) and five more samples that were welded according to approach 3 are even below the ideal limit. This indicates a higher amount of copper in the measured area between tips 2 and 4.

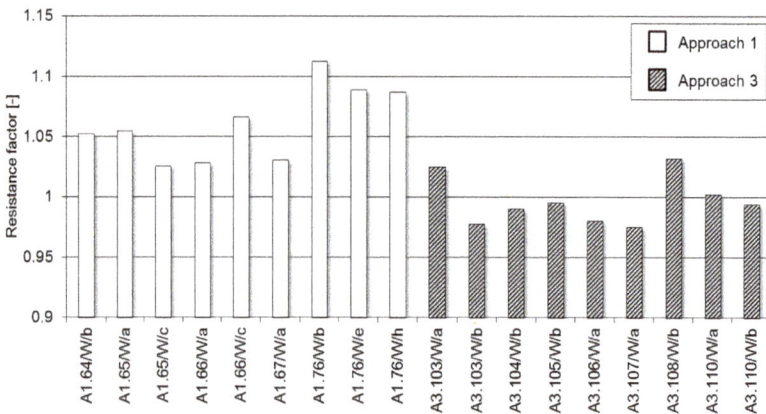

Figure 9. Resistance factors of the examined Al–Cu friction stir welds.

4. Conclusions

In this study, the mechanical and electrical properties of cross-section-adjusted friction stir butt welds were investigated. The applied materials were copper EN CW008A with a thickness of 3 mm and aluminum EN AW-1050A with a thickness of 4.7 mm. Therefore, three different approaches were tested with the aim to produce sound welds with properties similar to those of the used base materials.

1. Approach 1 included copper blanks that were edge bent on one side. These were positioned on the AS. The welds were produced with a welding speed of 450 mm/min.
2. Approach 2 included copper blanks that were edge bent on one side. These were positioned on the RS. Using this approach no sound welds could be achieved. This is the reason why samples that were welded according to this approach were ignored for further analysis.
3. Approach 3 included copper blanks that were edge bent on one side. These were positioned on the AS. Within this approach, the aluminum blanks were chamfered on one side to reduce the material deficit in the joining area. The welds were produced with a welding speed of 450 mm/min.
4. The results of the macrostructural analysis showed a lack of penetration for both approaches and inadmissible cavities for approach 1. In general, the underfill maximum was smaller for approach 3.
5. The average UTF for samples that were welded according to approach 1 was 7.78 kN while approach 3 achieved an average value of 8.89 kN. Compared to the aluminum base material, the joint efficiency was 57% for approach 1 and 65% for approach 3, respectively. The specimen with the best mechanical properties had a joint efficiency of 81%.
6. The average electrical resistance for approach 1 was 0.0075 mΩ and for approach 3 the electrical resistance was 0.0070 mΩ. The calculated resistance factor for both approaches indicated a long-term stable behavior of the joints.
7. In general, welds with good tensile properties lead to lower electrical resistances.

Especially by approach 3, promising results could be achieved. Considering the quality criteria from DIN EN ISO 25239-5, the lack of penetration has to be avoided. Therefore, in further experiments a longer tool pin will be used. The chamfer on the aluminum blank led to better mechanical and electrical joint properties due to the reduced material deficit in the joining area. However, exact positioning of the blanks was complicated by the chamfer with the result of a few welds showing insufficient scratching of the copper base material. Further investigations with standardized chamfer geometries and seam tracking could even improve the results.

Author Contributions: Conceptualization, N.E., A.H., and A.F.; Formal analysis, N.E. and A.H.; Investigation, N.E. and J.D.; Methodology, N.E.; Project administration, A.H. and A.F.; Supervision, A.F. and S.B.; Visualization, J.D.; Writing–original draft, N.E.; Writing–review and editing, A.H. and S.B.

Funding: This research was funded by Volkswagen AG corporate research.

Conflicts of Interest: The authors declare no conflict of interest.

References

1. Li, X.-W.; Zhang, D.-T.; Qiu, C.; Zhang, W. Microstructure and mechanical properties of dissimilar pure copper/1350 aluminum alloy butt joints by friction stir welding. *Trans. Nonferrous Met. Soc. China* **2012**, *22*, 1298–1306. [CrossRef]
2. Deutsches Kupfer-Institut e.V. *Kupfer in der Elektrotechnik—Kabel und Leitungen*; Deutsches Kupfer-Institut: Düsseldorf, Germany, 2000.
3. Bargel, H.-J.; Schulze, G. *Werkstoffkunde*; Springer: Düsseldorf, Germany, 2016.
4. Braunovic, M.; Myshkin, N.K.; Konchits, V.V. *Electrical Contacts: Fundamentals, Applications and Technology. Taylor and Francis Distributor*; CRC Press: Boca Raton, FL, USA, 2007.

5. Khodir, S.A.; Ahmed, M.M.Z.; Ahmed, E.; Mohamed, S.M.R.; Abdel-Aleem, H. Effect of Intermetallic Compound Phases on the Mechanical Properties of the Dissimilar Al/Cu Friction Stir Welded Joints. *J. Mater. Eng. Perform.* **2016**, *25*, 4637–4648. [CrossRef]

6. Xue, P.; Xiao, B.L.; Ni, D.R.; Ma, Z.Y. Enhanced mechanical properties of friction stir welded dissimilar Al–Cu joint by intermetallic compounds. *Mater. Sci. Eng. A* **2010**, *527*, 5723–5727. [CrossRef]

7. Xue, P.; Ni, D.R.; Wang, D.; Xiao, B.L.; Ma, Z.Y. Effect of friction stir welding parameters on the microstructure and mechanical properties of the dissimilar Al–Cu joints. *Mater. Sci. Eng. A* **2011**, *528*, 4683–4689. [CrossRef]

8. Barekatain, H.; Kazeminezhad, M.; Kokabi, A.H. Microstructure and Mechanical Properties in Dissimilar Butt Friction Stir Welding of Severely Plastic Deformed Aluminum AA 1050 and Commercially Pure Copper Sheets. *J. Mater. Sci. Technol.* **2014**, *30*, 826–834. [CrossRef]

9. Zhang, Q.-Z.; Gong, W.-B.; Liu, W. Microstructure and mechanical properties of dissimilar Al–Cu joints by friction stir welding. *Trans. Nonferrous Met. Soc. China* **2015**, *25*, 1779–1786. [CrossRef]

10. Al-Roubaiy, A.O.; Nabat, S.M.; Batako, A.D.L. Experimental and theoretical analysis of friction stir welding of Al–Cu joints. *Int. J. Adv. Manuf. Tech.* **2014**, *71*, 1631–1642. [CrossRef]

11. Thomas, W.M.; Nicholas, E.D.; Needham, J.C.; Murch, M.G.; Templesmith, P.; Dawes, C.J. Improvements Relating to Friction Welding. U.S. Patent 5,460,317, 10 June 1993.

12. Matthes, K.-J.; Schneider, W. *Schweißtechnik. Schweißen von Metallischen Konstruktionswerkstoffen*; Carl Hanser Verlag: München, Germany, 2016.

13. Celik, S.; Cakir, R. Effect of Friction Stir Welding Parameters on the Mechanical and Microstructure Properties of the Al-Cu Butt Joint. *Metals* **2016**, *6*, 133. [CrossRef]

14. Mishra, R.S.; De, P.S.; Kumar, N. *Friction Stir Welding and Processing: Science and Engineering*; Springer International Publishing: New York, NY, USA, 2014.

15. Ott, M. Neues Verfahren fügt Bleche besser zusammen. *Aluminium Praxis*, 3 February 2017.

16. Fratini, L.; Buffa, G.; Shivpuri, R. Improving friction stir welding of blanks of different thicknesses. *Mater. Sci. Eng. A* **2007**, *459*, 209–215. [CrossRef]

17. Sahu, P.K.; Pal, S. Mechanical properties of dissimilar thickness aluminium alloy weld by single/double pass FSW. *J. Mater. Process. Technol.* **2017**, *243*, 442–455. [CrossRef]

18. Ezumi, M. Friction Stir Welding Method of Two Members Having Different Thicknesses. Google Patent EP1825946A1, 29 August 2007.

19. Weigl, M.; Feineis, A.; Sascha, C.; Kunz, M. Verfahren und Vorrichtung zum Rührreibschweißen bei Materialien unterschiedlicher Dicke und bei Kehlnähten. Google Patent DE102014001050A1, 30 July 2015.

20. Werz, M.; Hoßfeld, M.; Volz, O. Verfahren zum Stumpfverschweißen sowie Bauteil und Rührreibschweißwerkzeug. Google Patent DE102013110034A1, 12 March 2015.

21. Deutsches Institut für Normung e.V. *Aluminium und Aluminiumlegierungen—Chemische Zusammensetzung und Form von Halbzeug—Teil 3: Chemische Zusammensetzung und Erzeugnisformen*; Beuth Verlag GmbH: Berlin, Germany, 2013; DIN EN 573-3.

22. Deutsches Institut für Normung e.V. *Kupfer und Kupferlegierungen—Platten, Bleche und Bänder aus Kupfer für die Anwendung in der Elektrotechnik*; Beuth Verlag GmbH: Berlin, Germany, 2014; DIN EN 13599.

23. Deutsches Institut für Normung e.V. *Rührreibschweißen—Aluminium—Teil 5: Qualitäts-und Prüfungsanforderungen*; Beuth Verlag GmbH: Berlin, Germany, 2012; DIN EN ISO 25239-5.

24. Deutsches Institut für Normung e.V. *Zerstörende Prüfung von Schweißverbindungen an Metallischen Werkstoffen—Querzugversuch*; Beuth Verlag GmbH: Berlin, Germany, 2013; DIN EN ISO 4136:2012.

25. Schmidt, P.A. *Laserstrahlschweissen Elektrischer Kontakte von Lithium-Ionen-Batterien in Elektro- und Hybridfahrzeugen*; Herbert Utz: München, Germany, 2015.

26. Essers, M.; Schiebahn, A.; Reisgen, U. Widerstandsbuckelschweißen von Al-Cu-Mischverbindungen zur Generierung elektrischer Kontaktierungen. Große Schweißtechnische Tagung, DVS-Studentenkongress, Nürnberg, Germany, 17 September 2015.

![metals logo] *metals*

MDPI

Article

Microstructure and Mechanical Characterization of a Dissimilar Friction-Stir-Welded CuCrZr/CuNiCrSi Butt Joint

Youqing Sun [1], Diqiu He [1,2], Fei Xue [2], Ruilin Lai [1,*] and Guoai He [1,2]

[1] State Key Laboratory of High Performance Complex Manufacturing, Central South University, Changsha 410083, China; sunyouqing@csu.edu.cn (Y.S.); 133701033@csu.edu.cn (D.H.); heguoai@csu.edu.cn (G.H.)

[2] Light Alloy Research Institute, Central South University, Changsha 410083, China; syqcsu@csu.edu.cn

* Correspondence: GuirongSu@csu.edu.cn, Tel.: +86-0731-8887-6230

Received: 11 April 2018; Accepted: 5 May 2018; Published: 8 May 2018

Abstract: Dissimilar CuNiCrSi and CuCrZr butt joints were successfully frictionstirwelded at constant welding speed of 150 mm/min and rotational speed of 1400 rpm with the CuCrZr alloy or the CuNiCrSi alloy located on the advancing side (AS). The microstructure and mechanical properties of joints were investigated. When the CuCrZr alloy was located on the AS, the area of retreating material in the nugget zone was a little bigger. The Cr solute-rich particles were found in the nugget zone on CuCrZr side (CuCrZr-NZ) while a larger density of solute-rich particles identified as the concentration of Cr and Si element was found in the nugget zone on CuNiCrSi side (CuNiCrSi-NZ). The Cr precipitates and δ-Ni$_2$Si precipitates were found in the base metal on CuNiCrSi side (CuNiCrSi-BM) but only Cr precipitates can be observed in the base metal on CuCrZr side (CuCrZr-BM). Precipitates were totally dissolved into Cu matrix in both CuCrZr-NZ and CuNiCrSi-NZ, which led to a sharp decrease in both micro-hardness and tensile strength from BM to NZ. When the CuNiCrSi was located on the AS, the tensile testing results showed the fracture occurred at the CuCrZr-NZ, while the fracture was found at the mixed zone of CuNiCrSi-NZ and CuCrZr-NZ for the other case.

Keywords: dissimilar joints; friction stir welding; microstructure; mechanical properties

1. Introduction

Copper alloys are widely used in industries for its high electrical conductivities. However, the strength of conventional copper alloys is usually limited in contrast with its high electrical conductivity [1]. CuCrZr and CuNiCrSi are excellent and unique copper alloys, which possess a desirable combination of high strength and good electrical conductivities after solution treatment and aging process. The ultimate tensile strength (UTS) and electrical conductivity of CuCrZr alloy can reach to more than 530 MPa and 80% IACS [1]. In comparison, the CuNiCrSi alloy shows a higher ultimate tensile strength of 600–800 MPa but a lower electrical conductivity of about 45% IACS due to different additions [2]. The dissimilar joints of CuCrZr and CuNiCrSi alloy with good balance of strength and electrical conductivity can be widely applied to many industries, such as large generator rotor and international thermo nuclear experiment reactor (ITER) components [3,4].

Owing to the impacts of oxygen and impurity, copper alloys are usually hard to be welded by conventional fusion welding techniques, such as the arc welding and the electrical beam welding. Zhang et al. [5] reported that the welding speed should be very slow in order to combat the high thermal diffusivity during the arc welding of copper joints, which contributed to the coarse microstructure in the heat affected zone (HAZ) and limited mechanical properties of joints. Kanigalpula et al. [6] reported that the mechanical properties of CuCrZr joints were limited by interdendritic micro-cracks and voids formed in the fusion zone during the electrical beam welding process. In contrast, friction stir welding (FSW), as a solid-state joining process, can overcome these metallurgical difficulties of conventional fusion welding [7]. Amarnath et al. [8] made a comparative study of gas tungsten art welding (GTAW) process and FSW process on electrolytic tough pitch (ETP) copper. They found that the mechanical property of FSW joints was superior to that of GTAW joints due to the generation of fine equiaxed grains. Jau-wen et al. [9] carried out a comparative study of tungsten inert gas welding (TIG) process and FSW process on pure copper. The result revealed that the mechanical property of FSW joints was better than that of TIG joints.

In recent years, substantial studies have been conducted on the microstructure and mechanical properties of friction-stir-welded copper joints. Mironov et al. [10] studied the relationship between microstructure evolutions and the rotational speeds ranging from 200 rpm to 1000 rpm in friction-stir-welded pure copper joints. The grain structure development was found to be predominantly controlled by continuous recrystallization in the welds produced at welding speeds of 200–500 mm/min, and fine-grained (1–2 µm) microstructure was obtained in the stir zone. Heidarzadeh et al. [11] found that the UTS of pure copper joints firstly increased up to maximum value (about 260 MPa) with the increasing of tool rotational speed, welding speed and tool axial force, and then decreased until the finish of the FSW process. Liu et al. [12] studied the effects of the tool rotational speeds (300, 400, 600, 800 and 1000 rpm) on microstructure and mechanical properties of friction-stir-welded copper joints. They observed that the grain size and the UTS were significantly affected by the tool rotational speeds. When the rotational speeds increased from 300 to 1000 rpm, the grain size increased while the UTS firstly increased to the maximum (277 MPa) and then decreased. Shen et al. [13] investigated the microstructure and mechanical properties of friction-stir-welded copper joints under different welding speeds (25, 50, 100, 150 and 200 mm/min). They found that defect-free joints were obtained at lower welding speeds and the maximum value of UTS of welded joints was 275 MPa. In order to improve the strength of copper joints, Xue et al. [14] used flowing water as additional rapid cooling method to lower peak temperature during the process of FSW. The joints using this method kept the microstructure with a high dislocation density so that the UTS of joints could reach to 340 Mpa, which was nearly equal to the strength of base metal.

Compared with substantial study on FSW of conventional copper joints, the reports on FSW of copper joints with high strength and good electrical conductivity were rather scarce. Sahlot et al. [15] studied the wear of the tool using in the FSW of CuCrZr, but he didn't discuss the characteristics of the joints. Jha et al. [16] and Lai et al. [17] investigated the characteristics of the CuCrZr joints welded by FSW. They both found that the properties of joints were primarily depended on Nano-precipitates. To the best of authors' knowledge, there have been no studies on microstructure and mechanical properties of the dissimilar CuCrZr and CuNiCrSi joints fabricated by FSW. Therefore, the present study mainly investigated the microstructure evolution and mechanical properties of dissimilar CuCrZr and CuNiCrSi joint. In particular, the relationship between microstructure and mechanical properties was studied as well.

2. Materials and Methods

In this study, the dimensions of CuCrZr and CuNiCrSi plates are 300 mm × 100 mm × 3 mm. The CuCrZr alloy was subjected to solution (920 °C for 0.5 h) and aged (440 °C for 2 h) treatments. The CuNiCrSi alloy was also treated through solution (800 °C for 2 h) and aged (450 °C for 5 h) process. The chemical compositions of these two alloys are listed in Table 1.

Table 1. The chemical compositions of CuCrZr alloy and CuNiCrSi alloy.

Alloy	Cu	Al	Mg	Cr	Ni	Zr	Fe	Si
CuCrZr	Bal	0.25	0.1	0.8	-	0.3	0.09	0.04
CuNiCrSi	Bal	-	-	0.5	2.0	-	≤0.15	0.5

2.1. Friction Stir Welding Process

Dissimilar CuCrZr and CuNiCrSi joints were friction stir welded by a tool at constant welding speed of 150 mm/min and constant rotation speed of 1400 rpm. The position-related microstructure and mechanical properties were also taken into consideration. The FSW tool had a concave shoulder 10 mm in diameter and a conical pin 2.8 mm in length, the pin with 4.5 mm in the top diameter and 3.5 mm in the bottom diameter was machined from H13 steel. The tool rotated in the clock wise direction with a fixed tool tilt angle of 2.5°. The plunging depth and plunging speed of the pin were selected as 0.1 mm and 0.05 mm/s. The laboratory equipment, experimental setup, schematic presentation of friction stir welding process and dimensions of FSW tool are shown in Figure 1.

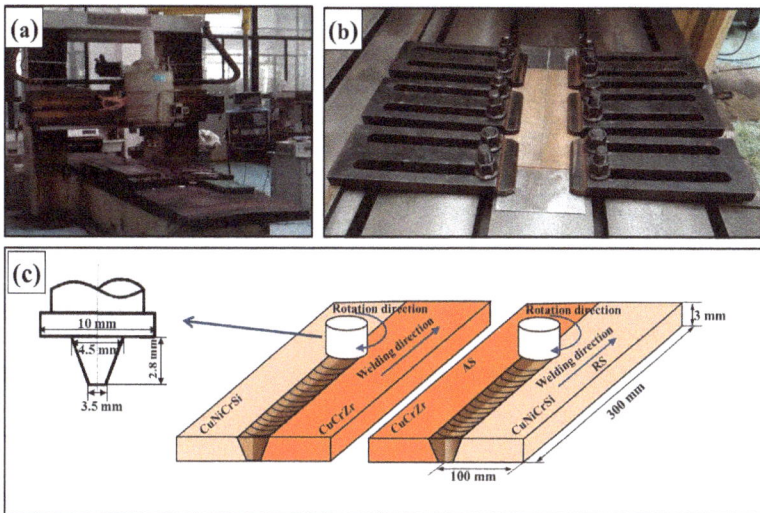

Figure 1. The laboratory equipment, experimental setup, schematic presentation of friction stir welding process and dimensions of FSW tool. (**a**) The laboratory equipment; (**b**) experimental setup; (**c**) schematic presentation of friction stir welding process and dimensions of FSW tool.

After the welding, these joints were cut into strips perpendicularly to the welding line. Specimens for metallographic observation and mechanical testing were made from these strips.

2.2. Microstructural Characterization

Specimens for microstructure analysis were etched in a solution of 40 mL H_2O, 10 mL HCl and 2 g Fe_3Cl after standard polishing process. The microstructure of different regions including the base metal on CuCrZr side (CuCrZr-BM), the nugget zone on CuCrZr side (CuCrZr-NZ), the base metal on CuNiCrSi side (CuNiCrSi-BM) and the nugget zone on CuNiCrSi side (CuNiCrSi-NZ) was observed on a confocal laser scanning microscope (CLSM) and a VHX-5000 3D microscope. To further study the microstructure of the nugget zone, a scanning electron microscope (SEM) was used. The SEM specimens were slightly etched to clarify CuCrZr-NZ and CuNiCrSi-NZ in the microscope. A transmission electron microscopy (TEM, Tecnai G2 F20) was also applied to analysis samples from CuCrZr-BM, CuCrZr-NZ, CuNiCrSi-BM and CuNiCrSi-NZ.

2.3. Mechanical Testing

Tensile tests were conducted on a universal electronic tensile testing machine (MTS Systems Corporation, Eden Prairie, MN, USA) with a testing speed of 2 mm/min. The tensile testing specimens have a gauge length of 150 mm and a width of 25 mm. Three tensile testing specimens were taken to repeat the experiments. The Vickers hardness was measured using a Vickers hardness machine (Huayin Testing Instrument Co., Ltd., Yantai, China) with a load of 100 g and a dwell time of 10 s. The measurement was carried out along the centerline and the distance between neighboring measured points was 0.5 mm.

3. Results

3.1. Microstructure of Dissimilar CuCrZr/CuNiCrSi Butt Joints

Figure 2 shows surface morphologies and cross-section macrographs of dissimilar CuCrZr/CuNiCrSi butt joints with the CuCrZr alloy located on the advancing side (AS) or the retreating side (RS), respectively. No defects were found on both the surface morphologies and the cross-section macrographs. From cross-section macrographs, it can be found that the area of retreating materials in the nugget zone (NZ) is a little bigger when the CuCrZr alloy is placed on the AS. Three distinct regions, the base metal (BM), the NZ, the thermo-mechanically affected zone (TMAZ), can be identified from macrographs.

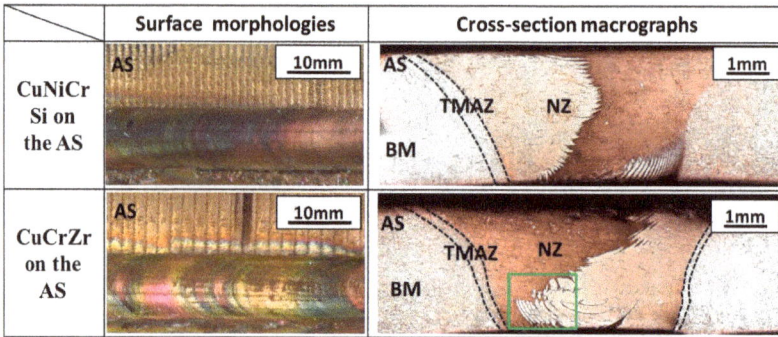

Figure 2. The surface morphologies and the cross-section macrographs of dissimilar CuCrZr/CuNiCrSi butt joints with the CuCrZr alloy or the CuNiCrSi alloy located on the advancing side (AS).

Figure 3 portrays magnified grain structures of the NZ, BM and TMAZ with the CuCrZr alloy or the CuNiCrSi alloy located on the AS. One thing to be noted is that the heat affected zone (HAZ) was not observed at the present study, which can be attributed to the high heat dissipation of copper alloy [18]. In Figure 3, it is seen that non-uniform distribution of coarse size grains exiting in the two base metals are completed transformed into even distribution, showing as equiaxed grains in the nugget zone after the FSW process. The average grain sizes of different zones measured by the mean linear intercept method are shown in Table 2. The grain size of CuNiCrSi-BM is a little bigger than that of CuCrZr-BM. Considerable numbers of coarse particles distribute randomly in both BM and NZ. Additionally, the density of coarse particles is larger in CuCrZr-NZ and CuCrZr-BM than that of CuNiCrSi-NZ and CuNiCrSi-BM. The chemical compositions of these coarse particles are analyzed by EDS which will be discussed in detail below.

Table 2. The average grain size of different zones.

Conditions	Grain Size/µm			
	CuNiCrSi-BM	CuNiCrSi-NZ	CuCrZr-BM	CuCrZr-NZ
CuNiCrSi on the AS	48 ± 2.5	1.5 ± 0.2	31 ± 4.1	0.8 ± 0.2
CuCrZr on the AS	47 ± 4.2	1.4 ± 0.5	30 ± 3.6	0.9 ± 0.3

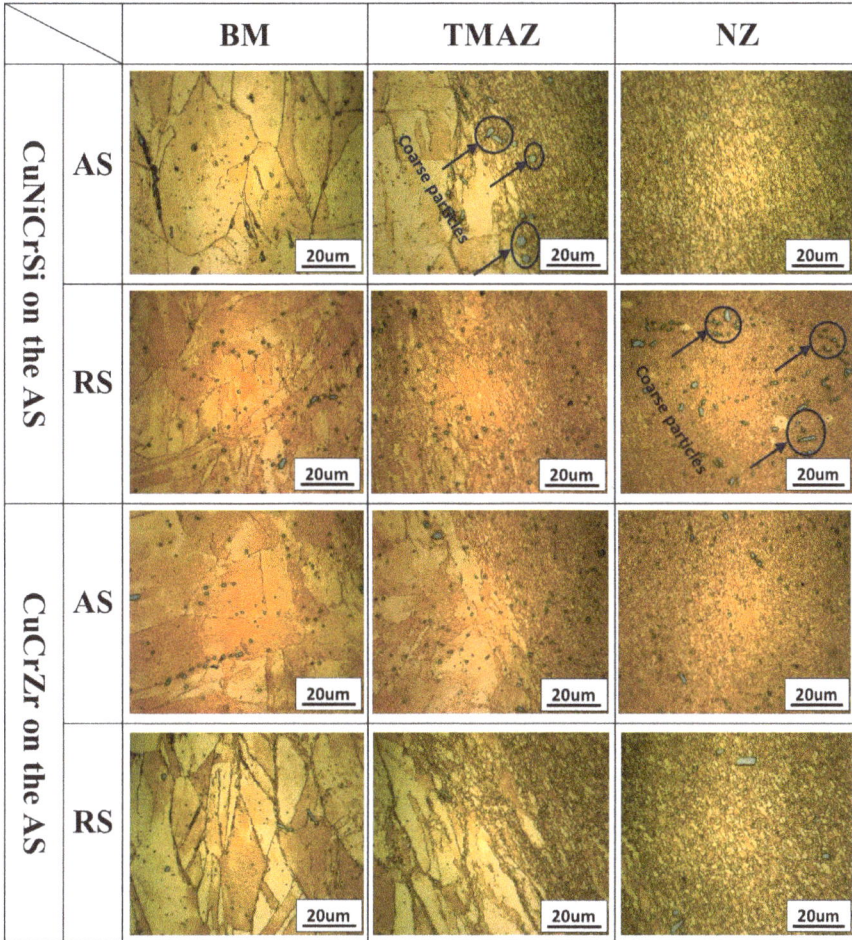

Figure 3. The magnified grain structures of the nugget zones (NZ), base metals (BM) and the thermo-mechanically affected zone (TMAZ) with the CuCrZr alloy or the CuNiCrSi alloy located on the AS.

Figure 4 shows the dislocation structures and sub-boundaries of grains in CuNiCrSi-NZ with the CuCrZr alloy on the AS. Figure 4a shows some grains are dislocation-free, while Figure 4b shows that there is a network structure containing high density of dislocations in some grains. This phenomenon means that the final grains in the NZ experiencing dynamic recovery at different degrees. As shown in Figure 4c, sub-grain boundaries are formed by absorbing dislocations. From Figure 4d, it can be found that some grains are passed through by sub-boundaries.

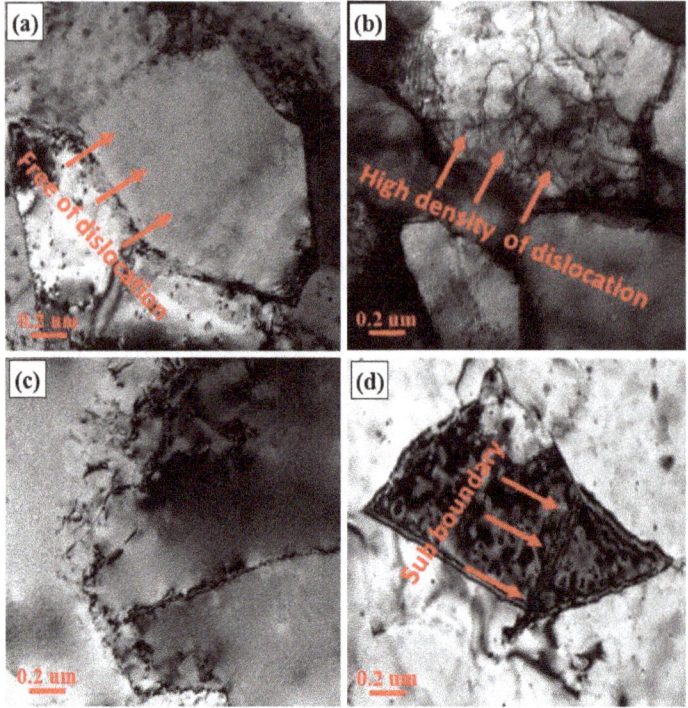

Figure 4. Dislocation structures and sub boundaries of grains in CuNiCrSi-NZ with the CuCrZr alloy on the AS. (**a**) Grains are free of dislocation; (**b**) Grains have high density of dislocation; (**c**) Dislocations are absorbed into sub-boundaries; (**d**) Grains are traversed by sub-boundaries.

Figures 5 and 6 are EDS images showing the analyses of elemental concentration in coarse particles distributing in CuCrZr-NZ and CuNiCrSi-NZ with the CuCrZr alloy on the AS. In Figure 5, it can be seen that the coarse particles in CuCrZr-NZ are composed of Cr element (green colour), the statistical result further reveals that these Cr element enriched particles contains 96.8 wt. % Cr and 3.2 wt. % Cu. In comparison, Cr element (green colour) and Si element (red colour) concentrate on coarse particles in CuNiCrSi-NZ. Besides, the EDS analysis (Figure 6) suggests that the atomic ratio of Cr and Si in these coarse particles is 3:1.

Figure 7 shows the bright field TEM micrographs as well as the relevant selected area electron diffraction (SAD) patterns of samples from CuCrZr-BM and CuCrZr-NZ with the CuCrZr alloy on the AS. The micrographs and SAD patterns are obtained with the incident beam parallel to the direction of [011]Cu. Figure 7a shows the lobe-lobe contrast precipitates homogeneously dispersing in Cu matrix are about 5–8 nm in length. The SAD pattern in Figure 7b reveals the reflections spots from Cu matrix and precipitates. The reflection spots of precipitates reveal that they are the FCC Cr precipitates. After the welding, no precipitates are detected in bright field TEM micrographs of samples from CuCrZr-NZ (Figure 7c), which is also confirmed by the corresponding SAD pattern (Figure 7d).

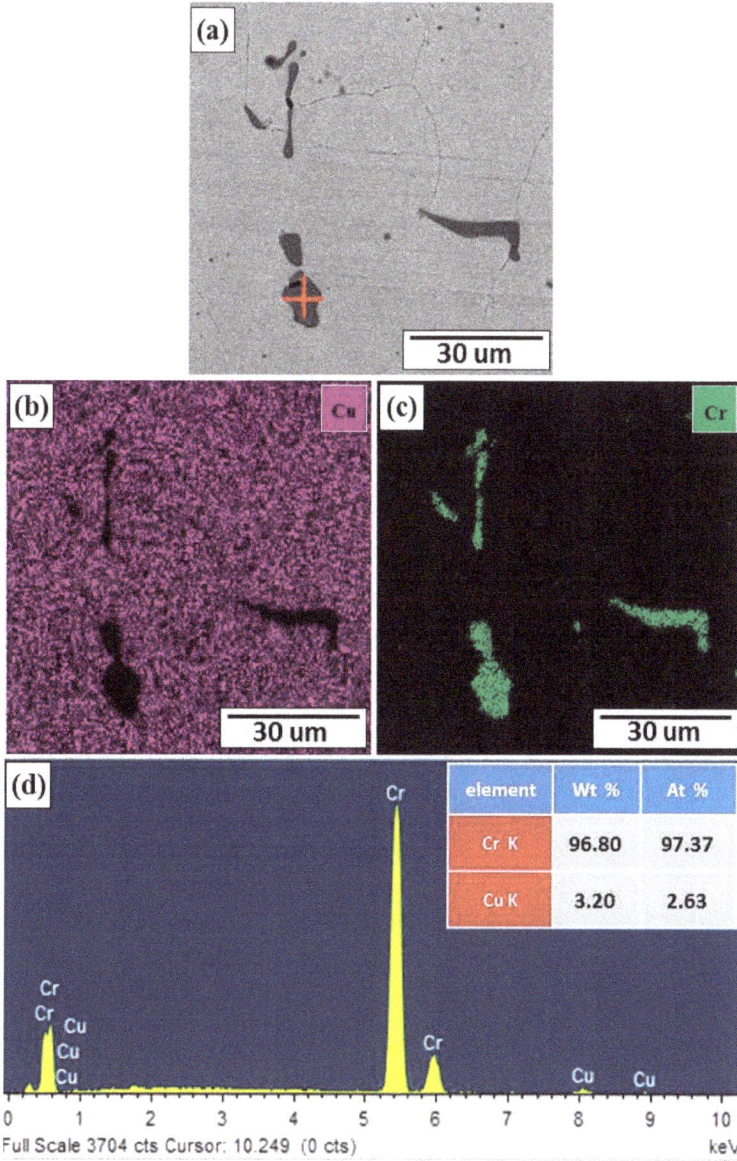

Figure 5. EDS images showing the analyses of elemental concentration in coarse particles distributing in CuCrZr-NZ with the CuCrZr alloy on the AS. (**a**) Micrograph of the surface revealing the presence of coarse particles; (**b**) copper; (**c**) chromium; (**d**) EDS analysis revealing the elements of coarse particles.

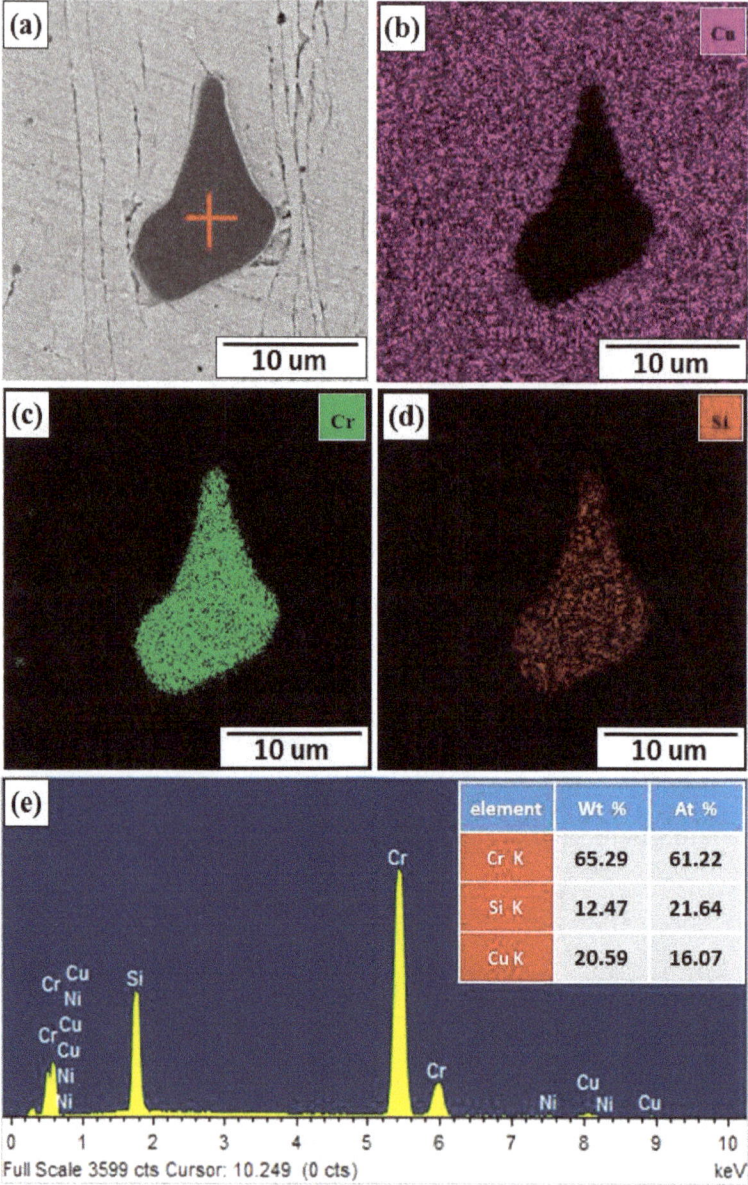

Figure 6. EDS images showing the analyses of elemental concentration in coarse particles distributing in CuNiCrSi-NZ with the CuCrZr alloy on the AS. (**a**) Micrograph of the surface revealing the presence of coarse particles; (**b**) copper; (**c**) chromium; (**d**) silicon; (**e**) EDS analysis revealing the elements of coarse particles.

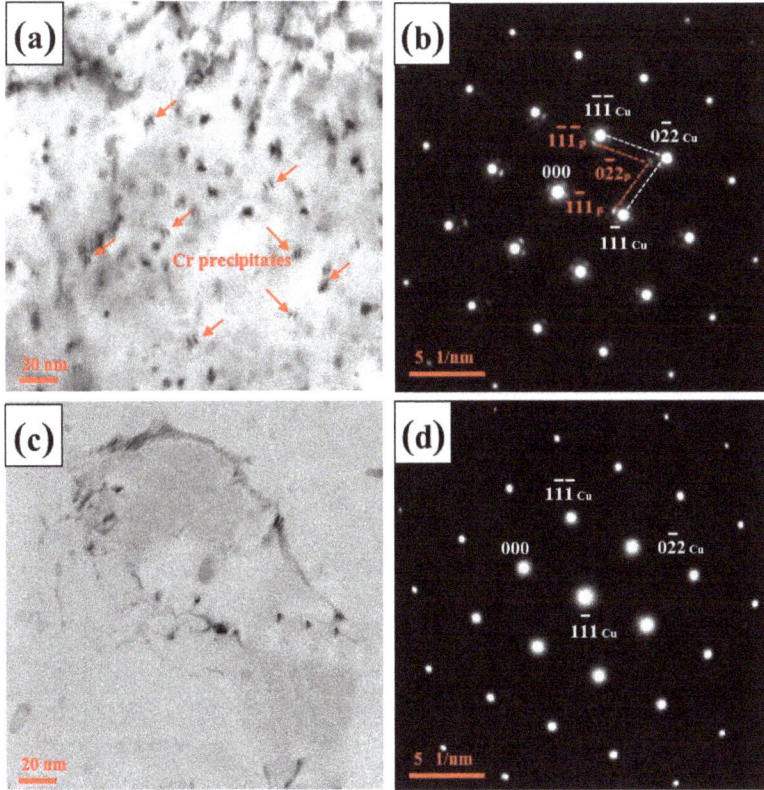

Figure 7. The bright field TEM micrographs as well as the relevant selected area electron diffraction (SAD) patterns of samples from CuCrZr-BM and CuCrZr-NZ with the CuCrZr alloy on the AS. (**a**) TEM micrographs showing the Cr precipitates in CuCrZr-BM; (**b**) SAD pattern of (**a**) in [011]Cu direction; (**c**) TEM micrographs showing the dissolution of precipitates in CuCrZr-NZ; (**d**) SAD pattern of (**c**) in [011]Cu direction.

Figure 8 reveals the bright field TEM micrographs with relevant SAD patterns of samples from CuNiCrSi-BM and CuNiCrSi-NZ with the CuCrZr alloy on the AS. Figure 8a shows that the lobe-lobe contrast Cr precipitates can also be defected in CuNiCrSi-BM. However, the reflection spots of Cr precipitates cannot be found from the relevant SDA pattern (Figure 8b) parallel to [011]Cu. Another type of precipitates which are rod-shaped in [011]Cu direction (Figure 8a) and disc-shaped in [111]Cu direction (Figure 8c) can also be found in CuNiCrSi-BM. Moreover, the corresponding SAD patterns parallel to [011]Cu direction (Figure 8b) and [111]Cu direction (Figure 8d) reveal the precipitates have a δ-Ni2Si crystal structure [2]. However, it is similar to samples from CuCrZr-NZ, no precipitates are detected in samples from CuNiCrSi-NZ based on bright field TEM micrographs (Figure 8e) and relevant SAD pattern (Figure 8f).

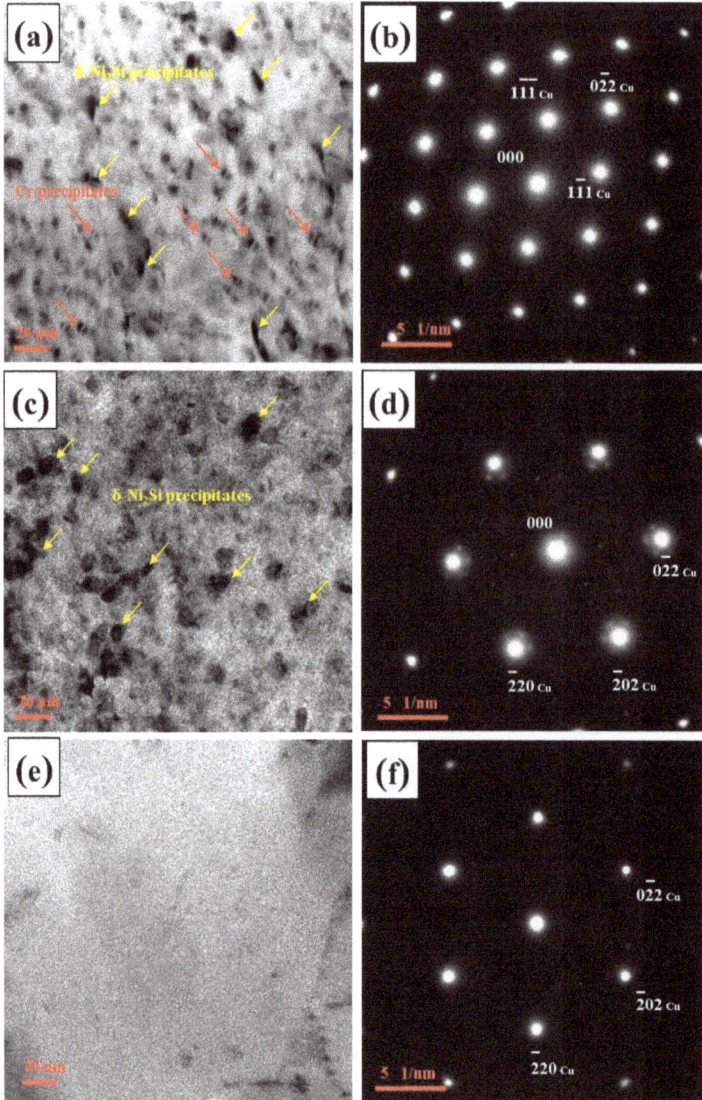

Figure 8. The bright field TEM micrographs with relevant SAD patterns of samples from CuNiCrSi-BM and CuNiCrSi-NZ with the CuCrZr alloy on the AS. (**a**) TEM micrographs showing the Cr and -Ni2Si precipitates in CuNiCrSi-BM; (**b**) SAD pattern of (**a**) in [011]Cu direction; (**c**) TEM micrographs showing the -Ni2Si precipitates in CuNiCrSi-BM; (**d**) SAD pattern of (**c**) in [111]Cu direction; (**e**) TEM micrographs showing the dissolution of precipitates in CuNiCrSi-NZ; (**f**) SAD pattern of (**e**) in [111]Cu direction.

Figure 9 reveals that the dislocations are pinned by Nano-precipitates in both CuNiCrSi-BM (Figure 9a) and CuCrZr-BM (Figure 9b) with the CuCrZr alloy on the AS, via which way the strengths of both CuNiCrSi alloy and CuCrZr alloy are improved.

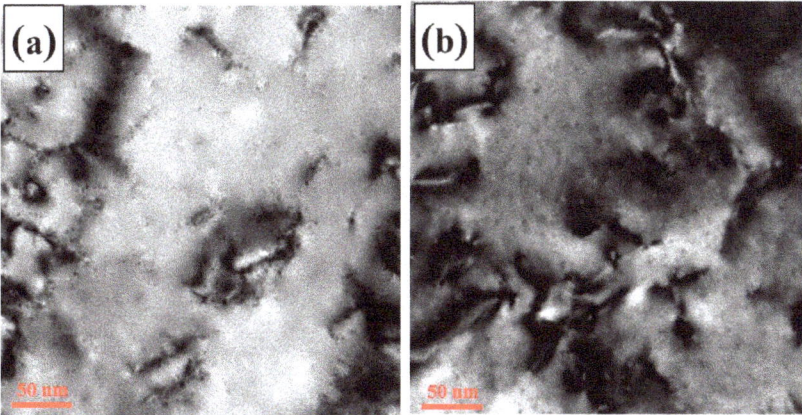

Figure 9. The dislocations are pinned by nano-precipitates in both CuNiCrSi-BM and CuCrZr-BM. (**a**) CuCrZr-BM; (**b**) CuNiCrSi-BM with the CuCrZr alloy on the AS.

3.2. Mechanical Characterization of CuCrZr/CuNiCrSi Butt Joints

Figure 10 shows the micro-hardness distribution of dissimilar CuCrZr/CuNiCrSi joints. The micro-hardness of CuNiCrSi-BM and CuCrZr-BM is 225 HV and 155 HV, respectively. However, it is seen that the micro-hardness decreases sharply to 150 HV and 125 HV in CuNiCrSi-NZ and CuCrZr-NZ. Besides, the micro-hardness profile distributes asymmetrically along the measuring line, which is a consequence of the different mechanical properties of CuCrZr and CuNiCrSi alloy.

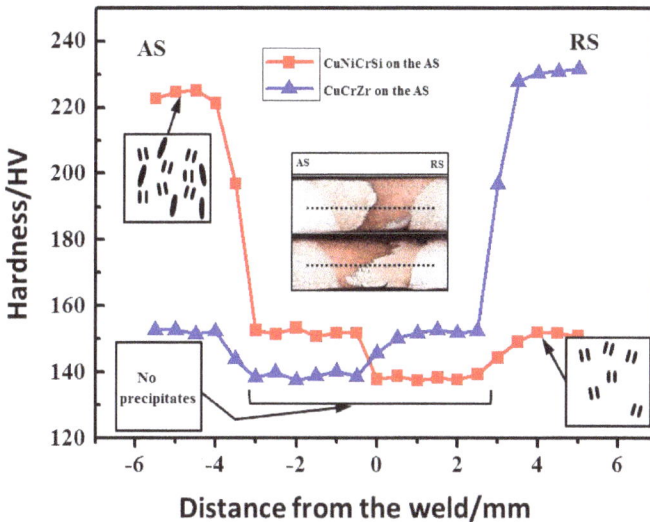

Figure 10. The micro-hardness of dissimilar CuCrZr/CuNiCrSi joints.

Figure 11 demonstrates the mechanical properties of dissimilar joints, whose detailed values are also listed in Table 3. It seems that the tensile strength of joints with CuCrZr located on the AS is a little bigger than that with the inverse material position. One thing to be noted is that the highest value of tensile strength is about 450 MPa in all welds, which indicates a decrease of mechanical properties compared with that of two aged base metals. However, the joints obtained at present study still exhibit a higher strength than most friction-stir-welded copper joints mentioned at previous study.

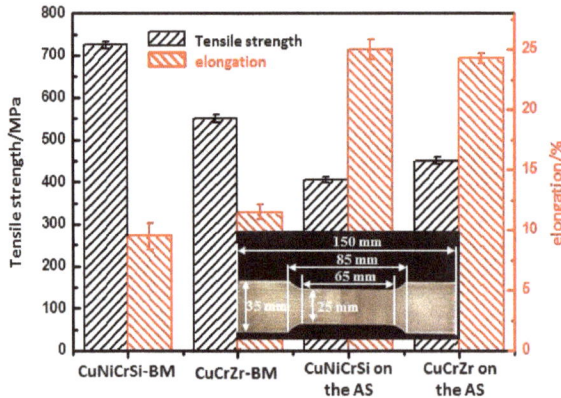

Figure 11. The mechanical properties of dissimilar joints.

Table 3. The mechanical properties of base metals and dissimilar joints.

Conditions	UTS (MPa)	ε (%)	Failure Location
CuNiCrSi-BM	725 ± 8	9.5 ± 1.1	-
CuCrZr-BM	550 ± 10	11.5 ± 0.6	-
CuNiCrSi on the AS	405 ± 7	25.0 ± 0.8	CuCrZr-NZ
CuCrZr on the AS	450 ± 8	24.3 ± 0.4	Mixed zone

Figure 12 shows the failure locations of tensile testing specimens. Obviously, all the welds failed at the stir zone. It is noted that the tensile specimens failed at the CuCrZr-NZ with CuNiCrSi located on the AS but at the mix zone of CuCrZr-NZ and CuNiCrSi-NZ with the inverse material position.

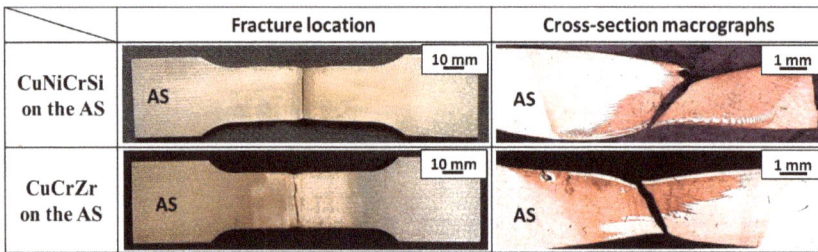

Figure 12. The failure locations and cross-section macrographs of tensile testing specimens with the the CuCrZr alloy or the CuNiCrSi alloy located on the AS.

Figure 13 shows the SEM micrographs of fracture surface of the dissimilar joints with the the CuCrZr alloy or the CuNiCrSi alloy located on the AS. Figure 13a shows the fracture surface of welds with the CuCrZr alloy on the AS while Figure 13c shows that with the inverse material position. Figure 13b,d show the magnified views of selected zones in Figure 13a,c, respectively. It can be found that the fracture surfaces both consist of fine populated dimples, which confirms that the dissimilar joints fail in the ductile mode of failure. No significant difference in the fracture mechanism with different material positions. Both selected samples experience extensive plastic deformation during the process of failure in these two conditions.

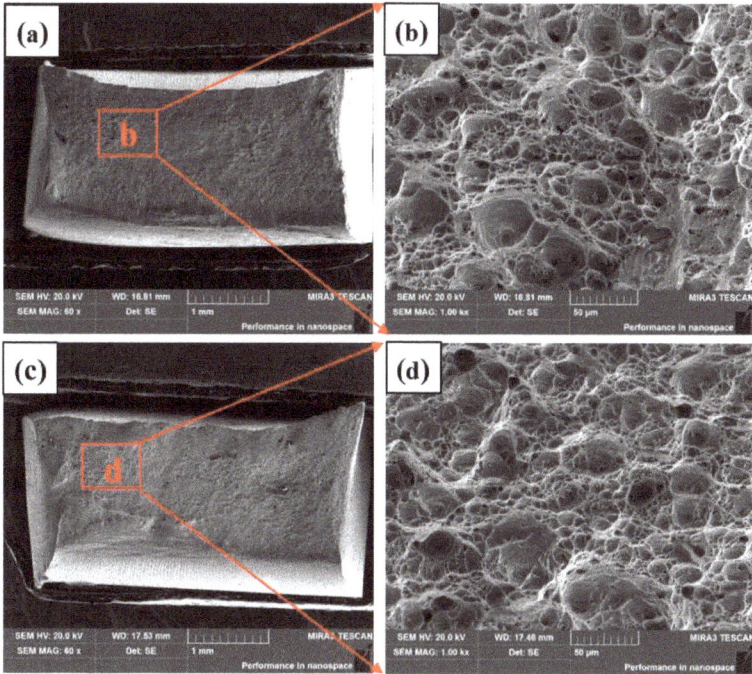

Figure 13. The SEM micrographs of fracture surface of the dissimilar joints with the the CuCrZr alloy or the CuNiCrSi alloy located on the AS. (**a**) the fracture surfaces of welds with CuCrZr alloy on the AS; (**b**) the magnified view of selected zone in (**a**); (**c**) the fracture surfaces of welds with CuNiCrSi alloy on the AS; (**d**) the magnified view of selected zone in (**c**).

4. Discussion

4.1. The Analysis of Microstructure of Dissimilar CuCrZr/CuNiCrSi Butt Joints

Figure 2 shows that the area of retreating materials in the NZ is seen to be a little bigger with the CuCrZr alloy placed on the AS. This is mainly because the higher flow stress of harder CuNiCrSi alloy. As the rotational pin starts to move through the base metals, the material in front of the pin is moved to rear side along the circular path. Zhu et al. [19] reported that both the material flow velocity and the friction force are reduced in this process. Moreover, the flow velocity and friction force of material on the rear advancing side come to the minimum. All these make it difficult for the material to move from the retreating side to the advancing side. When the CuNiCrSi alloy is located on the AS, the softer CuCrZr alloy needs to overcome bigger resistance as it is pushed to the advancing side by the rotational tool. Besides, both the velocity and friction force of CuCrZr alloy are reduced in this process. So it is difficult for CuCrZr alloy to penetrate into the advancing side. However, the results are reversed when the CuCrZr alloy is located on the AS. The area of retreating materials is larger in the NZ since it is easier for CuNiCrSi to move to the advancing side. Similar results can be found in other dissimilar friction-stir-welded joints, such as dissimilar joints of AA6061 and AA7075 [20]. They found that the material flow is more difficult when harder AA7075 was located on the AS.

Figure 3 shows the pancake-like grains in the BM are transformed into fine equiaxed grains in the NZ. It is because the dynamic recrystallization occurred in the NZ due to the thermal cycle and plastic deformation. Figure 4 shows the final grain structure of nugget zone is comprised of grains experiencing both dynamic recrystallization and partial dynamic recovery process. Su et al. [21] reported that repeated introduction of dislocations after dynamic recrystallization initiates partial recovery in the NZ, so the final grains are in state of dynamic recrystallization and partial dynamic recovery. What's more, it is observed that sub-grains grow with dislocations absorbed into sub-boundaries in the NZ from Figure 3c,d, which means the dynamic recrystallization process is a continuous dynamic recrystallization process [22].

In Figures 5 and 6, the coarse particles distributing in CuCrZr-NZ and CuNiCrSi-NZ with the CuCrZr alloy on the AS were analyzed. It is to be noted that the analyses are similar for welds with the inverse material location. From Figures 5 and 6, it is seen that the coarse particles in CuCrZr-NZ are primarily consisted of Cr element while Cr and Si element mainly concentrate on coarse particles in CuNiCrSi-NZ. Actually, these coarse particles are solute-rich particles which are formed during the solidification of the alloy due to the limited solubility of Cr and Si elements in Cu matrix [23]. Obviously, these coarse particles even cannot be solved during the process of FSW. No previously related study with similar experimental results can be compared by one to one but Jha et al. [16], who studied the FSW of aged CuCrZr alloy plates, reported the relevant results, and also found coarse particles in the NZ were identified as Cr concentration, However, these coarse particles are not effective on increasing strength any more [24]. In contrast, these coarse particles may act as internal stress concentrators which make the alloy easier to fracture [25].

In Figures 7 and 8, the Nano-precipitates distributing in welds with the CuCrZr alloy on the AS are studied, the results are the same in welds with the CuNiCrSi located on the AS. In Figure 7, Cr precipitates are observed in CuCrZr-BM. In fact, these Cr precipitates are found to generate from the aging process of CuCrZr alloy. These Cr precipitates can pin the movement of dislocations so that they are responsible for precipitation hardening of CuCrZr alloy [25]. From the previous studies, Jha et al. [16] also found Cr precipitates in the BM of 4-mm thick CuCrZr joints but he didn't offer the relevant SAD pattern. Lai et al. [17] found Nano-precipitates in the BM of butt-welded CuCrZr joints and offered the relevant SAD pattern, but the SAD pattern didn't reveal diffraction spots from fine precipitates. In contrast, Cr precipitates can also be defected in CuNiCrSi-BM but cannot be found in the relevant SAD pattern in Figure 8. It can be explained by the larger grain size of CuNiCrSi-BM as illustrated in Figure 3, which limit the choice of the diffraction conditions because only two or three grains could be examined in the visual field around the perforation [25]. Then the Cr precipitates coherent with Cu matrix are difficult to find in SAD pattern due to the limited specimen tilt. In addition to Cr precipitates, δ-Ni$_2$Si precipitates can also be found in CuNiCrSi-BM. Similar results can also be found in previous works about CuNiCrSi alloy [26,27]. All these precipitates can improve the strength of the CuNiCrSi alloy by pinning the movement of dislocations. Figures 7 and 8 shows the precipitates are dissolved in both CuNiCrSi-NZ and CuCrZr-NZ. It is because that high welding temperature and thermal cycling during FSW process can result in the dissolution of Nano-strengthening precipitates into the Cu matrix in the NZ. In general, there is always a higher demand on heat-input during the FSW process of Cu joints due to the high thermal conductivity of copper alloy. Jha et al. [16] found that the peak temperature was more than 800 °C in the friction stir welding of aged CuCrZr plates. The high heat-input in the NZ produces a supersaturated solution condition, which can result in the dissolution of strengthening precipitates into the Cu matrix in the NZ. Similar results can also be found in the friction stir welding of other precipitates hardening alloys, such as 6063 aluminum [28], 7075 aluminum [29], and thick CuCrZr plates [17].

4.2. The Relationship of Microstructure and Mechanical Characterization

From Figures 11 and 12, it can be concluded that the relative material position affected both failure location and tensile strength. Tensile specimens all failed at CuCrZr-NZ when the CuNiCrSi alloy was located on the AS, as shown in Figure 12. This phenomenon can be explained from two aspects. On the one hand, the micro-hardness is lowest in CuCrZr-NZ, which means CuCrZr-NZ is the softest zone in the whole joint. On the other hand, there is a larger density of coarse particles in CuCrZr-NZ, which may make CuCrZr-NZ an originating place to fracture by acting as internal stress concentrators. However, when the CuCrZr alloy is located at the AS, the CuNiCrSi alloy on the RS are stirred to the AS easily as before mentioned, so the specimens failed at the mixed zone of CuCrZr-NZ and CuNiCrSi-NZ with inverse material position. Since the strength of CuNiCrSi alloy is higher than CuCrZr alloy, the UTS of joints with the CuCrZr alloy located at AS is a little higher than that of joints with the opposite material position.

In Figures 10 and 11, both the micro-hardness and the tensile strength of joints are lower than those of two base metals. It is well-known that both the hardness distribution and tensile strength is directly dependent on the distribution of precipitates. The precipitates in CuNiCrSi-NZ and CuCrZr-NZ, which can pin the movement of dislocation and thus improve the strength of alloy, were both dissolved into Cu matrix. In this case, the dissolution of precipitates can lead to the decrease of both the micro-hardness and the tensile strength.

5. Conclusions

In the present study, the microstructure and mechanical properties of dissimilar CuNiCrSi and CuCrZr joints are investigated. The relationship between the microstructure and mechanical properties is studied as well. The following conclusions can be drawn:

(I) Defect-free joints are obtained under the constant welding speed of 150 mm/min and constant rotational speeds of 1400 rpm. The area of retreating materials in the NZ is bigger when the CuCrZr alloy was placed on the AS.

(II) Considerable numbers of coarse particles are found to distribute randomly in both BM and NZ. In CuCrZr-NZ, Cr solute-rich particles are detected, while concentration of Cr and Si element with larger density is found in CuNiCrSi-NZ.

(III) The Cr and δ-Ni_2Si precipitates are observed in CuNiCrSi-BM while only Cr precipitates are found in CuCrZr-BM. All these precipitates are dissolved in both CuCrZr-NZ and CuNiCrSi-NZ due to the high temperature, leading to the lower micro-hardness and tensile strength of joints when compared to base metals.

(IV) When the CuNiCrSi alloy was located on the AS, the fracture occurred in the CuCrZr-NZ due to the existence of larger density of coarse particles. However, failure was found at the mixed zone of CuNiCrSi-NZ and CuCrZr-NZ when the CuCrZr alloy was located on AS.

Author Contributions: Diqiu He was the principle investigator of the research. Youqing Sun, Fei Xue and Ruilin Lai carried out the welding tests and characterized the microstructure of the welded samples. Youqing Sun performed mechanical tastings, hardness, fractography, and wrote the paper. Guoai He provides tremendous help on analyzing and discussing the results as well as revising the manuscript.

Funding: This research was funded by National Basic Research Program of China (2014CB046605).

Acknowledgments: This work was supported by Science and Technology Innovation Projects of graduate students of central south university (2017zzts653) and the National Basic Research Program of China ("973 Program", 2014CB046605). Guoai He thank the supports from Scientific Research Foundation of Central South University (202045001).

Conflicts of Interest: The authors declare no conflict of interest.

References

1. Mishnev, R.; Shakhova, I.; Belyakov, A.; Kaibyshev, R. Deformation microstructures, strengthening mechanisms, and electrical conductivity in a Cu-Cr-Zr alloy. *Mater. Sci. Eng. A* **2015**, *629*, 29–40. [CrossRef]
2. Gholami, M.; Vesely, J.; Altenberger, I.; Kuhn, H.A.; Janecek, M.; Wollmann, M.; Wagner, L. Effects of microstructure on mechanical properties of CuNiSi alloys. *J. Alloy. Compd.* **2017**, *696*, 201–212. [CrossRef]
3. Shueh, C.; Chan, C.K.; Chang, C.C.; Sheng, I.C. Investigation of vacuum properties of CuCrZr alloy for high-heat-load absorber. *Nuclear Instrum. Methods Phys. Res.* **2017**, *841*, 1–4. [CrossRef]

4. Lipa, M.; Durocher, A.; Tivey, R.; Huber, T.; Schedler, B.; Weigert, J. The use of copper alloy CuCrZr as a structural material for actively cooled plasma facing and in vessel components. *Fusion Eng. Des.* **2005**, *75*, 469–473. [CrossRef]
5. Zhang, L.J.; Bai, Q.L.; Ning, J.; Wang, A.; Yang, J.N.; Yin, X.Q.; Zhang, J.X. A comparative study on the microstructure and properties of copper joint between MIG welding and laser-MIG hybrid welding. *Mater. Des.* **2016**, *110*, 35–50. [CrossRef]
6. Kanigalpula, P.K.C.; Pratihar, D.K.; Jha, M.N.; Derose, J.; Bapat, A.V.; Pal, A.R. Experimental investigations, input-output modeling and optimization for electron beam welding of Cu-Cr-Zr alloy plates. *International J. Adv. Manuf. Technol.* **2015**, *85*, 711–726. [CrossRef]
7. Thomas, W.M.; Nicholas, E.; Needham, J.; Murch, M.; Temple-Smith, P.; Dawes, C. Improvements Relating to Friction Welding. Patent EP 0,653,265, 17 May 1995.
8. Amarnath, V.; Karuppuswamy, P.; Balasubramanian, V. Comparative study of joining processes of high conductivity electrolytic tough pitch copper used in automotive industries. *Int. J. Veh. Struct. Syst.* **2017**, *9*, 1. [CrossRef]
9. Lin, J.W.; Chang, H.C.; Wu, M.H. Comparison of mechanical properties of pure copper welded using friction stir welding and tungsten inert gas welding. *J. Manuf. Process.* **2014**, *16*, 296–304. [CrossRef]
10. Mironov, S.; Inagaki, K.; Sato, Y.S.; Kokawa, H. Microstructural evolution of pure copper during friction-stir welding. *Philos. Mag.* **2015**, *95*, 367–381. [CrossRef]
11. Heidarzadeh, A.; Saeid, T. Prediction of mechanical properties in friction stir welds of pure copper. *Mater. Des.* **2013**, *52*, 1077–1087. [CrossRef]
12. Liu, H.J.; Shen, J.J.; Huang, Y.X.; Kuang, L.Y.; Liu, C.; Li, C. Effect of tool rotation rate on microstructure and mechanical properties of friction stir welded copper. *Sci. Technol. Weld. Join.* **2009**, *14*, 577–583. [CrossRef]
13. Shen, J.J.; Liu, H.J.; Cui, F. Effect of welding speed on microstructure and mechanical properties of friction stir welded copper. *Mater. Des.* **2010**, *31*, 3937–3942. [CrossRef]
14. Xue, P.; Xiao, B.L.; Zhang, Q.; Ma, Z.Y. Achieving friction stir welded pure copper joints with nearly equal strength to the parent metal via additional rapid cooling. *Scr. Mater.* **2011**, *64*, 1051–1054. [CrossRef]
15. Sahlot, P.; Jha, K.; Dey, G.K.; Arora, A. Quantitative wear analysis of H13 steel tool during friction stir welding of Cu-0.8%Cr-0.1%Zr alloy. *Wear* **2017**, *378–379*, 82–89. [CrossRef]
16. Jha, K.; Kumar, S.; Nachiket, K.; Bhanumurthy, K.; Dey, G.K. Friction Stir Welding (FSW) of Aged CuCrZr Alloy Plates. *Metall. Mater. Trans. A* **2018**, *49*, 223–234. [CrossRef]
17. Lai, R.; He, D.; He, G.; Lin, J.; Sun, Y. Study of the Microstructure Evolution and Properties Response of a Friction-Stir-Welded Copper-Chromium-Zirconium Alloy. *Metals* **2017**, *7*, 381. [CrossRef]
18. Akramifard, H.R.; Shamanian, M.; Sabbaghian, M.; Esmailzadeh, M. Microstructure and mechanical properties of Cu/SiC metal matrix composite fabricated via friction stir processing. *Mater. Des.* **2014**, *54*, 838–844. [CrossRef]
19. Zhu, Y.; Chen, G.; Chen, Q.; Zhang, G.; Shi, Q. Simulation of material plastic flow driven by non-uniform friction force during friction stir welding and related defect prediction. *Mater. Des.* **2016**, *108*, 400–410. [CrossRef]
20. Guo, J.F.; Chen, H.C.; Sun, C.N.; Bi, G.; Sun, Z.; Wei, J. Friction stir welding of dissimilar materials between AA6061 and AA7075 Al alloys effects of process parameters. *Mater. Des.* **2014**, *56*, 185–192. [CrossRef]
21. Su, J.Q.; Nelson, T.W.; Mishra, R.; Mahoney, M. Microstructural investigation of friction stir welded 7050-T651 aluminium. *Acta Mater.* **2003**, *51*, 713–729. [CrossRef]
22. Su, J.Q.; Nelson, T.W.; Sterling, C.J. Microstructure evolution during FSW/FSP of high strength aluminum alloys. *Mater. Sci. Eng. A* **2005**, *405*, 277–286. [CrossRef]
23. Pang, Y.; Xia, C.; Wang, M.; Li, Z.; Xiao, Z.; Wei, H.; Sheng, X.; Jia, Y.; Chen, C. Effects of Zr and (Ni, Si) additions on properties and microstructure of Cu-Cr alloy. *J. Alloy. Compd.* **2014**, *582*, 786–792. [CrossRef]
24. Holzwarth, U.; Pisoni, M.; Scholz, R.; Stamm, H.; Volcan, A. On the recovery of the physical and mechanical properties of a CuCrZr alloy subjected to heat treatments simulating the thermal cycle of hot isostatic pressing. *J. Nuclear Mater.* **2000**, *279*, 19–30. [CrossRef]
25. Holzwarth, U.; Stamm, H. The precipitation behaviour of ITER-grade Cu-Cr-Zr alloy after simulating the thermal cycle of hot isostatic pressing. *J. Nuclear Mater.* **2000**, *279*, 31–45. [CrossRef]
26. Lei, Q.; Xiao, Z.; Hu, W.; Derby, B.; Li, Z. Phase transformation behaviors and properties of a high strength Cu-Ni-Si alloy. *Mater. Sci. Eng. A* **2017**, *697*, 37–47. [CrossRef]

27. Lockyer, S.A.; Noble, F.W. Precipitate structure in a Cu-Ni-Si alloy. *J. Mater. Sci.* **1994**, *29*, 218–226. [CrossRef]
28. Sato, Y.S.; Kokawa, H.; Enomoto, M.; Jogan, S. Microstructural evolution of 6063 aluminum during friction-stir welding. *Metall. Mater. Trans. A* **1999**, *30*, 2429–2437. [CrossRef]
29. Rhodes, C.G.; Mahoney, M.W.; Bingel, W.H.; Spurling, R.A.; Bampton, C.C. Effects of friction stir welding on microstructure of 7075 aluminum. *Scr. Mater.* **1997**, *36*, 69–75. [CrossRef]

![metals logo] *metals*

MDPI

Article

Resistance Spot Welding of Aluminum Alloy and Carbon Steel with Spooling Process Tapes

Seungmin Shin, Dae-Jin Park, Jiyoung Yu and Sehun Rhee *

Department of Mechanical Engineering, Hanyang University, Seoul 133-791, Korea; glipzide@naver.com (S.S.); ew8270@naver.com (D.-J.P.); susagye@naver.com (J.Y.)
* Correspondence: srhee@hanyang.ac.kr

Received: 10 March 2019; Accepted: 1 April 2019; Published: 3 April 2019

Abstract: Many lightweight materials, including aluminum alloy, magnesium alloy, and plastic, have been used for automotives. Aluminum alloy—the most commonly utilized lightweight metal—has poor resistance spot weldability owing to its inherent properties, which demand the development of welding solutions. Various welding techniques are utilized to improve the resistance spot weldability of aluminum alloy, including DeltaSpot welding. However, the technological development for welding dissimilar metals (aluminum alloy and steel) required for vehicle body assembly is still in its nascent stages. This study proposes DeltaSpot welding (a resistance spot welding process with spooling process tapes) using the alloy combination of 6000 series aluminum alloy (Al 6K32) and 440 MPa grade steel (SGARC 440). The welding characteristics of the main process parameters in DeltaSpot welding were analyzed and the weldability of the combination of the aluminum alloy, Al 6K32, and 440 MPa grade steel was evaluated. In addition, the characteristics of the intermetallic compound layer between the 440 MPa grade steel and Al 6K32 sheets were identified via scanning electron microscopy/energy dispersive X-ray spectroscopy (SEM-EDS).

Keywords: DeltaSpot welding; spooling process tape; aluminum alloy; dissimilar metal welding; lobe curve

---

1. Introduction

The automotive industry has recently invested intensive and extensive research and development efforts to apply lightweight materials, such as high-strength steel, plastic, aluminum alloy, and magnesium alloy, to vehicle body structures for various purposes; for example, to reduce greenhouse gas and exhaust emissions to satisfy increasingly rigorous environmental regulations, to improve fuel efficiency in the face of rising oil prices due to energy resource depletion, to provide electronic equipment for user convenience, and to ensure durability and safety [1–10]. Aluminum, in particular, is about one-third the weight of steel and has excellent shock absorption, high specific strength, and high corrosion resistance owing to the passivity layer formed on the surface, and studies have attempted to apply it to vehicle body structures [11,12]. Aluminum alloys applied to vehicle body structures can be joined by several methods in mechanical joining, such as resistance spot welding (RSW), self-piercing rivet, friction stir welding, clinching, and adhesive spraying [13–20]. In the case of RSW, however, when copper (as an electrode material) is alloyed with aluminum alloy by RSW, the electrode is prone to contamination and a short service life, necessitating frequent electrode dressing. Furthermore, it is difficult to ensure resistance spot weldability owing to the low resistivity and high thermal expansion coefficient of aluminum. These aluminum-specific properties of high electrical and thermal conductivities make it a considerable challenge to secure appropriate RSW conditions for aluminum alloy. In general, welding is performed by applying a high heat input and short welding time. The oxide film formed on the surface of aluminum alloy causes welding defects, such as voids and cracks, in the joints [21–26].

Numerous studies have been conducted on the RSW of aluminum alloys. For instance, Thornton et al. studied the effects of the weld nugget diameter and the quality on weld strength in terms of the fatigue life of the joint in RSW [27]. Sun et al. examined the contact area change pattern and nugget formation process, focusing on the interfacial contact behavior during welding [28]. Senkara et al. analyzed the crack formation mechanism to investigate the causes of cracking in RSW and the effect of cracking on weld strength [29]. Subsequently, Browne et al. determined the process parameters influencing nugget formation during welding by performing a simulation considering electrical, thermal, and mechanical processes [30], and they presented the contact resistance values that facilitate RSW by analyzing the effect of the contact resistance on the base metal and estimating the contact resistance based on shunt resistance [31]. In their studies on RSW joining of two dissimilar metals, Qiu et al. examined the relationship between the thickness of the intermetallic compound (IMC) layer and the weld strength between aluminum alloy and steel [32], and the relationship between the IMC layer and weld zone location as well as welding current and material combination during the RSW of dissimilar metals using a cover plate [33]. Mortazavi et al. analyzed the fracture shape and IMC layer formation according to the welding current, as well as the failure modes and IMC layer formation depending on the welding current [34]. However, research on welding technology for dissimilar metals—steel and aluminum alloy—remains necessary, especially regarding solutions to improve the weldability of these metals.

In this study, a DeltaSpot welding machine using spooling process tapes was employed to improve the weldability of aluminum alloy and steel. The characteristics of the main parameters associated with weld strength and welding defects were investigated and the weld lobe curve was derived through welding experiments. A welding experiment was performed using a 6000 series aluminum alloy (main components: Magnesium and silicon) and steel. In addition, scanning electron microscopy/energy dispersive X-ray spectroscopy (SEM-EDS) analysis was performed to investigate the relationship between the weld strength and the IMC layer properties of coated vs. uncoated steel.

2. Experimental Procedure

2.1. Equipment

The DeltaSpot machine (Fronius, Wels, Austria) used in this study is mounted with a servogun to enable electrode force control. Furthermore, the DeltaSpot welding machine fundamentally differs from other spot welding machines in that a process tape runs between the base metal and the electrode. This process tape protects the electrode from contamination by the aluminum alloy, considerably extending the service life of the electrode. The process tape also increases weldability owing to its high electrical resistance, which compensates for the low resistivity of aluminum alloy and facilitates the resistance control of the weld zone. Figure 1 shows the schematic of a DeltaSpot welding process using process tapes. When the first welding is completed, the process tape between the electrode and the sheet advances. In the second welding, a new process tape is supplied and the welding proceeds. In the same way as the second welding, the third welding is supplied with a new process tape to proceed with the welding. This unique feature of DeltaSpot welding prevents electrode contamination and welding expulsion, and it increases the input heat by generating more resistance between the base metal and the electrode compared with other spot welding machines.

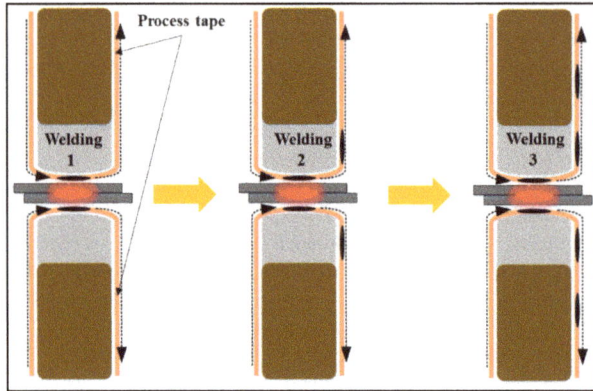

Figure 1. DeltaSpot welding process with process tape.

Figure 2 illustrates the resistance and temperature profiles in the spot welding area. Figure 2a shows the resistance and temperature profiles of a general spot welding area; Figure 2b shows those for a DeltaSpot welding area. As depicted in Figure 2b, significantly greater heat input can be obtained from the additional resistances occurring between the process tape and the electrode and between the process tape and the base metal.

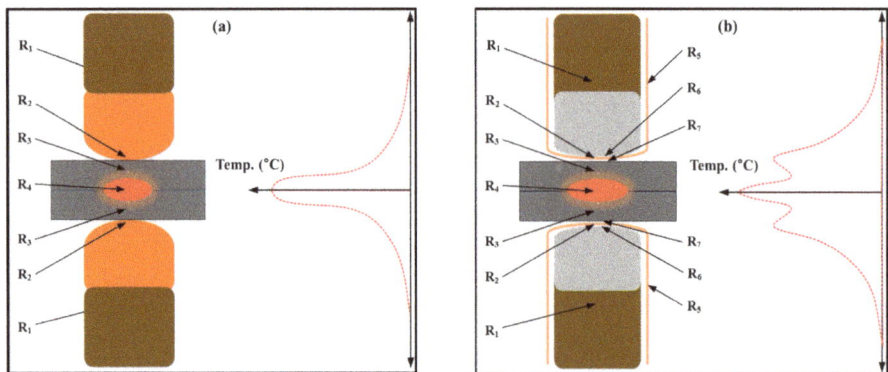

Figure 2. Spot welding resistance and temperature distribution: (**a**) General spot welding; (**b**) spot welding with process tape.

2.2. Materials and Experimental Method

The materials used in this study were 6000 series aluminum alloy (Al 6K32) and 440 MPa grade steel (SGARC 440). Their chemical compositions are presented in Table 1.

Table 1. Material compositions of Al 6K32 and SGARC 440 (wt %).

Al 6K32	Mg	Si	Fe	Cu	Mn	Cr	Zn	Ti
	0.02	1.0	0.13	0.01	0.07	0.01	0.01	0.01
SGARC 440	Si	Cu	Mn	Cr	Ni	Mo	V	C
	0.14	0.1	1.4	0.1	0.1	0.05	0.01	0.09

To compensate for the difference in resistivity between steel (higher) and aluminum (lower), two types of process tape were used: PT1407 with a lower resistance between the steel specimen and the electrode and PT3000 with a higher resistance between the aluminum specimen and the electrode. Table 2 summarizes the basic properties of these two types of process tapes.

Table 2. Process tape types.

Base Metal	Process Tape	Tape Material	Heat Input from Outside
Al 6K32	PT 3000	CrNi	High
SGARC 440	PT 1407	Steel	Medium

The tensile lap-shear test specimens were prepared in the shape illustrated in Figure 3, with a steel sheet fixed underneath an aluminum alloy sheet with an overlap length of 40 mm and the following sheet thicknesses: SGARC 440, 1.4 mm and 1.0 mm; Al 6K32, 1.6 mm and 1.0 mm. Three pairs of specimens for the tensile shear test, peel test, and cross-sectional examination were prepared. The experiment was repeated three times under the same welding conditions.

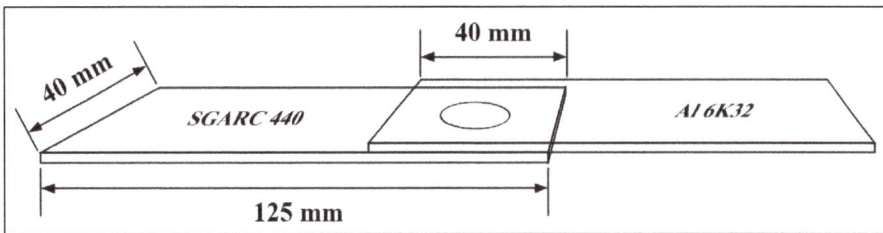

Figure 3. Spot welding specimen size and method.

Figure 4 shows an electrode (type R) with the Cr-Cu component used in the experiment. The radius of the electrode was 100 mm and the diameter of the electrode was 16 Ø (Figure 4a); Figure 4b shows the image of the electrode.

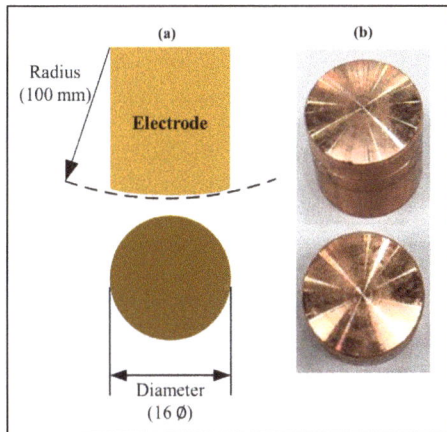

Figure 4. Electrode used in DeltaSpot: (**a**) electrode size; (**b**) electrode shape.

2.3. Weldability Evaluation Method

It is difficult to examine and measure an IMC layer, which is a significant factor in determining the weld quality of the dissimilar welding of aluminum alloy and steel, using non-destructive testing methods, such as X-ray testing, ultrasonic test, and computed tomography (CT). This is more difficult in RSW. Therefore, in this study, weldability was evaluated based on the magnitude of the tensile shear strength (TSS) measured by tensile shear testing. Furthermore, the nugget size of the weld zone was measured by examining its cross-section. For industrial applications, the lobe curve is used as the evaluation standard, as illustrated in Figure 5.

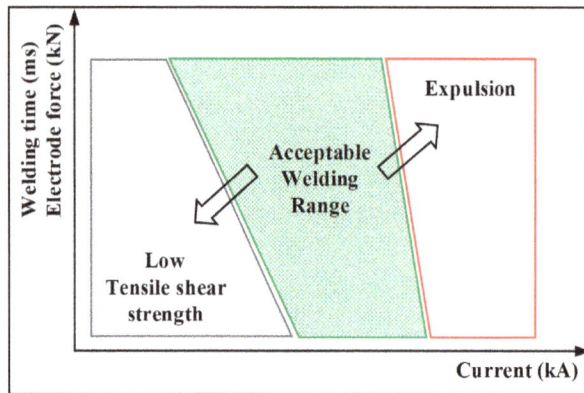

Figure 5. Schematic diagram illustrating the acceptable welding range (weld lobe). The vertical axis represents welding time or electrode force; the horizontal axis represents welding current.

As shown in Figure 5, the lobe curve is determined based on the permissible TSS for the minimum acceptable weld zone, whereby the weld strength fails to meet the strength requirement if it is smaller than the permissible TSS or if the nugget diameter measured by the peel test is $4\sqrt{t} - 5\sqrt{t}$ or smaller relative to the base metal thickness, t (mm). The maximum acceptable weld zone is determined by the occurrence of expulsion during welding by overheating. In general, the wider the acceptable weld zone, the better is the weldability. For the lobe curve used in this study, the horizontal axis represents the current and the vertical axis represents the current time and electrode force.

3. Results and Discussion

In our preliminary test, the up-slope time had significant effects on the weld strength and surface contamination of aluminum alloy sheets, and had a suitable condition to guarantee high tensile shear strength and improved surface quality of the weld below 150 ms. The down-slope time below the condition of 450 ms turned out to be a significant parameter for reducing weld defects of aluminum alloy sheets. It was also found that the pre-force time and hold time have significant effects on the weld surface quality and weld defects in certain test ranges: From −50 to 50 ms for the pre-force time and from 0 to 150 ms for the hold time. However, since force 2 did not have significant effects on both weld strength and weld defects below 5 kN, force 2 was fixed to 5 kN. Since the welding current, force 1, and main current time were found to be the most significant parameters among the eight parameters, and are usually adjusted in automotive assembly lines to improve welding quality, these parameters were selected and used for deriving suitable welding ranges. In addition, the welding schedules and specifications of actual automotive assembly lines were considered in order to select and set the test range of all eight process parameters.

3.1. Effect of DeltaSpot Welding in Improving Weldability

In this study, a weldability comparison was performed between inverter direct current (DC) spot welding and DeltaSpot welding. The materials used in this experiment are presented in Table 1; the sheet thickness was 1.4 mm for SGARC 440 and 1.6 mm for Al6K32. The basic properties of the DeltaSpot welding process tapes are presented in Table 2. Figure 6 shows the welding machines used for the comparison experiment: (a) and (b) are the inverter DC spot welding machine (Harm-Wende, Hamburg, Germany) and DeltaSpot welding machine (Fronius, Wels, Austria), respectively.

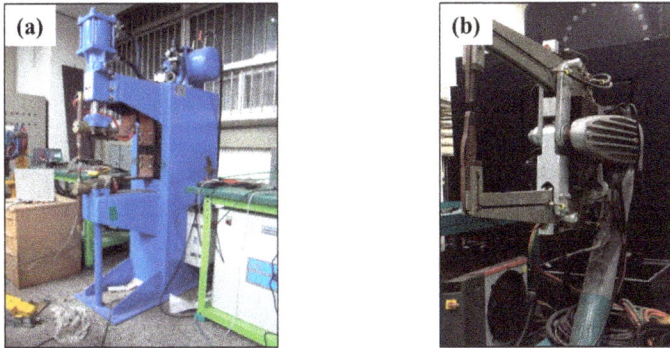

Figure 6. Experimental equipment: (**a**) Inverter direct current (DC) spot welding machine; (**b**) DeltaSpot welding machine.

The comparison experiment was performed under the welding conditions summarized in Table 3, which shows that the electrode force was fixed at 2 kN and the main current time and current were varied.

Table 3. Welding conditions.

Welding Conditions		Profiles of Welding Current and Electrode Force
Current (kA)	9, 11, 13	
Up slope time (ms)	0	
Main current time (ms)	166, 332, 500	
Down slope time (ms)	0	
Force 1 (kN)	2	
Force 2 (kN)	0	
Pre-force time (ms)	0	
Hold time (ms)	0	

Table 4 presents images of the fracture modes and cross-sections of the inverter DC spot weld zones of the test specimens welded under the conditions described in Table 3. Table 5 summarizes the fracture modes, expulsion, nugget diameter, and TSS.

In Table 4, when the main current time was 166 ms, interfacial fracture occurred at all current conditions and expulsion occurred at the current condition of 13 kA. At 332 ms, the plug fracture occurred at all current conditions, and the expulsion occurred at current conditions of 11 kA and 13 kA. At 500 ms, interfacial fracture occurred at all current conditions, and expulsion occurred at current conditions of 11 kA and 13 kA.

Table 6 shows images of the fracture modes and cross-sections of the DeltaSpot weld zones of the test specimens welded under the same welding conditions, and Table 7 presents the fracture modes, expulsion, nugget diameter, and TSS.

Table 4. DC spot weld fracture shape and cross-sectional image.

Main Current Time (ms)	Item	Current Level (kA)		
		9	11	13
166	Fracture mode			
	Cross Section			
332	Fracture mode			
	Cross Section			
500	Fracture mode			
	Cross Section			

Table 5. Weldability analysis of DC spot welding according to conditions.

Main Current Time (ms)	Item	Current Level (kA)		
		9	11	13
166	Fracture mode	Interfacial	Interfacial	Interfacial
	Expulsion	-	-	expulsion
	Nugget diameter (mm)	-	-	-
	TSS (kN)	1.2	2.7	2.4
332	Fracture mode	Plug	Plug	Plug
	Expulsion	-	expulsion	expulsion
	Nugget diameter (mm)	3.8	4.6	6.5
	TSS (kN)	2.1	2.9	3.8
500	Fracture mode	Interfacial	Interfacial	Interfacial
	Expulsion	-	expulsion	expulsion
	Nugget diameter (mm)	-	-	-
	TSS (kN)	2.8	3.4	3.4

Table 6. DeltaSpot weld fracture shape and cross-sectional image.

Main Current Time (ms)	Item	Current Level (kA)		
		9	11	13
166	Fracture shape			
	Cross Section			
332	Fracture shape			
	Cross Section			
500	Fracture shape			
	Cross Section			

Table 7. Weldability analysis of DeltaSpot welding according to conditions.

Main Current Time (ms)	Item	Current Level (kA)		
		9	11	13
166	Fracture mode	Interfacial	Interfacial	Interfacial
	Expulsion	-	-	-
	Nugget diameter (mm)	-	-	-
	TSS (kN)	1.2	1.9	5.2
332	Fracture mode	Plug	Plug	Plug
	Expulsion	-	-	expulsion
	Nugget diameter (mm)	7.4	7.2	7.5
	TSS (kN)	4.8	4.7	4.0
500	Fracture mode	Interfacial	Plug	Plug
	Expulsion	expulsion	expulsion	expulsion
	Nugget diameter (mm)	-	6.5	7.7
	TSS (kN)	2.5	4.6	4.5

In Table 6, when the main current time was 166 ms, interfacial fracture occurred at all current conditions, but no expulsion occurred. At 332 ms, plug fracture occurred at all current conditions and expulsion occurred at the current condition of 13 kA. At 500 ms, interfacial fracture occurred at the current condition of 9 kA, and expulsion occurred under all current conditions.

The experimental results in Tables 5 and 7 show that the DeltaSpot specimen was welded at a lower current and had a wider plug fracture case than that of the inverter DC specimen. Since DeltaSpot can obtain sufficient weld quality at a lower current range, the DeltaSpot equipment can apply a welding transformer with a smaller capacity compared to that of conventional inverter DC welding equipment, which leads to a cost reduction of the welding equipment and system. In addition, the nugget diameter of the welding condition under which the plug fracture occurred is larger than that for DC spot welding. Further, it can be confirmed that DeltaSpot welding achieves better weld strength than inverter DC spot welding even at a relatively low welding current. Thus, the comparative experiments confirm that DeltaSpot welding can be performed with a relatively larger suitable welding range and better weld quality. That is, DeltaSpot welding achieves a wider suitable welding range, which is attributed to the effect of compensating for the resistivity difference when welding dissimilar metals using a process tape with lower resistance for the steel electrode and that with a higher resistance for the aluminum alloy electrode.

3.2. Effects of DeltaSpot Welding Parameters

In this study, eight welding parameters, which are defined in Figure 7, were selected based on previous studies [35,36]. The up-slope time, main current time, down-slope time, and current level are related to the profile of the welding current. The profile of the electrode force was determined by setting the force 1, force 2, hold time, and pre-force time. In particular, the pre-force time is defined as the positive value when the magnitude of the electrode force changes from force 1 to force 2 during the main current time. Conversely, the pre-force time has a negative value when the magnitude of the electrode force changes from force 1 to force 2 after the main current time. To investigate the effects of the welding parameters, the basic welding conditions were set as indicated in Table 8, which presents the reference conditions for the experiment established based on the results of [35] regarding the DeltaSpot welding of aluminum alloy specimens with the same thickness. Force 2 was applied at the end of the welding process to reduce the resistivity loss through the high electrode force and to prevent welding defects during the current passage time, whereby force 2 should be at least 5 kN to have the effect of removing weld zone defects [35]. Therefore, force 2 was set at 5 kN.

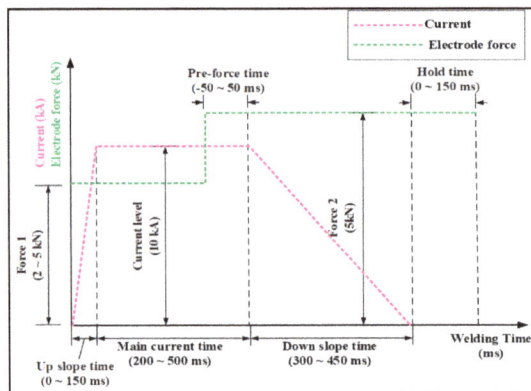

Figure 7. Schematic diagram illustrating the process parameters of DeltaSpot welding with test values.

Table 8. Reference conditions.

Parameter	Level
Current (kA)	10
Up-slope time (ms)	50
Main current time (ms)	300
Down-slope time (ms)	300
Force 1 (kN)	2
Force 2 (kN)	5
Pre-force time (ms)	0
Hold time (ms)	100

The characteristics of the process parameters associated with dissimilar metal welding were evaluated by welding aluminum alloy and carbon steel sheets of the same thickness (1.0 mm) using the DeltaSpot welding machine. In this study, based on the results of pre-tested aluminum alloy DeltaSpot welding, experiments were conducted under the welding conditions summarized in Table 9 to evaluate the characteristics of the process variables [35,36].

Table 9. Welding conditions according to parameters.

Parameter	Level
Current (kA)	10 (fixed)
Up-slope time (ms)	0, 50, 100, 150
Main current time (ms)	200, 300, 400, 500
Down-slope time (ms)	0, 150, 300, 450
Force 1 (kN)	2, 3, 4, 5,
Force 2 (kN)	5 (fixed)
Pre-force time (ms)	50, 0, −50
Hold time (ms)	0, 50, 100, 150

The experimental results for each process parameter are presented in Table 10. The weldability was evaluated in terms of expulsion, TSS, and fracture mode.

Table 10. Experimental results for DeltaSpot welding of 6000 series aluminum alloy (Al 6K32)/440 MPa grade steel (SGARC 440).

Parameter	Level	Expulsion	TSS (kN)	Fracture Mode
Up-slope time (ms)	0	X	3.66	Interfacial
	50	X	3.47	Plug
	100	X	3.23	Plug
	150	X	3.01	Interfacial
Main current time (ms)	200	X	2.79	Plug
	300	X	3.47	Plug
	400	O	3.53	Plug
	500	O	1.42	Plug
Down-slope time (ms)	0	X	2.93	Plug
	150	X	3.38	Plug
	300	X	3.47	Plug
	450	X	3.38	Plug
Force 1 (kN)	2	X	3.47	Plug
	3	X	2.39	Plug
	4	X	2.66	Interfacial
	5	X	2.41	Interfacial
Pre-force time (ms)	50	X	2.85	Interfacial
	0	X	3.47	Plug
	−50	X	3.44	Plug
Hold time (ms)	0	X	3.89	Plug
	50	X	3.19	Plug
	100	X	3.47	Plug
	150	X	3.02	Plug

The effects of the individual process parameters on the weldability of aluminum alloy with carbon steel via DeltaSpot welding were analyzed. The welding conditions were derived based on the analysis results; the up-slope time should be reduced to the minimum possible level because the weld strength decreases as the up-slope time increases. The experimental results reveal that a high weld strength and satisfactory button size were achieved at 50 ms. Given that the weld strength is determined in the initial welding phase, the main current time does not need to be long. The experimental results demonstrate that expulsion occurs when the main current time exceeds 400 ms. Accordingly, the optimum condition was set to 300 ms. The down-slope time should be maintained for at least 300 ms to ensure welding defect removal. At 450 ms, no button fracture and expulsion occurred; however, the reference weld strength was not achieved. Accordingly, the optimum condition was set to 300 ms. The optimum force 1 was set to 2 kN, because the highest weld strength and satisfactory button size were obtained at 2 kN. The application of the pre-force time was found to lower the weld strength. The hold time, which is the time to maintain the electrode force after the current has passed to prevent cracking during the cooling-induced contraction of the weld metal, was found to satisfy the weld strength and button size at 100 ms. Table 11 outlines the optimum conditions of the individual process parameters for DeltaSpot welding of dissimilar metals as derived from the analysis of the experimental results.

Table 11. Optimum welding conditions.

Parameter	Level
Up-slope time (ms)	50
Main current time (ms)	300
Down-slope time (ms)	300
Force 1 (kN)	2
Force 2 (kN)	5
Pre-force time (ms)	0
Hold time (ms)	100

3.3. Weldability Evaluation with Respect to the Main Current Time and Force 1

The main current time plays an important role in the RSW of aluminum alloy to steel. Its effect on weldability was analyzed, given its importance as a parameter, allowing sufficient time for the molten aluminum alloy to ensure good wetting of the surface of the heated carbon steel. The related experimental conditions are outlined in Table 12. All process parameters except for main current time, force 1, and current were set to the values listed in Table 11.

Table 12. Welding conditions according to main current time.

Parameter	Level
Current (kA)	5–14
Up-slope time (ms)	50
Main current time (ms)	300, 400, 500
Down-slope time (ms)	300
Force 1 (kN)	1–4.5
Force 2 (kN)	5
Pre-force time (ms)	0
Hold time (ms)	100

Tables 13–15 are lobe curves as functions of the main current time. The values within the green outline are those satisfying the standards for the permissible TSS (= no expulsion) and button size. No TSS values are provided for the areas affected by expulsion.

At a main current time of 300 ms (Table 13), the interfacial fracture was the main welding defect due to insufficient heat input in the low-current range. This characteristic made the weldability sensitive to the electrode force in the welding current range of 8 to 9 kA; however, a very stable weld zone was achieved at welding currents greater than 10 kA. An acceptable weld zone was achieved at currents exceeding 14 kA as well. In contrast, as the main current time increased, an acceptable weld zone was achieved in the low-current range, but expulsion occurred in the high-current range of 13 to14 kA due to excessive input heat. In addition, an overall tendency of the acceptable weld zone to move from the high- to low-current range was observed, whereby no acceptable weld zone was formed at currents less than 7 kA despite this change. These results indicate that the RSW of aluminum alloy to carbon steel is possible at a current ≥7 kA. Furthermore, increasing the main current time did not result in any significant increase in weld strength.

Table 13. Lobe curve according to main current time (300 ms).

						Current				
	4.5	-	-	-	-	-	3167	3791	3279	3460	3975
	4	-	-	-	-	2456	3167	3073	3132	3868	3051
	3.5	-	-	-	2074	2752	2863	3756	3213	3353	-
	3	-	-	2312	2317	2875	3144	3209	3065	3548	-
Force 1	2.5	-	-	2438	2831	3179	3684	3693	3025	-	-
	2	-	1349	2987	3784	3756	3467	4001	-	-	-
	1.5	-	2813	2961	3769	4295	-	-	-	-	-
	1	-	3052	-	-	-	-	-	-	-	-
		5	6	7	8	9	10	11	12	13	14

Table 14. Lobe curve according to main current time (400 ms).

						Current				
	4.5	-	-	-	2948	3203	3066	3237	3371	3440	-
	4	-	-	2033	3286	2988	3194	3493	3469	3716	-
	3.5	-	-	2292	2221	3148	3229	3379	-	-	-
	3	-	2008	3234	3569	3408	3510	3529	-	-	-
Force 1	2.5	-	2241	3212	3662	3514	-	-	-	-	-
	2	2078	2093	3580	3469	-	-	-	-	-	-
	1.5	2448	2771	3586	-	-	-	-	-	-	-
	1	3018	-	-	-	-	-	-	-	-	-
		5	6	7	8	9	10	11	12	13	14

Table 15. Lobe curve according to main current time (500 ms).

						Current				
	4.5	-	-	-	-	2750	3374	3575	3317	-	-
	4	-	-	-	2771	3031	3245	3386	3396	-	-
	3.5	-	-	2386	2923	3254	3371	3701	3957	-	-
	3	-	-	2023	2395	3293	3804	3711	4219	-	-
Force 1	2.5	-	-	2894	3321	3762	3467	-	-	-	-
	2	-	2705	3374	3504	3405	-	-	-	-	-
	1.5	-	2940	3228	3877	-	-	-	-	-	-
	1	537	3286	-	-	-	-	-	-	-	-
		5	6	7	8	9	10	11	12	13	14

3.4. Evaluation of Coating-Dependent Dissimilar Metal Welding Characteristics

In the RSW of aluminum alloy to carbon steel, the aluminum alloy melts and the carbon steel is heated at the aluminum alloy melting point or lower. The molten aluminum alloy provides wetting of the surface of heated carbon steel to form a new alloy layer, namely, an IMC layer, which acts as a bonding layer equivalent to a nugget in same-metal welding.

For comparison, the weld zone characteristics of dissimilar metal welding with zinc-coated and uncoated steel sheets, which are widely used in industrial settings, were evaluated. Weldability was evaluated under the experimental conditions outlined in Table 12. The lobe curve in Table 16 shows the resulting weldability characteristics.

Table 16. Lobe curve for non-coating steel.

							Current				
	4.5	-	-	-		651	976	1625	237	-	-
	4	-	-	-	95	445	1786	1172	263	-	-
	3.5	-	-	-	x	739	1119	721	195	-	-
	3	-	-	-	344	1220	513	1426	1177	-	-
Force 1	2.5	415	-	-	389	1553	336	839	-	-	-
	2	561	x	-	799	1093	145	1256	-	-	-
	1.5	477	x	x	-	-	-	-	-	-	-
	1	-	-	-	-	-	-	-	-	-	-
		5	6	7	8	9	10	11	12	13	14

Table 16 shows that the weld strength was very low and no button fracture appeared in the RSW of aluminum alloy to uncoated steel, demonstrating the existence of weldability problems without any regularity that would allow the identification of the effects of welding conditions on weldability. From this finding, it may be assumed that the zinc in the zinc-coating layer plays an important role in dissimilar metal welding of aluminum alloy onto carbon steel. This role may be explained by the fact that zinc, which has a melting point similar to that of aluminum alloy, melts during the welding process and forms an IMC layer, thus improving the bonding force; this effect cannot occur when aluminum alloy is welded onto uncoated steel without a zinc layer.

Figure 8a illustrates the results of the SEM-EDS analysis performed at different points in dissimilar metal RSW of zinc-coated steel: Figure 8b at point 1 inside the zinc-coated steel sheet (SGARC 440); Figure 8c at point 2 inside the IMC layer; and Figure 8d at point 3 inside the aluminum alloy sheet (Al 6K32). The results of component analysis at each point are summarized in Table 17.

Table 17. SEM-EDS components of zinc coating steel.

Position	Element (Wt %)		
	Al-k	Fe-k	Zn-k
Point 3	98.30	-	1.70
Point 2	69.04	30.30	0.65
Point 1	-	100.00	-

Figure 9a outlines the components of the aluminum alloy sheet, IMC layer, and uncoated carbon steel sheet as analyzed by SEM-EDS at three points in the two base metal sheets and the IMC layer: Figure 9b at point 1 inside the uncoated steel sheet (SPRC 440); Figure 9c at point 2 inside the IMC layer close to the uncoated steel sheet; Figure 9d at point 3 inside the IMC layer close to the aluminum alloy sheet; and Figure 9e at point 4 inside the aluminum alloy sheet (Al 6K32). The results of component analysis at each point are summarized in Table 18.

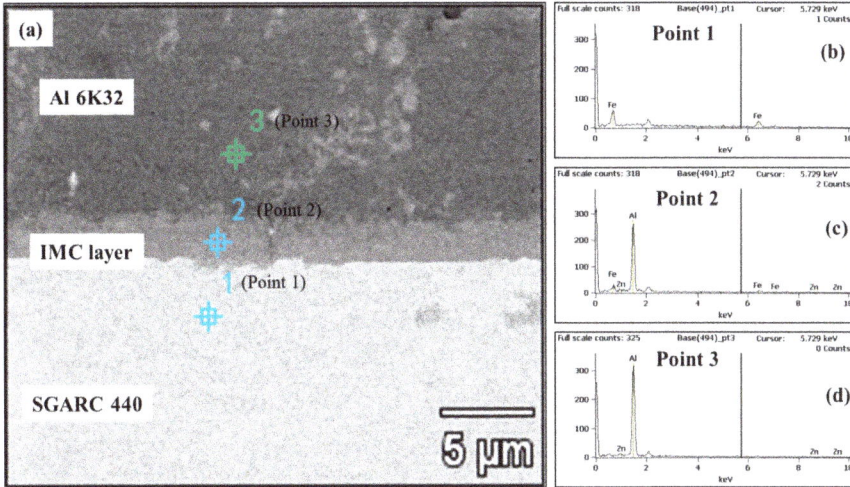

Figure 8. Energy dispersive X-ray spectroscopy (EDS) analysis locations in the welded section: (**a**) SEM image of dissimilar spot welds of aluminum alloy/zinc-coating steel.; (**b**) zinc-coating steel base metal point 1; (**c**) intermetallic compound (IMC) layer point 2; (**d**) aluminum alloy base metal point 3.

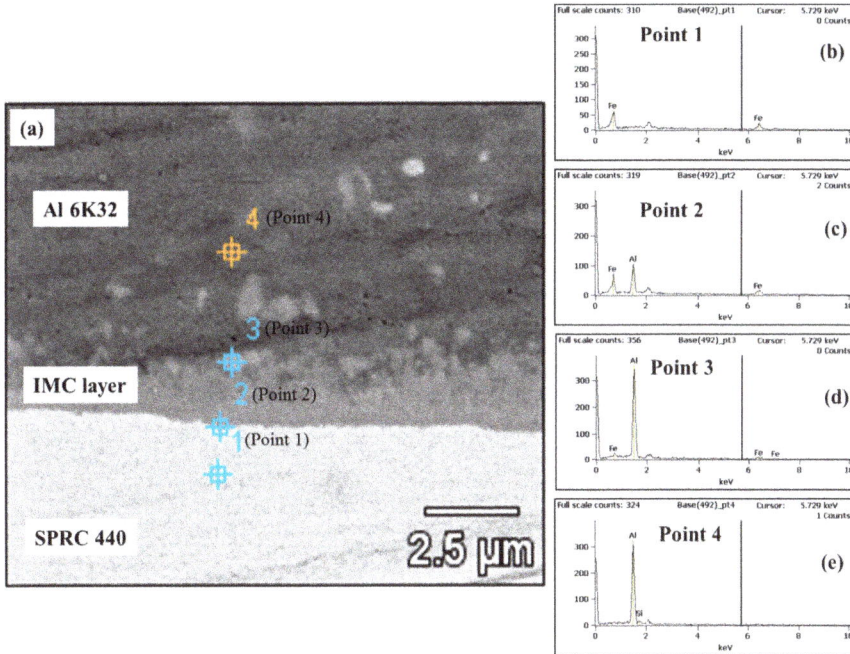

Figure 9. Energy dispersive X-ray spectroscopy (EDS) analysis locations in the welded section: (**a**) SEM image of dissimilar spot welds of aluminum alloy/non-coating steel. (**b**) Non-coating steel base metal point 1; (**c**) close to non-coating steel intermetallic compound (IMC) layer point 2; (**d**) close to aluminum alloy IMC layer point 3; (**d**) aluminum alloy base metal point 4.

Table 18. SEM-EDS components of non-coating steel.

Position	Element (Wt %)		
	Al-k	Si-k	Fe-k
Point 4	97.17	2.83	-
Point 3	74.36	-	25.64
Point 2	20.44	-	79.56
Point 1	-	-	100.00

The zinc-coated steel was analyzed to have the following components. The IMC layer zone (point 2) contained a compound with a mixing ratio of Al(69.04%):Fe(30.30%):Zn(0.65%). In the case of uncoated steel, a compound was extracted from the IMC layer closer to the aluminum alloy sheet (point 3), which was analyzed to have a mixing ratio closer to that of aluminum alloy, Al(74.36%):Fe(25.64%), and the compound extracted from the IMC layer closer to the uncoated steel sheet (point 2) was analyzed to have a mixing ratio of Al(20.44%):Fe(79.56%). These SEM-EDS analysis results suggest that the difference in the bonding force between coated and uncoated steel sheets is attributable to the presence of zinc in the IMC layer.

Figure 10 shows the difference in the thickness of the IMC layer between the zinc-coated steel (Figure 10a, SGARC 440) and uncoated steel (Figure 10b, SPRC 440) sheets: 3.264 µm vs. 2.390 µm. In general, a thinner IMC layer, which is highly brittle, has a higher strength. In the case of uncoated steel, however, a simple thickness comparison is not a good basis for the comparative evaluation of weld strength because of the different chemical compositions of the compounds extracted from the IMC layer [37].

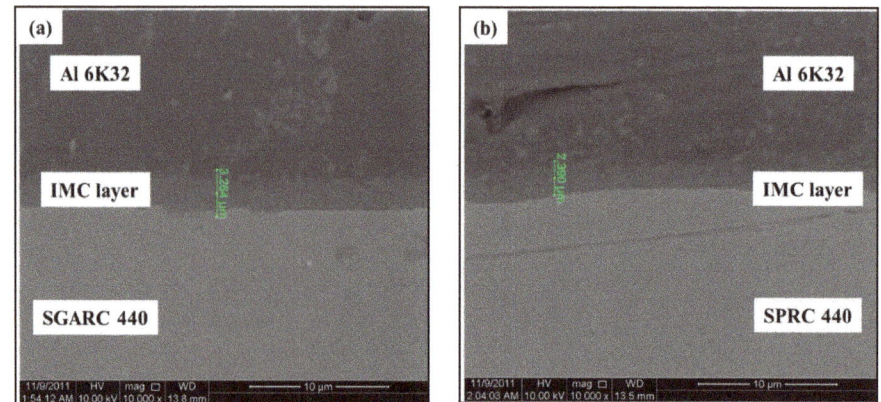

Figure 10. Intermetallic compound (IMC) layer thickness measurement: (a) Zinc-coating steel; (b) non-coating steel.

4. Conclusions

In this study, RSW between aluminum alloy (Al 6K32 1.0t) and carbon steel (SGARC 440 1.0t), which could not be solved using conventional RSW processes, was achieved by using a DeltaSpot welding system, and the RSW characteristics for different base metal combinations of aluminum alloy and various steel types were analyzed. The findings of this study can be summarized as follows:

(1)　In the RSW of aluminum alloy to steel, lobe curve comparison verified the superiority of DeltaSpot welding to inverter DC spot welding in terms of the acceptable weld zone and weld stability owing to the effects of the process tape.

(2)   The optimum values of six DeltaSpot welding process parameters were derived from the analysis results for the welding characteristics of eight process parameters selected based on the results of an earlier study on the RSW of an aluminum alloy (Al 6K32) of the same thickness.

(3)   Weldability was evaluated as a function of the main current time, a main parameter along with the current level and force 1. The results of the lobe curve analysis revealed an overall tendency of the acceptable weld zone to move from the high- to low-current range, whereby no acceptable weld zone was formed in the range lower than 7 kA.

(4)   Weldability of aluminum alloy to different steel types was evaluated in order to enhance its applicability to different welding process conditions based on the steel types used in vehicle body assembly. The comparison of the weldability of aluminum alloy to zinc-coated and uncoated steel sheets revealed its weldability to uncoated steel sheets to be very low. This low weldability seems to be attributable to the fact that zinc melted onto the surface of the zinc-coated steel sheet forms a compound in the IMC layer, thus improving the bonding force. The difference in the zinc content in the IMC layer was verified by SEM-EDS analysis.

The significance of this study lies in the fact that a method of improving the weldability in the dissimilar metal weld zone was developed for applications to aluminum alloy for industrial use as a lightweight material. Furthermore, the developed methodology can be used to analyze the weldability of various combinations of weld materials with the aim of establishing a database to strengthen DeltaSpot's applicability in the automotive industry.

Author Contributions: S.S. analyzed the data and wrote the paper; D.-J.P. performed the welding experiments; J.Y. analyzed the data and reviewed the paper; S.R. reviewed the paper.

Funding: This research was supported by the Ministry of Trade, Industry and Energy (MI, Korea).

Acknowledgments: This work was supported by the Industrial Technology Innovation Program (No. 10063421, 'Development of the in-line welds quality estimation system and network-based quality control technology in arc and spot welds of ultrahigh-strength steels for automotive parts assembly') funded By the Ministry of Trade, Industry and Energy (MI, Korea).

Conflicts of Interest: The authors declare no conflict of interest.

References

1.   Ramazani, A.; Mukherjee, K.; Abdurakhmanov, A.; Abbasi, M.; Prahl, U. Characterization of microstructure and Mechanical Properties of Resistance spot welded DP600 steel. *Metals* **2015**, *5*, 1704–1716. [CrossRef]

2.   Huin, T.; Dancette, S.; Fabrègue, D.; Dupuy, T. Investigation of the failure of advanced high strength steels heterogeneous spot welds. *Metals* **2016**, *6*, 111. [CrossRef]

3.   Yu, J. Adaptive Resistance Spot Welding Process that Reduces the Shunting Effect for Automotive High-Strength Steels. *Metals* **2018**, *8*, 775. [CrossRef]

4.   Džupon, M.; Kaščák, Ľ.; Spišák, E.; Kubík, R.; Majerníková, J. Wear of Shaped Surfaces of PVD Coated Dies for Clinching. *Metals* **2017**, *7*, 515. [CrossRef]

5.   Kazdal Zeytin, H.; Ertek Emre, H.; Kaçar, R. Properties of resistance spot-welded TWIP steels. *Metals* **2017**, *7*, 14. [CrossRef]

6.   He, X.; Deng, C.; Zhang, X. Fretting behavior of SPR joining dissimilar sheets of titanium and copper alloys. *Metals* **2016**, *6*, 312. [CrossRef]

7.   Fridlyander, I.N.; Sister, V.G.; Grushko, O.E.; Berstenev, V.V.; Sheveleva, L.M.; Ivanova, L.A. Aluminum alloys: promising materials in the automotive industry. *Met. Sci. Heat Treat.* **2002**, *44*, 365–370. [CrossRef]

8.   Kim, Y.; Park, K.Y.; Lee, K.D. Development of Welding Technologies for Lightweight Vehicle. *J. Weld. Join.* **2011**, *29*, 1–3. [CrossRef]

9.   Chang, W.S.; Choi, K.Y.; Kim, S.H.; Kweon, Y.G. Some Aspects of Friction Stir Welding and Its Application Technologies. *J. Weld. Join.* **2001**, *19*, 7–15.

10.   Yeon, Y.M.; Lee, W.B.; Jung, S.B. Microstructures and Characteristics of Friction-Stir-Welded Joints in Aluminum Alloys. *J. Weld. Join.* **2001**, *19*, 584–590.

11. Kim, H.T.; Kil, S.C. High Efficient Welding Technology of the Car Bodies. *J. Weld. Join.* **2016**, *34*, 62–66. [CrossRef]

12. Inaba, T.; Tokuda, K.; Yamashita, H.; Takebayashi, Y.; Minoura, T.; Sasabe, S. Wrought aluminum technologies for automobiles. *Kobelco Technol. Rev.* **2005**, *26*, 55–62.

13. Barnes, T.A.; Pashby, I.R. Joining techniques for aluminium spaceframes used in automobiles: Part II—Adhesive bonding and mechanical fasteners. *J. Mater. Process. Technol.* **2000**, *99*, 72–79. [CrossRef]

14. Yum, D.B.; Ko, J.B.; Choi, B.K.; Lee, S.G.; Kim, A.K. Evaluation of Resistance Spot Welding Weldability of Aluminum Alloy 5000 Series. *Trans. Kor. Soc. Mach. Tool Eng.* **2002**, *11*, 8–13.

15. Cai, W.; Wang, P.C.; Yang, W. Assembly dimensional prediction for self-piercing riveted aluminum panels. *Int. J. Mach. Tools Manuf.* **2005**, *45*, 695–704. [CrossRef]

16. Kim, J.Y.; Lee, C.J.; Lee, S.K.; Ko, D.C.; Kim, B.M. Effect of shape parameters of tool on improvement of joining strength in clinching. *Trans. Mater. Process.* **2009**, *18*, 392–400.

17. Rao, H.M.; Kang, J.; Huff, G.; Avery, K. Structural Stress Method to Evaluate Fatigue Properties of Similar and Dissimilar Self-Piercing Riveted Joints. *Metals* **2019**, *9*, 359. [CrossRef]

18. Peng, G.; Yan, Q.; Hu, J.; Chen, P.; Chen, Z.; Zhang, T. Effect of Forced Air Cooling on the Microstructures, Tensile Strength, and Hardness Distribution of Dissimilar Friction Stir Welded AA5A06-AA6061 Joints. *Metals* **2019**, *9*, 304. [CrossRef]

19. Patel, V.; Li, W.; Wang, G.; Wang, F.; Vairis, A.; Niu, P. Friction Stir Welding of Dissimilar Aluminum Alloy Combinations: State-of-the-Art. *Metals* **2019**, *9*, 270. [CrossRef]

20. Nakamura, T.; Obikawa, T.; Nishizaki, I.; Enomoto, M.; Fang, Z. Friction Stir Welding of Non-Heat-Treatable High-Strength Alloy 5083-O. *Metals* **2018**, *8*, 208. [CrossRef]

21. Li, Z.; Hao, C.; Zhang, J.; Zhang, H. Effects of sheet surface conditions on electrode life in resistance welding aluminum. *Weld. J.* **2007**, *86*, 81s–89s.

22. Park, S.H.; Park, B.C.; Kim, Y.G.; Beak, U.R. Fusion Zone Characteristics of Dissimilar Aluminum Alloys joining. In Proceedings of the KWS Conference, The Korean Welding and Joining Society, Jeonju, Korea, 15–16 November 2007; pp. 141–143.

23. Cho, S.M. Resistance welding and resistance joining technology to Fe-base material of Al-alloy. *J. Weld. Join.* **2001**, *19*, 14–22.

24. Rashid, M.; Fukumoto, S.; Medley, J.B.; Villafuerte, J.; Zhou, Y. Influence of lubricants on electrode life in resistance spot welding of aluminum alloys. *Weld. J.* **2007**, *86*, 62s–70s.

25. Yeon, Y.M.; Lee, W.B.; Lee, C.Y.; Jung, S.B.; Song, K. Joint Characteristics of Spot Friction Stir Welded A 5052 Alloy Sheet. *J. Weld. Join.* **2006**, *24*, 71–76.

26. Yeon, Y.M.; Lee, C.Y.; Lee, W.B.; Jung, S.B.; Chang, W.S. Spot friction stir welding and characteristics of joints in aluminum alloys. *J. Weld. Join.* **2005**, *23*, 16–20.

27. Thornton, P.H.; Krause, A.R.; Davies, R.G. Aluminum spot weld. *Weld. J. Res. Suppl.* **1996**, *75*, 101s–108s.

28. Sun, X.; Dong, P. Analysis of aluminum resistance spot welding processes using coupled finite element procedures. *Weld. J.* **2000**, *79*, 215s–221s.

29. Senkara, J.; Zhang, H. Cracking in spot welding aluminum alloy AA5754. *Weld. J.* **2000**, *79*, 194s–201s.

30. Browne, D.J.; Chandler, H.W.; Evans, J.T.; Wen, J. Computer simulation of resistance spot welding in aluminum: Part I. *Weld. J. Res. Suppl.* **1995**, *74*, 339s–344s.

31. Browne, D.J.; Chandler, H.W.; Evans, J.T.; James, P.S.; Wen, J.; Newton, C.J. Computer simulation of resistance spot welding in aluminum: Part II. *Weld. J. Res. Suppl.* **1995**, *74*, 417s–422s.

32. Qiu, R.; Iwamoto, C.; Satonaka, S. The influence of reaction layer on the strength of aluminum/steel joint welded by resistance spot welding. *Mater. Charact.* **2009**, *60*, 156–159. [CrossRef]

33. Qiu, R.; Iwamoto, C.; Satonaka, S. Interfacial microstructure and strength of steel/aluminum alloy joints welded by resistance spot welding with cover plate. *J. Mater. Process. Technol.* **2009**, *209*, 4186–5193. [CrossRef]

34. Mortazavi, S.N.; Marashi, P.; Pouranvari, M.; Masoumi, M. Investigation on joint strength of dissimilar resistance spot welds of aluminum alloy and low carbon steel. *Adv. Mater. Res.* **2011**, *264–265*, 384–389. [CrossRef]

35. Yu, J.; Choi, Y.; Shim, J.; Cho, Y.; Rhee, S. A study on resistance spot welding for aluminum alloys with spooling process tapes. In Proceedings of the International Conference on Advances in Welding Science and Technology for Construction, Energy and Transportation, AWST, held in Conjunction with the 63rd Annual Assembly of the International Institute of Welding, IIW 2010, Istanbul, Turkey,, 15–16 July 2010; pp. 679–684.
36. Yeom, J. A Study on Resistance Spot Welding of Aluminum Alloys Based on the Current and Electrode Force Characteristics Analysis. Master's Thesis, Hanyang University, Seoul, Korea, 2007.
37. Ikeuchi, K.; Yamamoto, N.; Takahashi, M.; Aritoshi, M. Effect of interfacial reaction layer on bond strength of friction-bonded joint of Al alloys to steel. *Trans. JWRI* **2005**, *34*, 1–10.

metals

MDPI

Article

Development of Low Silver AgCuZnSn Filler Metal for Cu/Steel Dissimilar Metal Joining

Peng Xue [1], Yang Zou [1], Peng He [2,*], Yinyin Pei [3], Huawei Sun [3], Chaoli Ma [4] and Jingyi Luo [5]

[1] School of Materials Science and Engineering, Nanjing University of Science and Technology, Nanjing 210094, China; xuepeng@njust.edu.cn (P.X.); zouyang@njust.edu.cn (Y.Z.)

[2] State Key Laboratory of Advanced Welding and Joining, Harbin Institute of Technology, Harbin 150001, China

[3] State Key Laboratory of Advanced Brazing Filler Metals & Technology, Zhengzhou Research Institute of Mechanical Engineering, Zhengzhou 450001, China; peiyy@zrime.com (Y.P.); sunhuawei@zrime.com (H.S.)

[4] College of Materials Science and Technology, Nanjing University of Aeronautics & Astronautics, Nanjing 210016, China; machaoli@nuaa.edu.cn

[5] Jinhua Jinzhong Welding Materials Co., Ltd., Jinhua 321016, China; ljy00630@163.com

* Correspondence: hepeng@hit.edu.cn; Tel.: +86-0451-8641-8146

Received: 20 January 2019; Accepted: 6 February 2019; Published: 8 February 2019

Abstract: The microstructure and properties of a Cu/304 stainless steel dissimilar metal joint brazed with a low silver Ag16.5CuZnSn-xGa-yCe braze filler after aging treatment were investigated. The results indicated that the addition of Ce could reduce the intergranular penetration depth of the filler metal into the stainless steel during the aging process. The minimum penetration depth in the Ag16.5CuZnSn-0.15Ce brazed joint was decreased by 48.8% compared with the Ag16.5CuZnSn brazed joint. Moreover, the shear strength of the brazed joint decreased with aging time while the shear strength of the AgCuZnSn-xGa-yCe joint was still obviously higher than the Ag16.5CuZnSn joint after a 600 h aging treatment. The fracture type of the Ag16.5CuZnSn-xGa-yCe brazed joints before aging begins ductile and turns slightly brittle during the aging process. Compared to all the results, the Ag16.5CuZnSn-2Ga-0.15Ce brazed joints show the best performance and could satisfy the requirements for cost reduction and long-term use.

Keywords: Ag-Cu-Zn; Rare earth; aging treatment; microstructure; mechanical properties

---

1. Introduction

The increasing demand for advanced manufacturing has required equipment with complex structures and diversified properties. This demand has made the use of mixed-materials an urgent need in various fields. For example, brass has excellent electrical and thermal conductivity, but the low strength of brass joints can't satisfy the requirements of industrial applications, such as bearing compressing loads. Meanwhile, stainless steel posseses high strength but has poor thermal performance [1–3]. Therefore, a combination of brass and stainless steel could possibly satisfy requirements for both conductivity and mechanical properties. Currently, the composite structure of brass and stainless steel has been widely used in the fabrication of cooling equipment, pressure vessels, liquid cryogenic storage tanks, and heat exchangers [4–6].

Brazing has been proved to be an efficient way to achieve the joining of dissimilar metals by using appropriate filler metals. Currently, the Ag-Cu-Zn series filler metal possess a number of advantages: excellent wetting performance, excellent mechanical properties, outstanding conductivity, and corrosion resistance. Therefore, Ag-Cu-Zn series filler metals have been commonly used in brazing various ferrous metals and most non-ferrous metals, including steel to steel, copper to steel, copper to copper, and titanium alloy to steel [5–8]. To further improve the performance of a brazed joint, a series of trace additions including Ca [9], In [10], Ni [11], and rare earth elements [12] has been used to improve the mechanical properties and reliability of a Ag-Cu-Zn brazed joint. However, with increasing competitive pressures in the manufacturing industry, silver-based filler metals are unable to satisfy present requirements, due to the high cost of Ag (>30 wt.%). Therefore, the research of low silver filler metal without sacrificing the properties of Ag-Cu-Zn based filler metal has become an urgent issue.

Present research mainly focuses on improving the properties of Ag-Cu-Zn based filler metals and the microstructure of joints through micro alloying [13–15], while research on the brazed joints of brass/stainless steel, especially on their mechanical properties during the aging process, is still rare. Considering the reliability of a brass/stainless steel joint in long-term service, filler metals with designed additions of Ga and Ce were prepared in this paper. The effect of Ga and Ce addition on the microstructure, properties, and fracture morphology of a brass/stainless steel joint after aging treatment was investigated, and the relationship between the microstructure and its mechanical properties was analyzed.

2. Experimental Procedure

All alloys were made from pure Ag (99.95 wt.%), pure Cu (99.95 wt.%), pure Zn (99.95 wt.%), pure Sn (99.95 wt.%), pure Ga (99.95 wt.%), and pure Ce (99.5 wt.%). Raw materials of Ag, Cu, Zn, and Ga were first melted in a medium induction furnace. Then, Sn-5Ce master alloy ingots were added into the melted liquid alloy, which was held for 15 min, and mechanical stirring was performed every 5 min. Based on the previous research [16,17], the designed additions of Ga and Ce in the filler metal are listed in Table 1.

Table 1. Compositions of filler metals (wt.%).

No.	Ag	Cu	Zn	Sn	Ga	Ce
1	16.5	Bal.	35.44	2	0	0
2	16.5	Bal.	34.60	2	2	0
3	16.5	Bal.	34.26	2	0	0.15
4	16.5	Bal.	32.70	2	2	0.15

Figure 1 shows the geometry and dimensions of the brazing specimens for the shear strength test. According to China's National Standard, GB/T 11363-2008 [18], H62 brass and 304 stainless steel were used as base metals and were cut into plates with dimensions of 80 mm × 25 mm × 3 mm. The brazed joint specimens were prepared through automatic torch brazing using FB102 flux (B_2O_3(35 wt.%) + KBF_4(23 wt.%) + KF(42 wt.%)). All specimens were cleaned in an ultrasonic batch using DI water for 20 min after brazing. The strength of the brazed joints was tested on a SANS-CMT5105 electromechanical universal testing system (MTS, Minnesota, MN, USA) at room temperature, with a constant crosshead speed of 2 mm/min. To ensure the accuracy of the results, each test was conducted five times under the same conditions, and the average value was taken as the final result.

According to the test method, the military standard for microelectronics, MIL-STD-883 [19], the specimens were placed in an oven maintained at a constant temperature for performing a high temperature storage (HTS) test to determine the reliability of the brass/304 stainless steel brazed joints. The aging temperature was 150 °C and the storage time was 0 h, 200 h, 400 h, and 600 h. The cross sections of the brass/304 stainless steel joints were prepared by standard polishing techniques and subsequently etched with etchant solutions ((NH_4)$_2S_2O_8$(15 g) + H_2O (100 mL) + $NH_3 \cdot H2O$ (2 mL)). Microstructure observations were conducted with an optical microscope and a thermal field emission scanning electron microscope (SEM) (Hitachi, Tokyo, Japan) equipped with energy dispersive X-ray spectroscopy (EDS) (Hitachi, Tokyo, Japan).

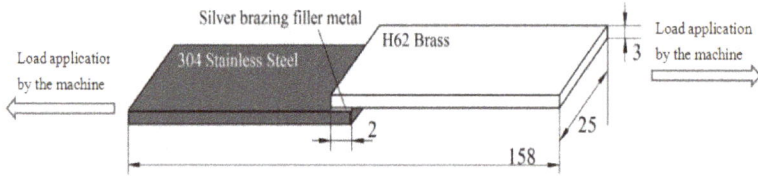

Figure 1. Brazing specimens for shear strength test.

3. Results and Discussion

3.1. Effect of the Aging Treatments on the Interfacial Microstructure of the Brazed Joints

Figures 2 and 3 illustrate the interfacial microstructure of the brazed joints near the stainless steel side after various aging times. The major composition of the brazing seams consisted of a Cu based solution, a Cu-Zn phase, and a Ag-rich phase. It can be seen that the major composition in the interfacial microstructure and morphology of the brazing seams had no significant change after aging treatment. Moreover, the intergranular penetration phenomenon from the filler metal into the stainless steel could be evidently observed, and the penetration depth increased with aging time. Moreover, the white phase appeared in the brazing seam, and the interfacial microstructure became more bulky over time.

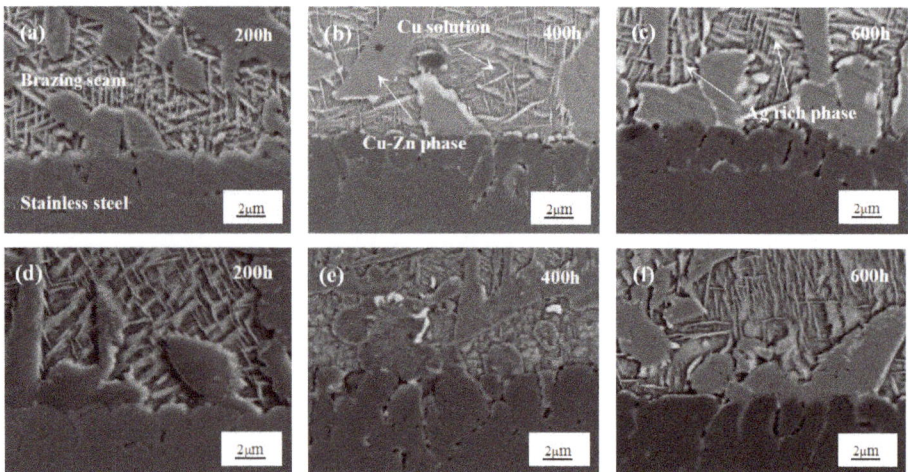

Figure 2. Interfacial microstructure of the Ag16.5CuZnSn-xGa brazed joints after aging treatment: (a–c) Ag16.5CuZnSn, (d–f) Ag16.5CuZnSn-2Ga.

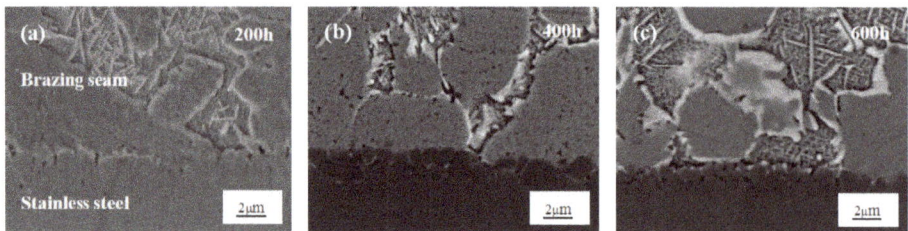

Figure 3. *Cont.*

Figure 3. Interfacial microstructure of the Ag16.5CuZnSn-xGa-yCe brazed joints after aging treatment: (**a–c**) Ag16.5CuZnSn-0.15Ce, (**d–f**) Ag16.5CuZnSn-2Ga-0.15Ce.

Figure 4 illustrates the EDS results of the Ag16.5CuZnSn-2Ga brazed joints near the stainless steel side after a 400 h aging treatment. As a result of intergranular penetration of the filler metal into the stainless steel, it can be found that Ag accumulated at the grain boundary, but no obvious accumulation of elements Fe and Cr was observed near the stainless steel side. Intergranular penetration has proven to be the key factor which affects the microstructure and mechanical properties of a brazed joint. Extensive intergranular penetration may significantly deteriorate the strength and plasticity of the brazed joint, resulting in fractures and connection failures [20,21]. Hence, the intergranular penetration depth during the aging process was calculated to further determine the reliability of the brazed joints after aging treatment.

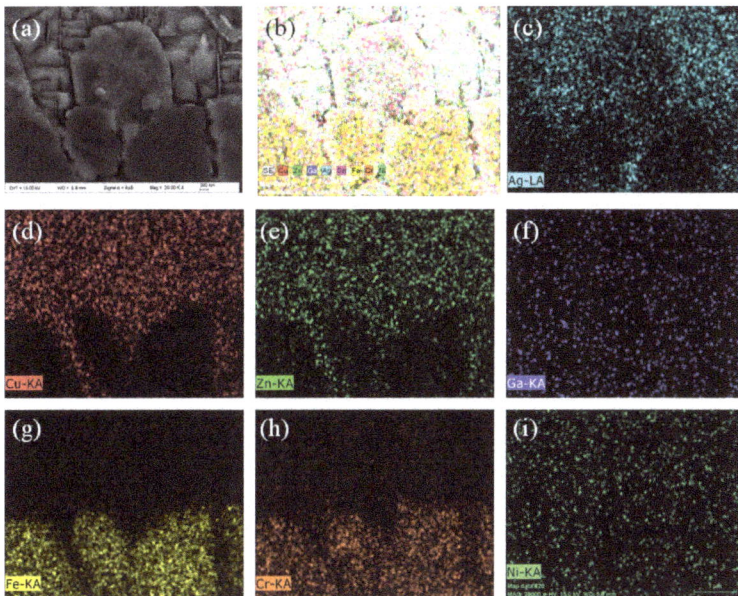

Figure 4. EDS results of the Ag16.5CuZnSn-2Ga brazed joints after 400 h: (**a**) Cross section morphology of brazed joints, (**b**) surface scanning, (**c**) Ag, (**d**) Cu, (**e**) Zn, (**f**) Ga, (**g**) Fe, (**h**) Cr, (**i**) Ni.

3.2. Effect of the Aging Treatment on Intergranular Penetration Depth in the Brazed Joints

Figure 5 shows a diagram of intergranular penetration depth calculation. The intergranular penetration depth H from the filler metal into the stainless steel was calculated by $H = S/d$, where S represents the penetration area, which was analyzed by image process software, and d was measured as the width of the interface. Figure 6 shows the average intergranular penetration depth in the Ag16.5CuZnSn-xGa-yCe brazed joints during aging. It can be seen that the intergranular penetration depth from the filler metal into the stainless steel increased with aging time. Compared to the original Ag16.5CuZnSn filler metal, the intergranular penetration depth was greatly affected by the addition of Ga/Ce during the aging process. After a 600 h aging treatment, the maximum intergranular penetration depth was obtained in the Ag16.5CuZnSn-2Ga brazed joint, with a value of 5.258 μm, while the minimum penetration depth in Ag16.5CuZnSn-0.15Ce was about 1.6 μm. The penetration depth in the Ag16.5CuZnSn-2Ga-0.15Ce brazed joint reached 2.044 μm after a 600 h aging treatment, which decreased by 48.8% compared to the original Ag16.5CuZnSn brazed joint. From the results above, it can be concluded that the addition of Ce could reduce the intergranular penetration depth of filler metal into stainless steel during the aging process. Previous studies have proven that the pining effect caused by moderate intergranular penetration could be beneficial to the mechanical properties of brazed joints [22]. However, excessive intergranular penetration would destabilize the interfacial layer in the brazed joint near the stainless side, resulting in the formation of voids and fractures in the non-planar interfacial layer, which would greatly reduce the mechanical properties of the brazed joint. Therefore, the penetration depth is expected to be well controlled.

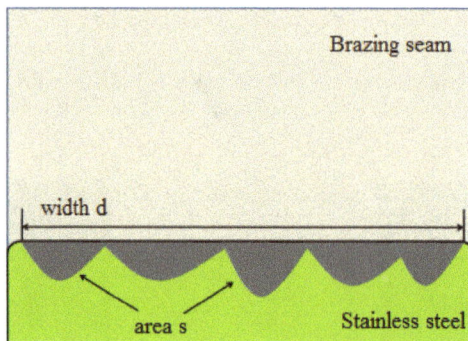

Figure 5. Diagram of intergranular penetration depth calculation.

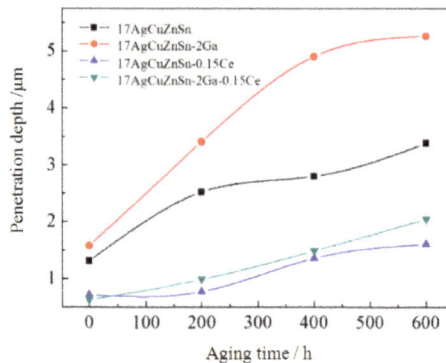

Figure 6. Effect of the aging treatment on intergranular penetration depth in the Ag16.5CuZnSn-*x*Ga-*y*Ce brazed joints.

3.3. Effect of the Aging Treatment on the Microstructure of the Brazing Seam

Figure 7 illustrates the microstructure of the Ag16.5CuZnSn-2Ga brazing seam after aging for 600 h. After a long time aging, a white phase appeared in the brazing seam; the composition of this white phase consisted of Cu: 11.01 at.%, Zn: 26.05 at.%, and Ag: 60.40 at.%, based on the EDX result in Figure 7c. Therefore, it can be concluded that the white phase is a Ag-rich phase. Moreover, a bulk-like phase was found in the Ag16.5CuZnSn-0.15Ce brazing seam after 600 h of aging, as shown in Figure 8a. These intermetallic compounds(IMCs) were surrounded by a white phase. From the EDX results in Figure 8b–f, the white phase consisted of a Ag-rich phase, while the composition of the bulk-like phase, which can be interpreted as an RE-phase, consisted of elements Ce, Sn, and Ag. According to the Hume-Rothery theory, the element Ce possess a large atomic radius and prefers to form an intermetallic compound, instead of a solution, with Sn, Ag, and Cu, due to the large deviation of its atomic radius [23]. Furthermore, formations of these IMCs tend to accumulate ahead of the solid liquid interface during the brazing process, resulting in an increase of the under-cooling rate of the filler metals; hence, the microstructure of the brazing seam is refined.

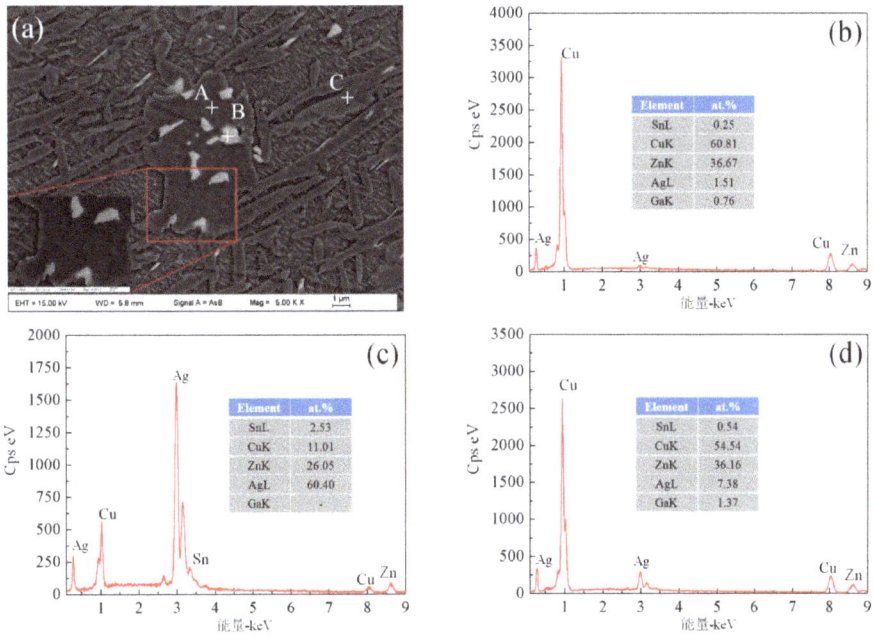

Figure 7. Microstructure of the Ag16.5CuZnSn-2Ga brazing seam after aging treatment for 600 h: (**a**) microstructure of Ag16.5CuZnSn-2Ga brazing seam, (**b**) EDX result of spot A, (**c**) spot B, (**d**) spot C.

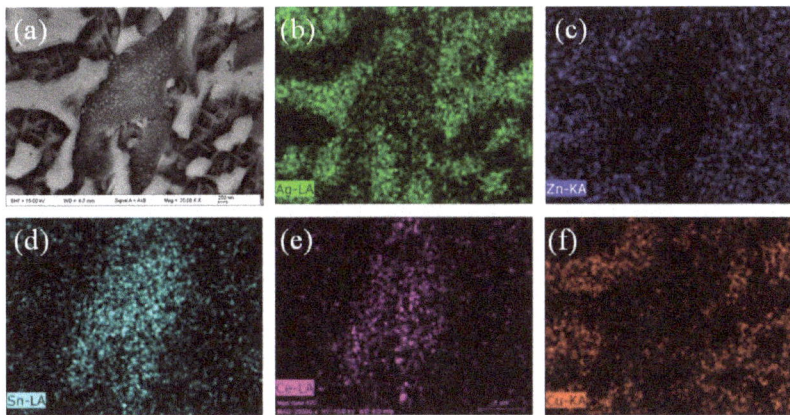

Figure 8. EDX result of the RE-phase in the Ag16.5CuZnSn-0.15Ce brazed joint after aging treatment for 600 h: (**a**) microstructure of the RE-phase, (**b**) Ag, (**c**) Zn, (**d**) Sn, (**e**) Ce, (**f**) Cu.

3.4. Effect of the Aging Treatment on the Mechanical Properties of Brazed Joints

Figure 9 illustrates the effect of an aging treatment on the mechanical properties of the brazed joints; it can be seen that the shear strength of the brazed joints decreases with aging time. After an aging treatment for 600 h, the shear strength of Ag16.5CuZnSn, Ag16.5CuZnSn-2Ga, Ag16.5CuZnSn-0.15Ce, and Ag16.5CuZnSn-2Ga-0.15Ce was lowered by 16.5%, 18.99%, 13.1%, and 15.0% respectively. Based on the results from Figures 7 and 8, since the Ag-rich phase possessed higher hardness compared to the phases in the brazing seam, the detached Ag-rich phase during the aging process caused defects due to the difference of physical properties with the surrounding microstructure [24], which led to the deterioration of the mechanical properties of the brazed joints.

Moreover, the shear strength of the brazed joints bearing Ga/Ce was still obviously higher than the original Ag16.5CuZnSn joint. It can be concluded that the addition of Ga and Ce could significantly improve the mechanical properties of Ag16.5CuZnSn brazed joints after an aging treatment. Therefore, brazed joints could satisfy the requirements for long-term use.

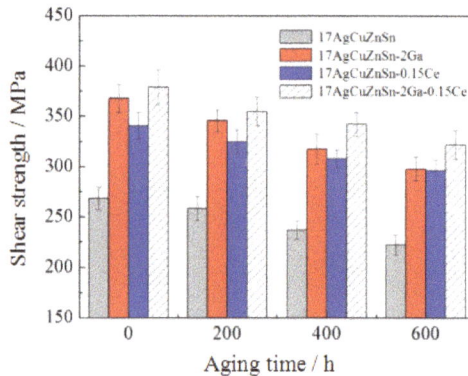

Figure 9. Shear strength of Ag16.5CuZnSn-xGa-yCe after the aging treatment.

3.5. Effect of the Aging Treatment on the Fracture Morphology of Brazed Joints

The fracture morphology of brazed joints can describe the effect of Ga and Ce additions on the mechanical properties of filler metals. Figure 10 illustrates the fracture morphology of Ag16.5CuZnSn-2Ga, Ag16.5CuZnSn-0.15Ce, and Ag16.5CuZnSn-2Ga-0.15Ce brazed joints with and without aging treatment. It can be seen that the fracture morphology of brazed joints before aging showed a dimpled structure, which means that the fracture type was ductile. Meanwhile, the fine and uniform dimples in the fracture of the Ag16.5CuZnSn-2Ga-0.15Ce brazed joint showed excellent performance compared to the other joints. Moreover, the microstructure of the fractures in the brazed joints became coarse; there were particles and cracks observed in the fracture after aging treatment for 600 h, resulting in a slightly more brittle fracture type. The EDX results of spot A and B illustrate that the main composition of the fracture microstructure consists of Ag, Cu, and Zn, which demonstrates that the fracture occurred in the brazing seam. Based on the results in Figures 6–8, the transformation of the microstructure, including intergranular penetration and the detached Ag-rich phase, could be the main factor that deteriorated the mechanical properties of the brazed joints during the aging process.

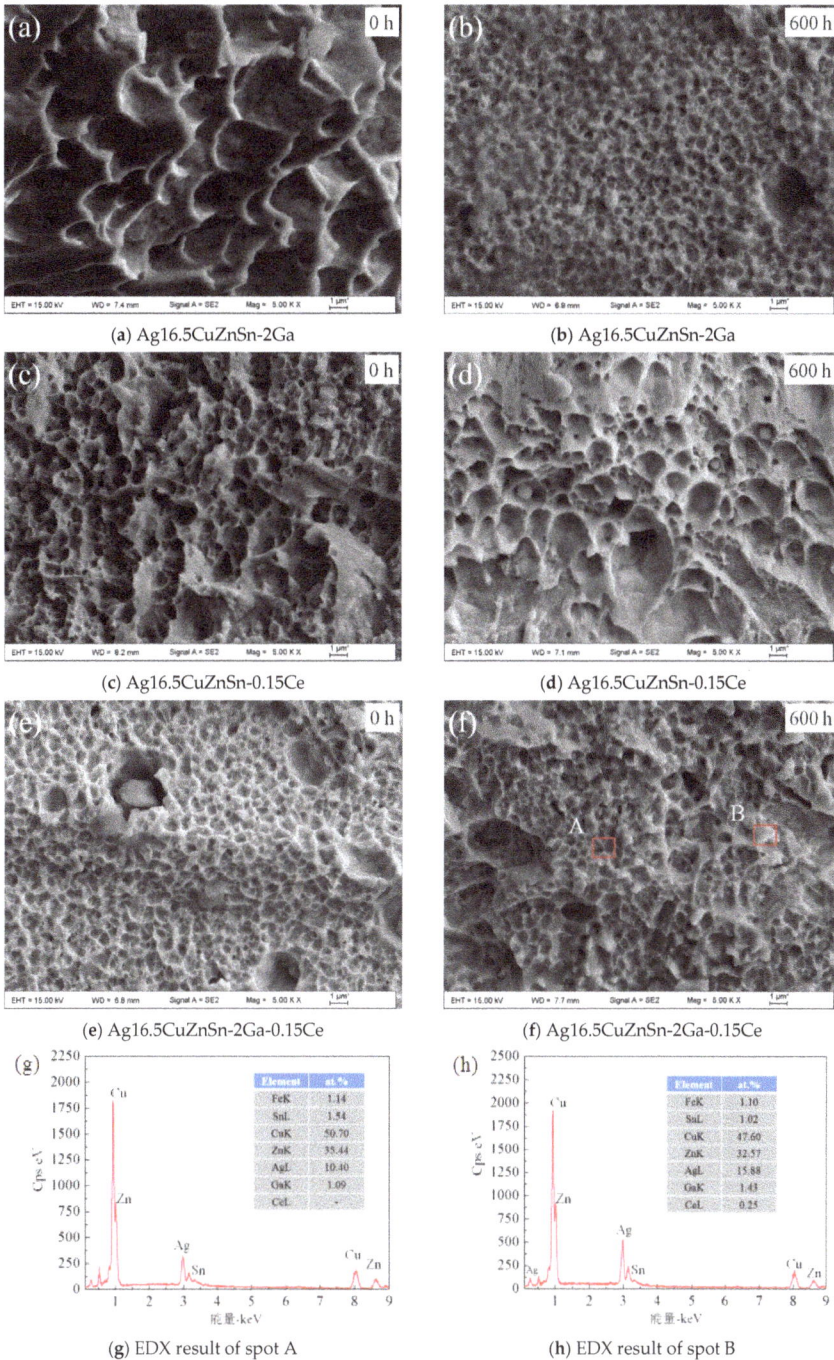

(a) Ag16.5CuZnSn-2Ga

(b) Ag16.5CuZnSn-2Ga

(c) Ag16.5CuZnSn-0.15Ce

(d) Ag16.5CuZnSn-0.15Ce

(e) Ag16.5CuZnSn-2Ga-0.15Ce

(f) Ag16.5CuZnSn-2Ga-0.15Ce

(g) EDX result of spot A

(h) EDX result of spot B

Figure 10. Fracture morphology of Ag16.5CuZnSn-xGa-yCe brazed joints after a 600 h aging treatment.

4. Conclusions

(1) The addition of Ce could reduce the intergranular penetration depth from filler metal to stainless steel during the aging process. The minimum penetration depth in the Ag16.5CuZnSn-0.15Ce brazed joint was 1.6 μm after a 600 h aging treatment, which was decreased by 48.8% compared to the original Ag16.5CuZnSn brazed joint.

(2) The shear strength of brazed joints decreased with aging time, and the brazed joints bearing Ga/Ce possessed notably higher mechanical properties than the original Ag16.5CuZnSn joint after a 600 h aging treatment. The optimum shear strength was obtained in the Ag16.5CuZnSn-2Ga-0.15Ce brazed joint.

(3) The fracture type of the Ag16.5CuZnSn-xGa-yCe brazed joints before aging began ductile and turned slightly brittle during the aging process. The Ag16.5CuZnSn-2Ga-0.15Ce brazed joint showed optimum performance compared to the other joints.

Author Contributions: Conceptualization, P.X. and Y.Z.; methodology, P.X. and Y.Z; software, P.X.; validation, P.X., Y.Z and P.H.; formal analysis, P.X. and Y.P.; investigation, P.X. and H.S.; resources, Y.P., H.S. and J.L.; data curation, P.X., C.M. and J.L.; writing—original draft preparation, P.X.; writing—review and editing, P.X. and P.H.; visualization, P.X.; supervision, P.H.; project administration, P.X.; funding acquisition, P.X.

Funding: This work is supported by the National Natural Science Foundation of China (Grant No.51605226) and the Zhejiang Postdoctoral Science Foundation. The authors would like to express gratitude to the support from the Chinese Scholars Council (No.201706845020).

Conflicts of Interest: The authors declare no conflict of interest.

References

1. Phanikumar, G.; Manjini, S.; Dutta, P.; Chattopadhyay, K.; Mazumder, J. Characterization of a continuous CO_2 laser-welded Fe-Cu dissimilar couple. *Metall. Mater. Trans. A* **2005**, *36*, 2137–2147. [CrossRef]

2. Jafari, M.; Abbasi, M.; Poursina, D.; Gheysarian, A.; Bagheri, B. Microstructures and mechanical properties of friction stir welded dissimilar steel-copper joints. *J. Mech. Sci. Technol.* **2017**, *31*, 1135–1142. [CrossRef]

3. Yao, C.; Xu, B.; Zhang, X.; Huang, J.; Fu, J.; Wu, Y. Interface microstructure and mechanical properties of laser welding copper-steel dissimilar joint. *Opt. Lasers Eng.* **2009**, *47*, 807–812. [CrossRef]

4. Chen, S.; Zhai, Z.; Huang, J.; Zhao, X.; Xiong, J. Interface microstructure and fracture behavior of single/dual-beam laser welded steel-Al dissimilar joint produced with copper interlayer. *Int. J. Adv. Manuf. Technol.* **2016**, *82*, 631–643. [CrossRef]

5. Leon-Patino, C.A.; Aguilar-Reyes, E.A.; Braulio-Sanchez, M.; Rodríguez-Ortiz, G.; Bedolla-Becerril, E. Microstructure and shear strength of sintered Cu-Al₂O₃ composite joined to Cu using Ag-Cu and Cu-Zn filler alloys. *Mater. Des.* **2014**, *54*, 845–853. [CrossRef]

6. Zhang, L.; Feng, J.; Zhang, B.; Jing, X. Ag-Cu-Zn Alloy for Brazing TiC Cermet/Steel. *Mater. Lett.* **2005**, *59*, 110–113. [CrossRef]

7. Venkateswaran, T.; Xavier, V.; Sivakumar, D.; Pant, B.; Janaki Ram, G.D. Brazing of stainless steels using Cu-Ag-Mn-Zn braze filler: Studies on wettability, mechanical properties, and microstructural aspects. *Mater. Des.* **2017**, *121*, 213–228.

8. Alexander, V.; Joerg, B.; Sigurd, J.; Dieter, K.; Georg, P.; Wolfgang, K. Use of Silver Alloys as Cadium-Free Brazing Solder. U.S. Patent 5352542, 4 October 1994.

9. Sui, F.; Long, W.; Liu, S.; Zhang, G.; Bao, L.; Li, H.; Chen, Y. Effect of Calcium on the Microstructure and Mechanical Properties of Brazed Joint Using Ag-Cu-Zn Brazing Filler Metal. *Mater. Des.* **2013**, *46*, 605–608. [CrossRef]

10. Ma, X.; Li, L.-F.; Zhang, Z.-H.; Wang, H.; Wang, E.-Z.; Qiu, T. Microstructure and Melting Properties of Ag-Cu-In Intermediate Temperature Brazing Alloys. *Rare Met.* **2015**, *34*, 324–328. [CrossRef]

11. Lai, Z.; Xue, S.; Han, X.; Gu, L.; Gu, W. Study on Microstructure and Property of Brazed Joint of AgCuZn-X (Ga, Sn, In, Ni) Brazing Alloy. *Rare Metal. Mater. Eng.* **2010**, *39*, 397–400.

12. Li, Z.; Jiao, N.; Feng, J.; Chen, Y. Effect of P and rare-earth La on microstructure and property of AgCuZnSn brazing alloy. *Trans. Chin. Weld. Inst.* **2007**, *28*, 1–4.

13. Wang, X.; Yu, D.; He, Y.; Huang, S.; Chen, R. Effect of Sn Content on Brazing Properties of Ag Based Filler Alloy. *Mater. Sci.* **2013**, *3*, 16–21.
14. Ma, C.; Xue, S.; Zhang, T.; Jiang, J.; Long, W.; Zhang, G.; Zhang, Q.; He, P. Influences of In on the Microstructure and Mechanical Properties of Low Silver Ag-Cu-Zn Filler Metal. *Rare Met. Mater. Eng.* **2017**, *46*, 2565–2570.
15. Cao, J.; Zhang, L.X.; Wang, H.Q.; Wu, L.Z.; Feng, J.C. Effect of Silver Content on Microstructure and Properties of Brass/steel Induction Brazing Joint Using Ag-Cu-Zn-Sn Filler Metal. *J. Mater. Sci. Technol.* **2011**, *27*, 377–381. [CrossRef]
16. Ma, C.; Xue, S.; Wang, B. Study on novel Ag-Cu-Zn-Sn brazing filler metal bearing Ga. *J. Alloys Compd.* **2016**, *688*, 854–862. [CrossRef]
17. Lai, Z.M. Effects of Ga/In and rare earth Ce on microstructures and properties of brazed joint of 30AgCuZn filler metal. Ph.D. Thesis, Nanjing University of Aeronautics and Astronautics, Nanjing, China, 1 March 2011.
18. GB/T 11363-2008. *Test Method of the Strength for Brazed and Soldered Joint*; Standardization Administration of China: Beijing, China, 2008.
19. *Test Method Standard Microcircuits, MIL-STD-883G, METHOD 2019.7, Die Shear Strength*; Military and Government Specs & Standards (Naval Publications and Form Center) (NPFC): Englewood, CO, USA, 2018.
20. Ina, K.; Koizumi, H. Penetration of liquid metals into solid metals and liquid metal embrittlement. *Mater. Sci. Eng. A* **2004**, *387–389*, 390–394. [CrossRef]
21. Lian, X.; Qu, W.; Li, H.; Wang, G. Grain boundary penetration behavior analysis of OFC brazed with AgCu28 brazing filler. *J. Beijing Univ. Aeronaut. Astronaut.* **2014**, *40*, 717–720.
22. Beura, V.K.; Xavier, V.; Venkateswaran, T.; Kulkarni, K.N. Interdiffusion and microstructure evolution during brazing of austenitic artensitic stainless steel and aluminum-bronze with Ag-Cu-Zn based brazing filler material. *J. Alloys Compd.* **2018**, *740*, 852–862. [CrossRef]
23. Troparevsky, M.C.; Morris, J.R.; Daene, M.; Wang, Y.; Lupini, A.R.; Stocks, G.M. Beyond Atomic Sizes and Hume-Rothery Rules: Understanding and Predicting High-Entropy Alloys. *JOM* **2015**, *67*, 2350–2363. [CrossRef]
24. Chi, C.-T.; Chao, C.-G.; Liu, T.-F.; Lee, C.-H. Aluminum element effect for electron beam welding of similar and dissimilar magnesium-aluminum-zinc alloys. *Scr. Mater.* **2007**, *56*, 733–736. [CrossRef]

![metals logo] **metals** — MDPI

Article

Microstructural and Mechanical Characterization of Dissimilar Metal Welding of Inconel 625 and AISI 316L

Fatih Dokme [1,2,*], Mustafa Kemal Kulekci [2] and Ugur Esme [2]

[1] Sisecam Chemicals Group Soda Sanayii A.S. Kromsan Chromium Chemicals Plant, 33003 Mersin, Turkey
[2] Tarsus University Faculty of Technology, 33400 Mersin, Turkey; mkkulekci@yahoo.com (M.K.K.);
     uguresme@gmail.com (U.E.)
*   Correspondence: fdokme@gmail.com; Tel.: +90-324-241-6810

Received: 4 September 2018; Accepted: 2 October 2018; Published: 4 October 2018

Abstract: This study investigated the microstructure of the dissimilar metal welding of Inconel 625 and AISI 316L using Continuous Current Gas Tungsten Arc Welding (CCGTAW) and Pulsed Current Gas Tungsten Arc Welding (PCGTAW) processes with ERNiCr-3, TIG 316L and twisted (ERNiCr-3 and TIG 316L) fillers. Microstructure examinations were carried out using an optical microscope and Scanning Electron Microscopy (SEM)/Energy Dispersive X-Ray (EDAX). The results of the study showed the existence of a partially melted zone (PMZ) on the AISI 316L side. Weld zone (WZ) analysis showed the existence of a multi-directional grain growth on the 316L side in all specimens, although less growth was found on the Inconel 625 side. Grain growth almost disappeared using PCGTAW with twisted fillers. SEM/EDAX investigations indicated that secondary deleterious secondary phases were tiny and white in five experiments. However, a meager amount of precipitates occurred in PCGTA welding with twisted fillers. Moreover, these were particularly innocent precipitates, represented by black dots in images, whereas other tiny white secondary phases are known to be brittle. As a result, PCGTA welding with twisted fillers exhibited the best metallurgical properties.

Keywords: dissimilar metal welding; Inconel 625; AISI 316L; microstructure; filler metals

1. Introduction

Inconel 625 is a nickel-based super alloy, which has good corrosion resistance properties at cryogenic as well as very high temperatures. Thus, it is the preferred material in many heavy industries. It is used in the reactor-core of nuclear power plants, boilers, piping, exhaust systems of racing cars, impellers of chemical vessels, power plant turbine blades, aerospace engine components, etc. [1–3]. It is often used as dissimilar joints, with several metals in field applications. For example, bimetallic joints of Inconel 625 and AISI 304 L are used in high temperature media in thermal power and nuclear power plants to cryogenic applications [4]. Bimetallic joints of Inconel 625 and duplex stainless steel pipes are used in sub-sea manifolds [5]. Bimetallic combinations of Inconel 625 and austenitic stainless steels were used by NASA for manufacturing of sub-scale boilers [6]. The bimetallic joints of Inconel 625 and austenitic stainless steel AISI 316 L are used in the chemical process at very high temperatures, especially the chromic acid calcining process, which contains a diluted acidic environment. Therefore, many researchers have studied it during last decade [7–11]. One of the main reasons is that it reduces cost of materials, and the other is that it improves design as a result of operating environment requirements [12]. Different metals have different chemical and physical properties; hence, the dissimilar metal welding process has some disadvantages. A weld pool consists of a molten metal mixture, which has unknown chemical and mechanical properties [13,14]. Thus, some specific parameters, such as proper filler metal and current type selection, could minimize weld defects.

Gas Tungsten Arc Welding (GTAW) is one the preferred dissimilar welding technique in the manufacturing industry due to its high development, ease of use, low investment value and low operating cost. The main disadvantage of GTAW is that the welding of thick parts requires a multi-pass, which causes a higher heat input. This causes a Laves phase formation as a result of the segregation of Nb and Mo at a high temperature. This leads to a change in the microstructure and mechanical properties of both the heat affected zone (HAZ) and weld zone, leading to a decreased corrosion resistance [15].

Because of the content of Nb and Mo elements, Inconel 625 exhibits a precipitation disposition in welding treatment. The solidification of Inconel 625 starts at nearly 1368 °C. During fusion weld overlay solidification, NbC carbides (LE → γ + NbC) and austenite forms occur at 1250 °C, and an austenite and brittle intermetallic Laves phase (LE → γ + Laves phase) occurs at 1150 °C. If the alloy has a high Si and low C content, an additional carbide (M6C) could occur during solidification [16].

Ramkumar et al. [10] studied the corrosion behavior of the similar and dissimilar TIG welding of Inconel 718 and SS 316L. Both similar weldments of Inconel 718 and SS 316L possessed a lower hardness value than that of dissimilar weldments. A higher hardness occurred as a result of the brittle phases of Cr and Ni precipitation in HAZ. Corrosion tests showed that the maximum weight loss was found in dissimilar weldments, which backed up the thesis that carbides decrease corrosion resistance. In some studies, various improvements were recommended for minimizing micro-segregations and precipitates in a liquid nitrogen aided cooling system for GTAW [17], double shot laser shock peening on the fusion zone [18], and a post weld heat treatment (PWHT) [16,19]. PWTH reduces residual stresses, although it causes increasing precipitations [20] in some cases.

One of the major parameters for dissimilar welding is the selection of appropriate filler metals for both base metals. Prabu et al. [21] studied the effects of two filler materials, ErNiCrMo-4 and ERNiCrCoMo-1, on the dissimilar welding of Inconel 625 and AISI 904L using GTAW. Researchers found that the segregated zone is wider and more visible in ERNiCrCoMo-1 welding compared with ERNiCrMo-4 welding, because of the migration of C. Ni-based filler metal and AISI 904 L have different chemical compositions, and a wider unmixed zone occurred on the AISI 904 L side rather than on the Inconel 625 side. This is evidence of the importance of the similarity between base and filler metals in the welding process. Moosavy et al. [22] investigated the microstructural evolutions of the dissimilar welding of AISI 310 and Inconel 657 using four types of filler metals: Inconel 82, Inconel A, Inconel 617 and type 310 SS. Thus, the presence of a high Mo content of Inconel 617 filler generated brittle phases. Likewise, 310 SS led to continuous precipitates. Inconel 82 generated NbC precipitates, whereas Inconel A did not show the formation of NbC in the interdentric region. PWHT ensured the removal of the unmixed zone of Inconel 657 on the base metal side. Mortezaie and Shamanian [9] studied the dissimilar welding of Inconel 718 and AISI 310S using GTAW with three different filler metals. According to their results, Inconel 82 exhibited the best corrosion resistance and highest energy impact in a V-Charpy test. Ramkumar et al. [23] researched the influence of filler metal and welding techniques on the dissimilar welding of Inconel 718 and AISI 316L. The authors used two filler materials, ER2553 and ERNiCu-7, and two welding techniques, Continuous Current Gas Tungsten Arc Welding (CCGTAW) and Pulse Current Gas Tungsten Arc Welding (PCGTAW). They controlled the precipitations through PCGTAW. Furthermore, the Nb free ErNiCu-7 filler did not cause precipitations. Hejripour and Aidun [24] conjoined Inconel 718 and AISI 410 by GTAW. Researchers selected the twisted filler metals of 718-410 and 82-410. Secondary phase formations formed a higher volume in the 718-410 weld zone than in the 82-410 weld zone because of the higher Nb content.

Kumar et al. [25] investigated metallurgical and mechanical properties of bimetallic joints of Inconel 625 and AISI 316L materials. Researchers used CCGTA and PCGTA welding processes employing ERNiCr-3 and ER2209 filler metals. Microstructural investigations showed that secondary phases were occurred at the HAZ of Inconel 625 employing ERNiCr-3 filler metal.

However, PCGTA welding minimized secondary phases formation. Tensile tests present all fractures occurred at the parent metal of AISI 316L. And both welding techniques employing ERNiCr-3 exhibited better mechanical properties. Researchers recommend PCGTA welding employing ERNiCr-3 for bimetallic joints of Inconel 625 and AISI 316L. In another study focused on dissimilar welding of Inconel 625 and AISI 316 L, Kournadi and Haghighi [26] investigated the effects of different welding methods on metallurgical and mechanical properties. Authors used shield metal arc welding (SMAW) and GTA welding employing ERNiCrMo-3 and TIG 316L filler metals and an electrode NiCrMo-3. Results showed that SMA welding reduced metallurgical and mechanical properties. GTA welding employing ERNiCrMo-3 exhibited the best properties for welding Inconel 625 and AISI 316L.

There are several welding methods, such as laser beam (LB), electron beam (EB), friction stir, explosion welding (EW), etc., for joining various dissimilar metals. Xie et al. [27] investigated interface characteristics of explosive welded and hot rolled TA1/X65 bimetallic plate. In the vortex zone brittle intermetallic compounds and cracks were detected. Topolski et al. [28] studied the properties of Ti6Al4V/Inconel 625 Bimetal obtained by EW. Results showed that chemical composition was clearly changed in the fusion line and elemental transfer was obtained between the metals. The main disadvantage of EW is its use of thin sheets. Mohammed et al. [29] studied the fiber laser welding of duplex stainless steel 2205 and austenitic stainless steel 304. From the SEM results, formation of micro-voids confirmed the ductile mode of fracture. Small HAZ obtained between metals and rapid solidification in the fusion zone improved the mechanical properties of the weldments. Shakil et al. [30] investigated electron beam welding of dissimilar Inconel 625 and AISI 304L. Columnar and cellular dendritic microstructures were observed in the fusion zone. Ni and Fe elements transferred from Inconel 625 to FZ and from FZ to AISI 304L. These technological welding types provide some advantages on metallurgical and mechanical properties of weldments. However, there are similarly some problems, such as undesirable elemental transfers, presence of brittle structures, etc. On the other hand, they have weaknesses such as thickness limits, as well as investment and operating costs. Therefore, GTA welding is commonly used with cheap costs and wider thickness limits.

Although bimetallic joints of Inconel 625 and AISI 316L metals used several applications, limited studies focused on the dissimilar metal welding of Inconel 625 and AISI 316L. Furthermore, studies that focused on the dissimilar metal welding of Inconel and stainless steel experimented with thin plates in general. In this investigation, 20 mm thickness Inconel 625 and AISI 316L, conjoined with CCGTA and PCGTA welding processes with ERNiCr-3, TIG 316L and twisted of both filler metals, was studied. TIG 316L is a Nb free filler, which is expected to prevent the Laves phase. Metallurgical characterizations were carried out using Optical Microscopy (OM) and Scanning Electron Microscopy (SEM), and compositional analysis was carried out using Energy Dispersive Spectroscopy (EDS). The results of microstructural tests are presented in corresponding sections.

2. Materials and Methods

The chemical composition of the as-received solution of annealed Inconel 625 and AISI 316L base metals and filler metals are given in Table 1. Base metals were machined to rectangular samples with the following dimensions: 200 mm × 60 mm × 20 mm. Standard double V-Butt configurations (with a root face thickness of 1 mm, a 2 mm gap, and with a 60° angle) were employed for the samples. Two filler wires were selected: ERNiCr-3 and TIG316L (thickness 3.2 mm). However, both wires combined and created a third filler metal for welding. Three of the samples were welded using CCGTA welding, whereas the other three samples were welded using the PCGTA welding process.

Welding parameters are given in Table 2. High purity argon gas (99.9%) was used as a shielding gas at 15 L per minute. Samples were welded using the semi-automatic GTAW (automatic welding speed with manual filler feeding) setting of the Fronius Magicwave Synergic 2200 (Fronius International GmbH, Wels/Austria) welding machine (shown in Figure 1) in continuous and pulsed current modes. After welding, samples were sliced to several coupons for mechanical and metallurgical tests.

Figure 1. Semi-automatic Gas Tungsten Arc Welding (GTAW) setting.

Table 1. Chemical composition of the base and filler metals (wt %).

Base/Filler	C	Cr	Si	Ni	P	Mn	Nb + Ta	Mo	Fe	Other
Inconel 625	0.04	22.12	0.15	Rem.	0.012	0.09	3.38	8.32	4.48	Cu—0.05 Al—0.11 Ti—0.21 Co—0.3
AISI 316L	0.02	17.25	0.04	10.38	0.039	1.78	-	2.15	Rem.	S—0.004
ERNiCr-3	0.04	20.0	0.09	73.00	0.003	2.8	2.4	-	1.5	Ti—0.40 S—0.002 Cu—0.03 Other < 0.5
TIG316L	0.02	18	0.45	12	0.04	1.70	-	2.3	Rem.	Cu—0.75 S—0.03

Table 2. Process parameters employed in Continuous Current Gas Tungsten Arc Welding (CCGTAW) and Pulse Current Gas Tungsten Arc Welding (PCGTAW).

Welding	Filler	Pass	Current (A)		Voltage (V)	Duty Cycle	Frequency (Hz)	Welding Speed (mm/s)
			Peak	Back				
CCGTAW	ERNiCr3	Root 1	110	-	14	-	-	0.56
		Pass 1	110	-	17	-	-	0.71
		Cap 1	110	-	18	-	-	0.95
		Root 2	110	-	15	-	-	0.65
		Pass 2	110	-	17	-	-	0.82
		Cap 2	110	-	20	-	-	1.09
	TIG 316L	Root 1	110	-	15	-	-	0.62
		Pass 1	110	-	17	-	-	0.79
		Cap 1	110	-	19	-	-	1.02
		Root 2	110	-	13	-	-	0.78
		Pass 2	110	-	16	-	-	0.92
		Cap 2	110	-	18	-	-	1.18
	TWISTED	Root 1	120	-	16	-	-	0.81
		Pass 1	120	-	19	-	-	1.02
		Cap 1	120	-	20	-	-	1.15
		Root 2	120	-	14	-	-	0.95
		Pass 2	120	-	16	-	-	1.13
		Cap 2	120	-	18	-	-	1.29
PCGTAW	ERNiCr3	Root 1	170	100	12	50%	6	0.56
		Pass 1	170	100	15	50%	6	0.71
		Cap 1	170	100	16	50%	6	0.95
		Root 2	160	90	13	50%	6	0.65
		Pass 2	150	85	14	50%	6	0.82
		Cap 2	150	85	16	50%	6	1.09
	TIG 316L	Root 1	170	100	11	50%	6	0.62
		Pass 1	170	100	15	50%	6	0.79
		Cap 1	170	100	17	50%	6	1.02
		Root 2	160	90	10	50%	6	0.78
		Pass 2	150	85	14	50%	6	0.92
		Cap 2	140	80	16	50%	6	1.18
	TWISTED	Root 1	180	110	13	50%	6	0.81
		Pass 1	170	100	15	50%	6	1.02
		Cap 1	170	100	17	50%	6	1.15
		Root 2	170	100	12	50%	6	0.95
		Pass 2	160	90	14	50%	6	1.13
		Cap 2	150	85	15	50%	6	1.29

Each sample has been studied with respect to both parent metals, both heat affected zones (HAZ), and a weld zone in five sections. Specimens were polished using the sandpaper sheets of silicon carbide (SiC) with various grit sizes from 150 to 1000, then by a final mechanical polishing with 2 µm polishing paste. To display the microstructures at the specified zones of weldments, a mixture of 10% oxalic acid was used for electrolytic etching, with a 6 DC supply for 20 s at room temperature. Optical investigations were carried out using Carl Zeiss brand Axio Imager M2M model optical microscope (Carl Zeiss, Cambridge, UK). The microstructural and element distribution analysis were carried out using Carl Zeiss brand Evo MA10 model Scanning Electron Microscope (SEM) (Carl Zeiss, Cambridge, UK) equipped with an Ametek brand Apollo X model Energy Dispersive Spectrometer (EDS) (Ametek, Mahwah, NJ, USA). Both point and line mapping analyses were implemented to reveal elemental displacements across weldments.

3. Results

3.1. Macrostructure of the Weldments

The photographs of both the CCGTAW and PCGTAW dissimilar weldments of Inconel 625 and AISI 316L metals are shown in Figure 2. Weld seams are free of slags. The cross-section macrographs of the CCGTA and PCGTA welded samples are shown in Figure 3.

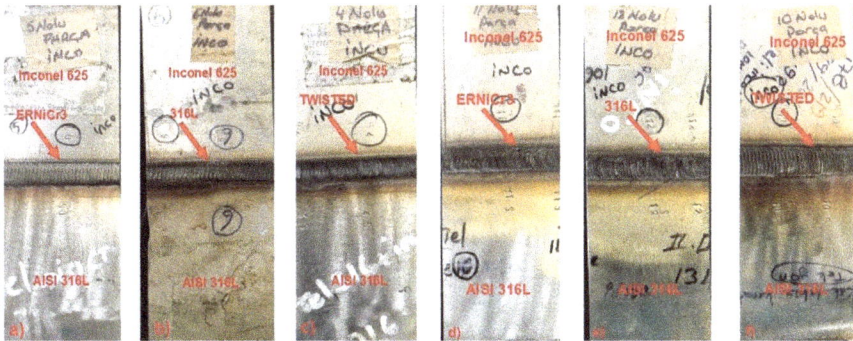

Figure 2. Photographs of the CCGTA (**a–c**) and PCGTA (**d–f**) weldments of Inconel 625 and AISI 316 L.

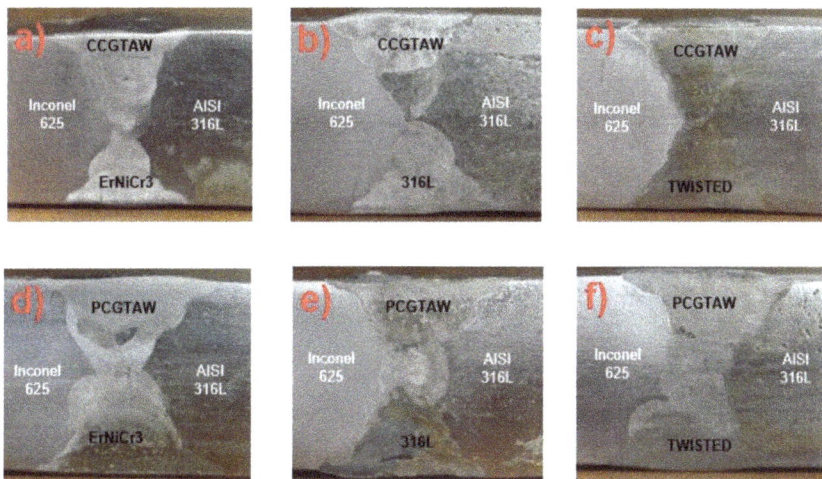

Figure 3. Cross-section macrographs of the CCGTA (**a–c**) and PCGTA (**d–f**) weldments of Inconel 625 and AISI 316 L.

According to visual testing (VT) and ultrasonic testing (UT), all experiments present full penetration without any lack, porosities, spatters, etc.

3.2. Microstructural Investigation

3.2.1. Microstructure of the CCGTA Welding

Interfacial microstructures of the CCGTA welded samples are shown in Figure 4. Investigations revealed that secondary phases prominently present as an unmixed zone (UZ) on all Inconel 625 sides (Figure 4b,d,f). The UZ on the Inconel 625 side welded with an ERNiCr3 filler has disappeared, although no aging treatment was performed. A partially melted zone (PMZ) was observed in the interface on the AISI 316L side with TIG 316L and twisted fillers (Figure 4c,e). Additionally, equi-axed and columnar dendrites were observed in the weld zone (WZ) microstructure on the AISI 316L sides.

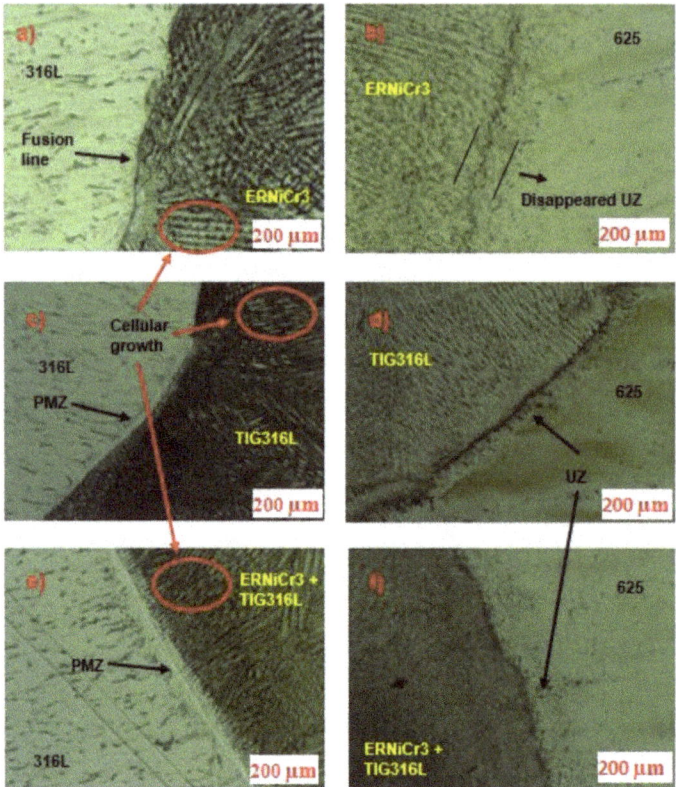

Figure 4. Microstructures of bimetallic joints of Inconel 625 and AISI 316L obtained by CCGTA welding, employing ERNiCr3 (**a**,**b**), TIG 316L (**c**,**d**), Twisted Fillers (**e**,**f**).

3.2.2. Microstructure of the PCGTA Welding

Interfacial microstructures of the PCGTA welded samples are shown in Figure 5. Secondary phases of UZ were also obtained in PCGTA welding. Particularly on the Inconel 625 side of 316L welding with twisted fillers (Figure 5d,f), UZ is clearer. On the AISI 316L side, equi-axed and columnar dendrites were observed in WZ. Twisted fillers presented PMZ on the AISI 316L side, and UZ disappeared on the Inconel 625 side. However, minimum grain coarsening was observed in WZ, which indicates improved metallurgical properties.

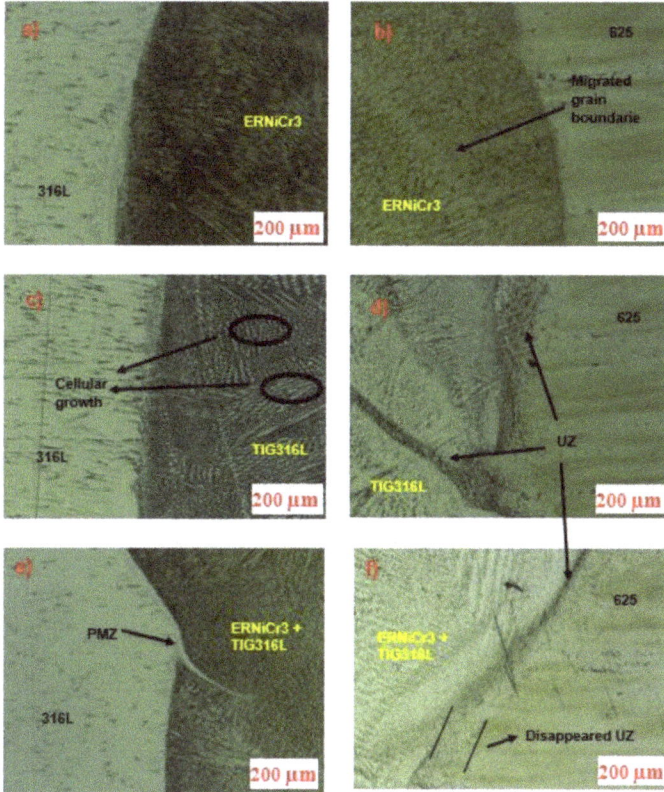

Figure 5. Microstructures of bimetallic joints of Inconel 625 and AISI 316L obtained by PCGTA welding, employing ERNiCr3 (**a**,**b**), TIG 316L (**c**,**d**), Twisted Fillers (**e**,**f**).

3.2.3. Line Mapping Analysis of the CCGTA Welding

Line mapping analysis of the CCGTA welding process is represented in Figure 6. Major elemental transfers has been detected as Ni, Fe and Cr. As can be seen in Figure 6a–c, Fe moved from AISI 316L to WZ or vice versa, and Ni moved from WZ to AISI 316L, in all filler types. Nb and Mo transfers occurred from Inconel 625 to WZ relatedly. Cr migration is meager in all experiments.

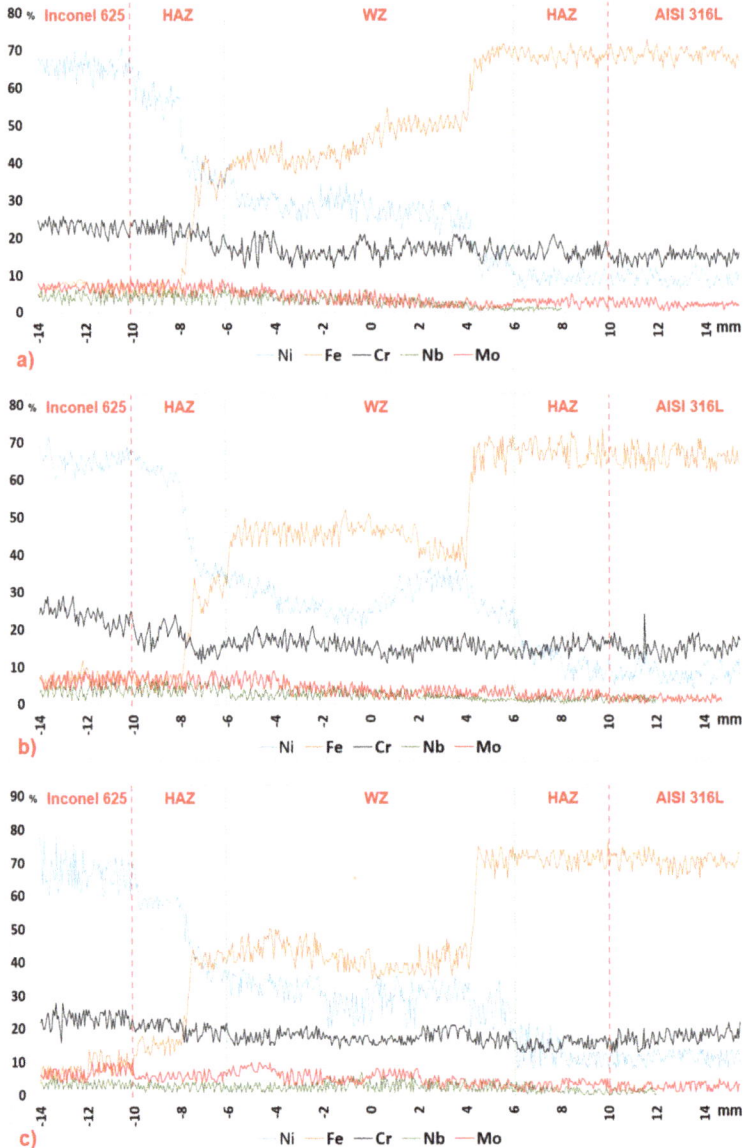

Figure 6. Line mapping analysis of bimetallic joints of Inconel 625 and AISI 316L obtained by CCGTA welding, employing ERNiCr3 (**a**), TIG 316L (**b**), Twisted Filler (**c**).

3.2.4. Line Mapping Analysis of the PCGTA Welding

Line mapping analysis of the PCGTA welding process is represented in Figure 7. In all cases, the element Fe migrated from AISI 316L to WZ and vice versa, and the element Ni migrated from WZ to AISI 316L, as shown in Figure 7d,e,f. A meager amount of Cr migration from both Inconel 625 to WZ and from AISI 316L to WZ was observed for all fillers. Maximum amount of Nb migration was observed from Inconel 625 to WZ employing TIG 316L.

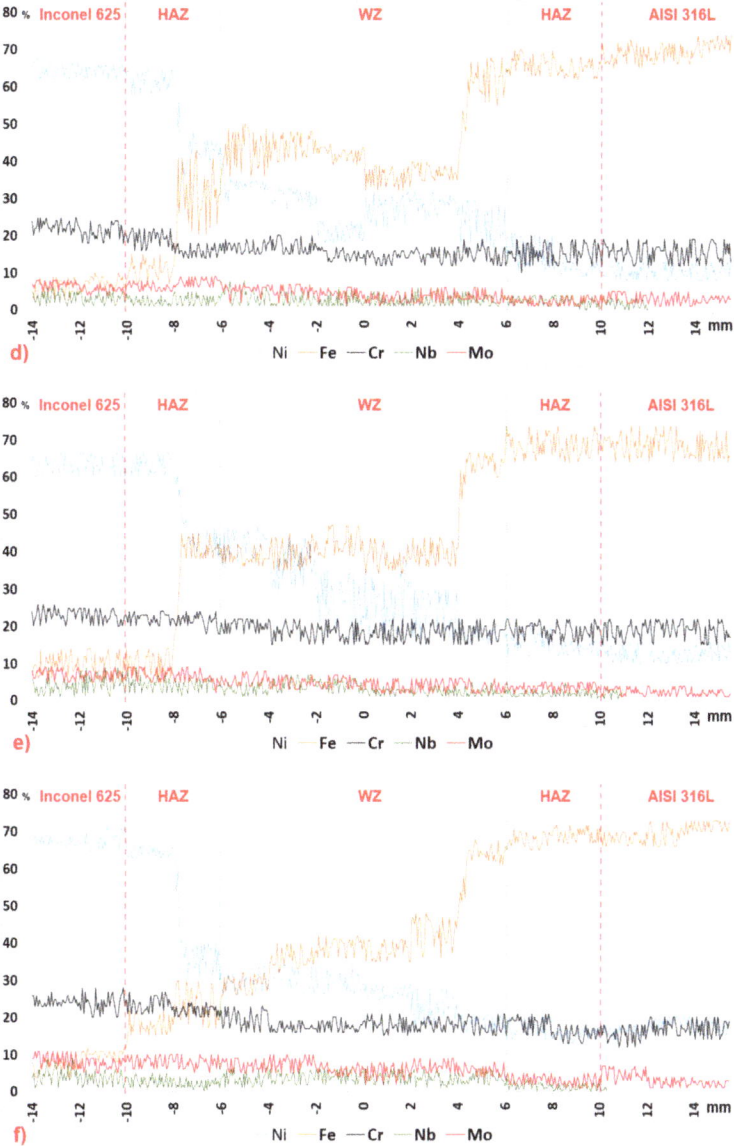

Figure 7. Line mapping analysis of bimetallic joints of Inconel 625 and AISI 316L obtained by PCGTA welding, employing ERNiCr3 (**d**), TIG 316L (**e**), Twisted Filler (**f**).

3.2.5. SEM/EDAX Analysis of CCGTA Welding

The presence of tiny secondary phases, which are represented in white in Figure 8, was observed in the SEM/EDAX spot analysis of the weld zones of CCGTA welding. Analysis showed that these secondary phases in the fusion zone consisted of an enriched amount of Nb, Cr, Mo, Ni and Nb. The pitting points, Spot 4 and Spot 6, contain a restricted element, whereas other shiny white spots contain a great number of elements. Secondary phases were found to be continuous in all weldments.

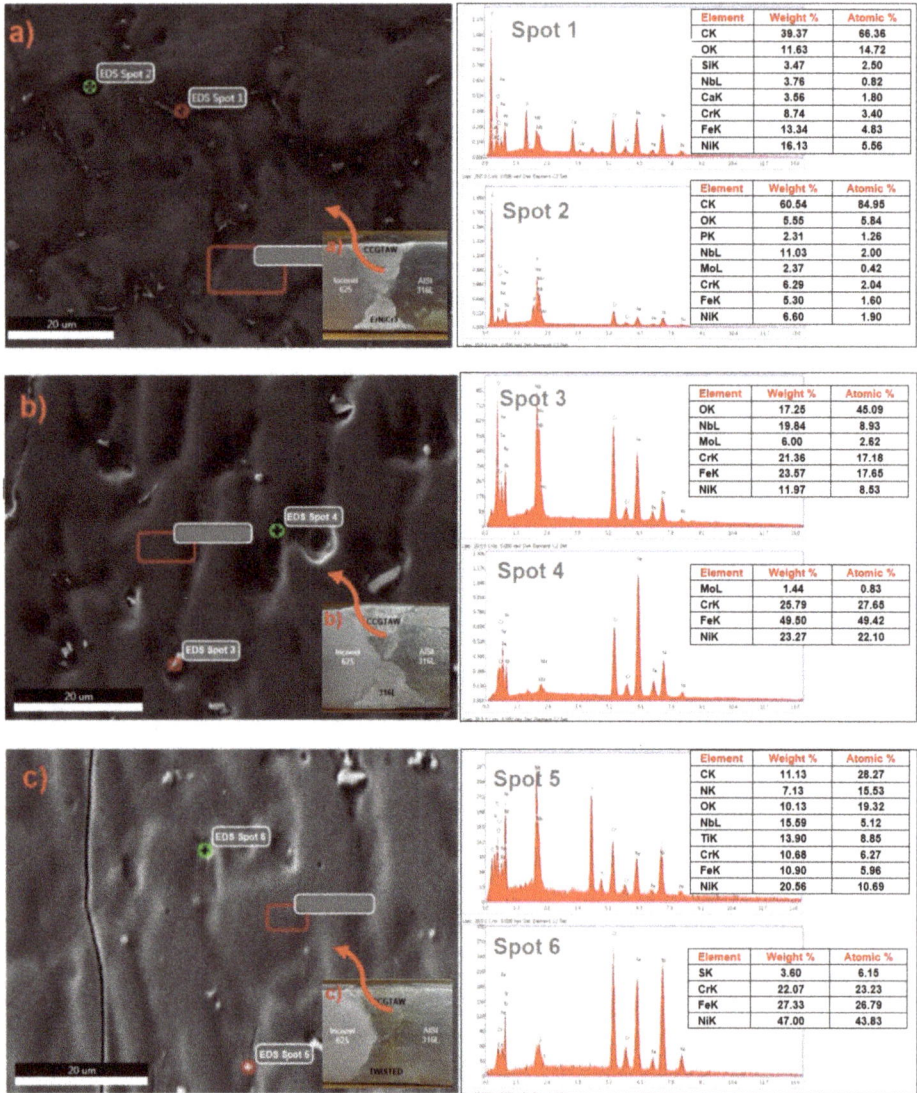

Figure 8. SEM/EDAX point analysis of the dissimilar CCGTA welding of Inconel 625 and AISI 316L, employing ERNiCr-3 (**a**), TIG 316L (**b**), Twisted Filler (**c**).

3.2.6. SEM/EDAX Analysis of PCGTA Welding

SEM/EDAX point analysis results of PCGTA welding zones are given in Figure 9. Secondary phases contain an enriched amount of Al, Nb, Ni, Cr and Ti, and they are found to be continuous. In ERNiCr3 and TIG 316L filler results, only a specific element was determined. The twisted filler created more element movement at WZ.

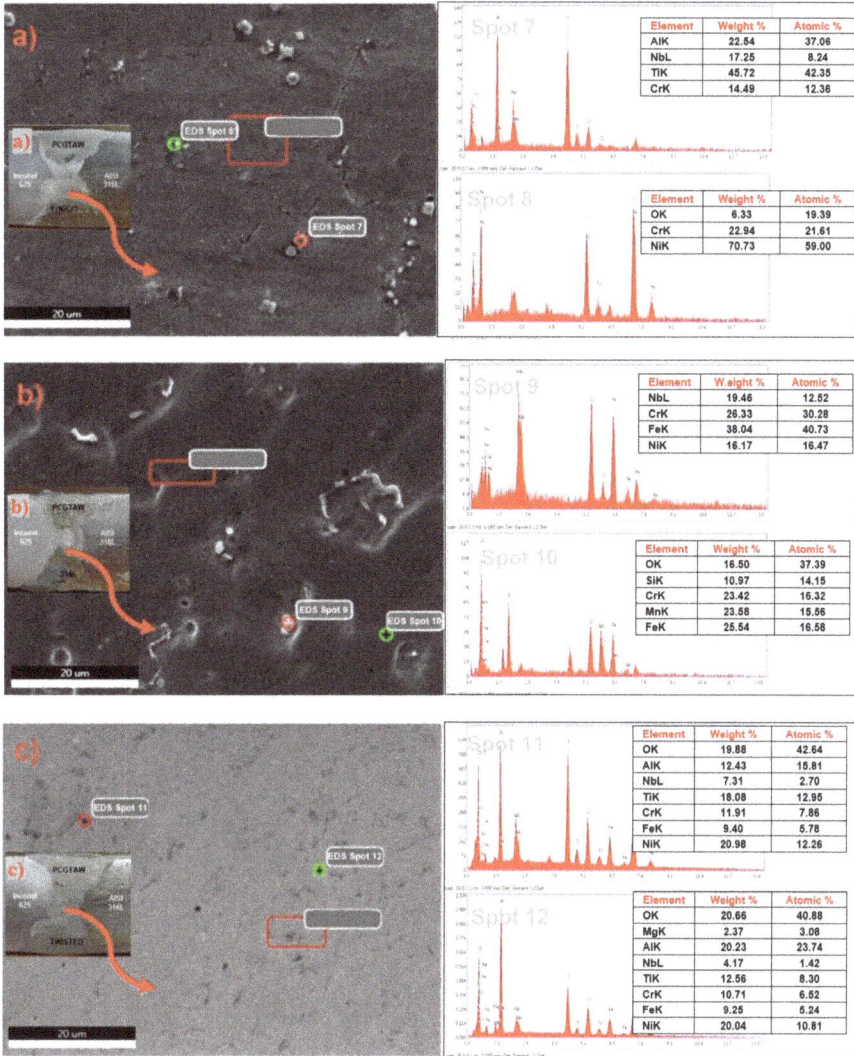

Figure 9. SEM/EDAX point analysis of the dissimilar PCGTA welding of Inconel 625 and AISI 316L, employing ERNiCr-3 (**a**), TIG 316L (**b**), Twisted Filler (**c**).

3.3. Mechanical Investigation

3.3.1. Hardness Tests

The micro-hardness measurements were made by using Vickers hardness tester. Hardness studies were carried out across the entire width of the weldments vis-à-vis cap, filler and root passes. Results are given in Figure 10. The average weld hardness for CCGTA welding was found minimum in the root pass 265.16 HV, 245.10 HV, 250.85 HV employing ERNiCr-3, TIG 316L and twisted fillers respectively. Peak hardness value was measured (385.30 HV) at the cap pass employing ERNiCr-3 filler metal. The average weld hardness for whole weldment was found to be its maximum (281.72 HV) employing ERNiCr-3 filler metal.

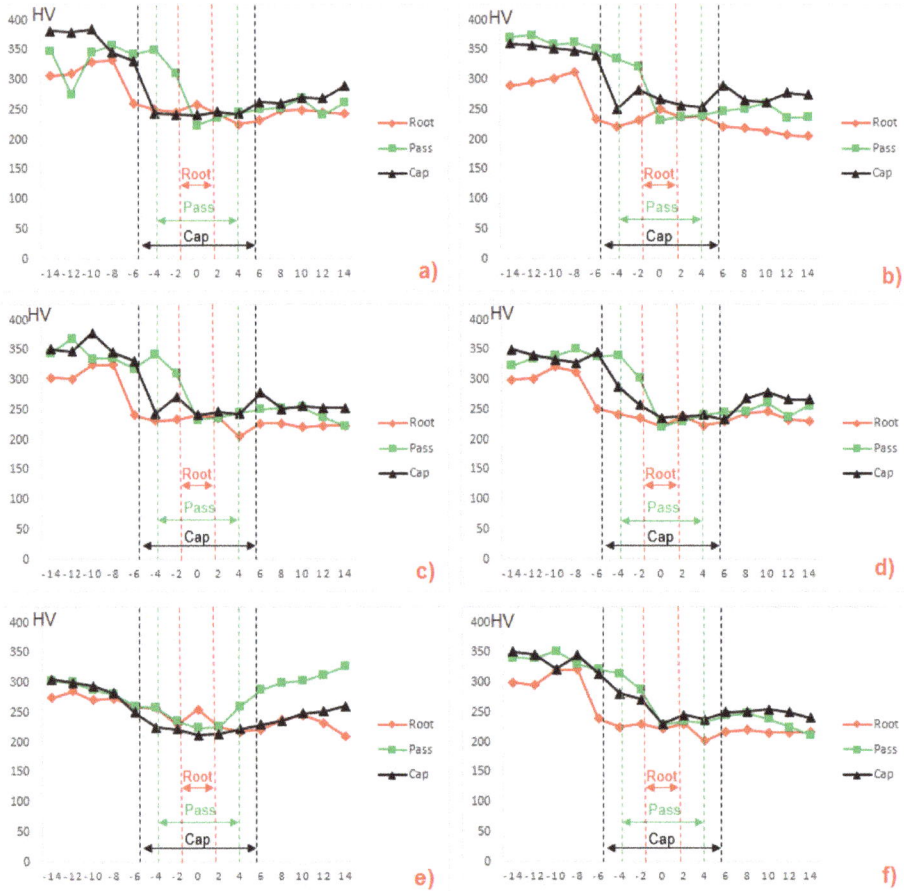

Figure 10. Micro hardness results for CCGTA welding employing ERNiCr-3 (**a**), TIG 316L (**b**), Twisted Filler (**c**), and PCGTA welding employing ERNiCr-3 (**d**), TIG 316L (**e**), Twisted Filler (**f**).

The average weld hardness for PCGTA welding was found minimum in the root pass similar with CCGTAW, with values 255.34 HV, 246.33 HV, 244.78 HV employing ERNiCr-3, TIG 316L and twisted fillers respectively. Peak hardness value was measured (389.63 HV) at the filler pass employing ERNiCr-3 filler metal. The average weld hardness for whole weldment was found maximum (275.08 HV) employing ERNiCr-3 filler metal.

3.3.2. Tensile Test

Tensile test was carried out on the dissimilar weldments obtained from the CCGTA and PCGTA welding techniques employing ERNiCr-3, TIG 316L and twisted fillers, and are shown in Figure 11. Results of CCGTA weldments clearly depicted that the fractures occurred at WZ employing ERNiCr-3 and TIG 316L, whereas fracture occurred at parent metal of AISI 316L side employing twisted fillers. Results of PCGTA weldments have shown that the fracture occurred at WZ employing ERNiCr-3, whereas fractures occurred at parent metal of AISI 316L side employing TIG 316L and twisted fillers.

CCGTAW

PCGTAW

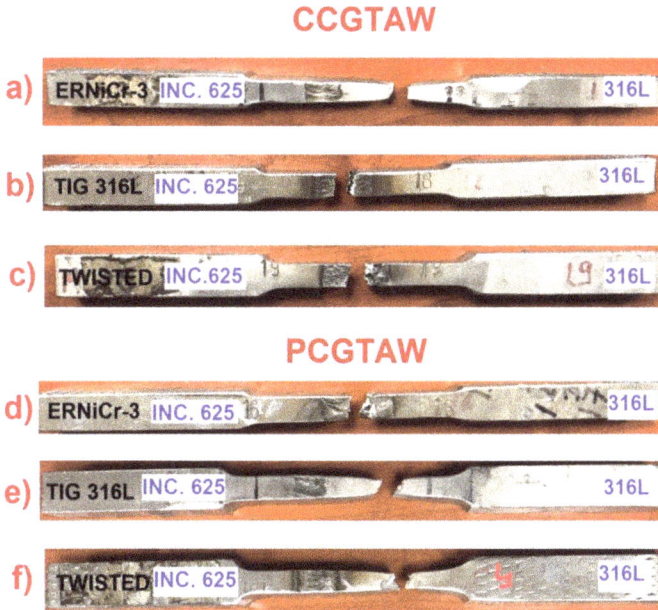

Figure 11. Tensile test results CCGTAW employing ERNiCr-3 (**a**), TIG 316L (**b**), Twisted Filler (**c**), and PCGTA welding employing ERNiCr-3 (**d**), TIG 316L (**e**), Twisted Filler (**f**).

The average tensile test properties of the CCGTA and PCGTA weldments are represented in Table 3. As it seen from the results, ductility measured in terms of percentage elongation at the break load 34%, 32% and 36% (CCGTAW emp. ERNiCr-3, PCGTAW emp. TIG 316L, PCGTAW emp. twisted, respectively) which fractures occurred at parent metal of AISI 316L.

Table 3. Tensile Test Results of CCGTAW and PCGTAW

Weldment	Yield Strength (MPa)	Tensile Strength (MPa)	Elongation (%)	Fracture Zone
CCGTAW emp. ERNiCr-3	442	674	34	AISI 316L
CCGTAW emp. TIG 316L	389	529	13	WZ
CCGTAW emp. Twisted	437	567	19	WZ
PCGTAW emp. ERNiCr-3	401	532	17	WZ
PCGTAW emp. TIG 316L	446	661	32	AISI 316L
PCGTAW emp. Twisted	451	687	36	AISI 316L

4. Discussion

Non-Destructive Testing (NDT) analysis and macro-photograph observations show that all weldments were free from gap, lack of penetration, spatter, defects, etc. This indicated that all techniques and filler metals were proper for the dissimilar welding of Inconel 625 and AISI 316L.

The interfacial and weld zone microstructures of CCGTA and PCGTA welding are given in Figures 4 and 5, respectively. Employing TIG 316L and twisted fillers, clear PMZs were observed in the HAZ of CCGTA welding (Figure 4c,e). PMZ generally formed right next to the exterior of the fusion zone, where the material is over-heated by the eutectic temperature [31]. PMZ could cause cracking, and this is an undesirable result. Figure 8c depicts an obvious crack. It is clear that widest PMZ's are concretized cracking. Figures 4 and 5 show CCGTA and PCGTA welding microstructures; widest PMZ's were determined in CCGTA welding employing twisted fillers (Figure 4e,f). Employing twisted filler by using CCGTA welding might cause unknown chemical composition at WZ, which leads to PMZ, hence cracking. However, UZs were obviously determined at the HAZ of Inconel 625 with TIG 316L and twisted fillers (Figure 4d,f). In the dissimilar welding process, UZ formation is an expectation when the melting range of base and filler metals is largely varying. In this connection, the heat input was not able to provide a proper welding pool, leading to the formation of an unmixed zone [8,32]. Researchers observed that the UZ area is greater on the nickel-based Inconel 625 side compared with that on the AISI 316L side. Dev et al. [8] exhibited similar results in their study, and they attributed them to the melting-solidification rate of the nickel element. UZs were also confirmed in the HAZ on the PCGTA welding Inconel 625 side, as shown in Figure 5d,e. It is confirmed that the current type could not prevent the formation of UZ. Multi-pass welding might cause a multi directional grain growth, however, PCGTA welding could control grain growth at WZ [23]. In PCGTA welding, peak and back currents provide a fluctuation of temperature in the weld pool, which effects the grain configuration during solidification. It can be seen in Figure 5 that PCGTA welding controls the grain growth, excepting the TIG 316L welded specimen. Naffakh et al. [22] subjected dissimilar weldments of Inconel 657 and AISI 310 to heat treatment after welding. UZs disappeared as a consequence of the heat treatment. CCGTA welding with ERNiCr-3, shown in Figure 4a, and PCGTA welding with the twisted filler, shown in Figure 5f, indicate that UZs also disappeared without any heat treatment. Researchers tend to agree that a homogeneous cooling rate might occur in these regions. In PCGTA welding with ERNiCr-3, shown in Figure 5b, Migrated Grain Boundaries (MGBs) were clearly observed at WZ. According to DuPont et al. [15], an Ni-rich filler ensures a fully austenitic phase in WZ in addition to multi-pass welding, which would likely cause MGBs.

Line mapping analysis demonstrated a migration of elements on both sides. In CCGTA welding, Fe movement from AISI 316L to WZ or vice versa, and Ni movement from WZ to AISI 316L, was observed. Due to similar chemical compositions (Table 1), Ni and Cr movements from Inconel 625 to WZ were minimal, as shown in Figure 6. A meager amount of Cr migration from AISI 316L to WZ for all fillers ERNiCr3 and TIG 316L fillers has been observed. Due to the difference in the chemical composition of the metals, Ni migrated from Inconel 625 to WZ (Figure 6d). In PCGTA welding, a miserable Cr movement was observed with ERNiCr-3 and twisted fillers, as shown in Figure 7d,f. Employing twisted fillers reveals the dramatically edged migrations of Fe and Ni elements, which tend to mix the two metals.

SEM/EDAX analysis indicated that the formation of Nb-, Ti-, Cr-, Mo-, Mn- and Ni-rich phases occurred in the WZs. Similar precipitations were observed at the HAZ of Inconel 625 and HAZ of AISI 316L for all specimens. It is well-known that deleterious secondary phases may reduce the corrosion resistance of weldment [24]. The formation of Nb-rich secondary phases was observed on the GTA welds with ERNiCr-3 because of its Nb content. As expected, the formation of Nb-rich phases in CCGTA welding with an ERNiCr-3 filler can clearly be seen in Figure 8a. However, secondary phase formation was slightly reduced with PCGTA welding, as shown in Figure 9a. Kumar et al. [25] also found similar microstructural results in their study. Nevertheless, Nb, Ti, Ni and Cr precipitates were located, which also reduces the corrosion resistance [10]. The TIG 316L filler material does not contain Nb in the chemical composition, although an Nb-rich formation was confirmed. Nb migrated from Inconel 625 to WZ. This caused the formation of a Laves phase, as can be seen in Spot 3. In Figure 8c, an enriched amount of Nb, Ti, Cr and Ni precipitates occurred in CCGTA welding with twisted fillers. Ti and Ni elements at a high temperature using the welding process are prone to form TiN [33]. The TiN phase was inert when the Laves phases occurred. Nb-, Cr-, Ni- and Mn-rich precipitation is illustrated in Figure 9b. The analysis of PCGTA welding with twisted fillers is demonstrated in Figure 9c. According to the figure, the absence of white secondary phases was observed. However, differently viewed, infrequent secondary phases represented by black dot shapes also occurred. These phases contain a rich amount of Al, Ti, Cr and Ni. This is the lowest density formation compared with the other experiments. The main welding parameter was current type. In the experiments, pulse current leads to control solidification phases. Therefore, PCGTA welding ensures the minimization of secondary phase formation. These results support the study of Kumar et al. [25]. In sum, the OM results also support the SEM results that PCGTA welding with twisted fillers has the best metallurgical properties.

When welded metals cool slowly, they cause the extension of micro-segregation. To minimize micro-segregation heat input should be reduced. In other words, minimum heat input would increase cooling rate, and this helps to reduce segregation of Nb and Laves phase [26]. CCGTA and PCGTA welding processes have nearly the same heat input with stationary welding parameters as base metal filler, current etc. Researchers observed Laves Phases PCGTA welded specimens. Authors suppose that high current range could cause high heat input and so form Laves phases.

It is evident from the hardness tests that maximum WZ hardness magnitudes were detected at ERNiCr-3 welded specimens for both CCGTA and PCGTA welding (Figure 10). This could be formation of secondary phases. PCGTA welding hardness results are lower that CCGTA welding results. It is clear from the micro hardness results that PCGTA welding reduces the formation of secondary phases, which improves the properties of weldments. Fracture occurred at WZ in three tests. The obvious crack, which can be seen at SEM analysis, was cause to reduced weld tensile strength. The researchers observed micro-cracks in the WZ which was due to the segregation of Nb. These micro cracks were caused due to fractures at WZ during tensile test of PCGTA welding employing ERNiCr-3. This result is conflict with the study of Kumar et al. [25], which could be peak current differences between the studies. Authors commented that peak-base current for ERNiCr-3 was selected high. Tensile test results show that sample PCGTA welding employing twisted fillers exhibits the highest tensile strength and elongation.

5. Conclusions

This study investigates the microstructural properties of the dissimilar metal welding of a nickel-based super alloy, Inconel 625, and an austenitic stainless steel, AISI 316L, obtained by CCGTA and PCGTA welding with ERNiCr-3, TIG 316L and twisted (ERNiCr-3 and TIG 316L) fillers. The results of the study can be summarized as follows:

- Successful weldments obtained for all specimens in terms of macrostructure, without any lack of penetration, crack, spatter, etc.
- Microstructural investigations illustrated that in CCGTA welding with TIG 316L and twisted fillers, PMZs were obtained, which may cause cracking. TIG 316L filler metal weldments showed grain coarsening and twisted fillers could produce a laxer chemical composition with CCGTA welding process. However, CCGTA welding with ERNiCr-3 demonstrated almost regular boundaries on both sides. PCGTA welding with twisted fillers showed PMZ on the AISI 316L side and caused UZ to disappear on the Inconel 625 side. PCGTA welding reduced PMZ and UZ employing twisted fillers and minor grain coarsening.
- WZ microstructural analysis exhibited the existence of multi-directional grain growth in all specimens on the AISI 316L side. Grain growth was less on the Inconel 625 side and almost disappeared on the Inconel 625 side in PCGTA welding with twisted fillers.
- It was observed that the formation of Ni-, Ti-, Nb-, Mo-, Mn- and Cr-rich secondary precipitates appeared in all specimens. However, in PCGTA welding with twisted fillers, a meager amount of precipitates occurred. Moreover, these are particular precipitates that are represented by black dots in images, whereas others are tiny white secondary phases.
- Hardness and tensile test results show that PCGTA welding improve mechanical properties. Twisted fillers exhibit the best mechanical properties for bimetallic joint for Inconel 625 and AISI 316L.
- According to the study results, for the dissimilar metal welding of Inconel 625 and AISI 316L, the best process is PCGTA with twisted fillers (ERNiCr-3 and TIG 316L).

The results of the study will be very useful for the Original Equipment Manufacturers (OEM) in producing equipment with bimetallic joints.

Author Contributions: Conceptualization, F.D. and M.K.K.; methodology, F.D., M.K.K. and U.E.; formal analysis, F.D.; investigation, F.D. and M.K.K.; resources, F.D.; data curation, F.D., M.K.K. and U.E.; writing-original draft preparation, F.D.; writing-review and editing, F.D.; visualization, M.K.K. and U.E.; supervision, M.K.K.; project administration, M.K.K. and U.E.; funding acquisition, F.D.

Funding: This work was supported by Sisecam Chemicals Group Soda Sanayii A. S. Kromsan Chromium Chemicals Plant.

Acknowledgments: The authors would like to thank Sisecam Chemicals Group Soda Sanayii A. S. in Mersin, Turkey for their support of this study. Special appreciation is also expressed to Kromsan Chromium Chemical Plant mechanical maintenance personnel who helped to prepare experiments.

Conflicts of Interest: The authors declare no conflicts of interest.

References

1. Moosavy, H.N.; Aboutalebi, M.R.; Seyedein, S.H.; Mapelli, C. A Solidification Model for Prediction of Castability in the Precipitation-strengthened Nickel-based Superalloy. *J. Mater. Process. Technol.* **2013**, *213*, 1875–1884. [CrossRef]
2. Korrapati, P.K.; Avasarala, V.K.; Bhushan, M.; Ramkumar, K.D.; Arivazhagan, N. Narayanan, S. Assessment of Mechanical Properties of PCGTA Weldments of Inconel 625. *Procedia Eng.* **2014**, *75*, 9–13. [CrossRef]
3. Shankar, V.; Sankara, K.B.; Mannan, S.L. Microstructural and Mechanical Properties of Inconel 625 Superalloy. *J. Nucl. Mater.* **2001**, *288*, 222–232. [CrossRef]
4. Ramkumar, K.D.; Mithilesh, P.; Varun, D.; Reddy, A.R.G.; Arivazhagan, N.; Narayanan, S.; Kumar, K.G. Characterization of Microstructure and Mechanical Properties of Inconel 625 and AISI 304 Dissimilar Weldments. *J. Int. Ste. Inst. Jap.* **2014**, *54*, 900–908. [CrossRef]
5. American Welding Society. Available online: https://app.aws.org/wj/2002/06/feature2/ (accessed on 23 September 2018).
6. Dreshfield, R.L.; Moore, T.J.; Bartolotta, P.A. Post-Test Examination of a Pool Boiler Receiver. DOE/NASA/33408-6; U.S Department of Energy Conservation and Renewable Energy Office of Solar Heat Technologies: Washington, WA, USA, 1992; pp. 1–4.
7. Mitilesh, P.; Varun, D.; Reddy, A.R.G.; Ramkumar, K.D.; Arivazhagan, N.; Narayanan, S. Investigations on Dissimilar Weldments of Inconel 625 and AISI 304. *Procedia Eng.* **2014**, *75*, 66–70. [CrossRef]
8. Dev, S.; Ramkumar, K.D.; Arivazhagan, N.; Rajendran, R. Investigations on the Microstructure and Mechanical Properties of Dissimilar Welds of Inconel 718 and Sulphur Rich Martensitic Stainless Steel AISI 416. *J. Manuf. Process.* **2018**, *32*, 685–689. [CrossRef]
9. Mortezaie, A.; Shamanian, M. An Assessment of Microstructure, Mechanical Properties and Corrosion Resistance of Dissimilar Welds between Inconel 718 and 310S Austenitic Stainless Steel. *Int. J. Press. Vessel. Pip.* **2014**, *116*, 37–46. [CrossRef]
10. Ramkumar, T.; Selvakumar, M.; Narayanasamy, P.; Begam, A.A.; Mathavan, P.; Raj, A.A. Studies on the Structural Property, Mechanical Relationships and Corrosion Behaviour of Inconel 718 and SS 316L Dissimilar Joints by TIG welding without Using Activated Flux. *J. Manuf. Process.* **2017**, *30*, 290–298. [CrossRef]
11. Ahmad, H.W.; Hwang, J.H.; Lee, J.H.; Bae, D.H. An Assessment of the Mechanical Properties and Microstructural Analysis of Dissimilar Material Welded Joint between Alloy 617 and 12Cr Steel. *Metals* **2016**, *6*, 242. [CrossRef]
12. Kim, J.S.; Park, Y.I.; Lee, H.W. Effects of Heat Input on the Pitting Resistance of Inconel 625 Welds by Overlay Welding. *Met. Mater. Int.* **2015**, *21*, 350–355. [CrossRef]
13. Knapp, S. Mechanical Properties of an Inconel Dissimilar Metal Weld. Master's Thesis, University of Ottowa, Ottawa, ON, Canada, 2013.
14. Kulekci, M.K.; Esme, U.; Kahraman, F.; Ocalir, S. An examination of the mechanical properties of dissimilar steel welds. *Mater. Test.* **2016**, *58*, 362–370. [CrossRef]
15. Badiger, R.I.; Narendranath, S.; Srinath, M.S. Microstructure and Mechanical Properties of Inconel-625 Welded Joint Developed Through Microwave Hybrid Heating. *J. Eng. Manuf.* **2017**. [CrossRef]
16. DuPont, J.N.; Lippold, J.C.; Kiser, S.D. *Welding Metallugy and Weldability of Nickel-Base Alloys*; John Wiley & Sons, Inc.: Hoboken, NJ, USA, 2009; pp. 82–100.
17. Manikandan, S.G.K.; Sivakumar, D.; Rao, K.P.; Kamaraj, M. Microstructural Characterization of Liquid Nitrogen Cooled Alloy 718 fusion zone. *J. Mater. Process. Technol.* **2014**, *214*, 3141–3149. [CrossRef]

18. Ramkumar, K.D.; Kumar, P.S.G.; Krishna, V.R.; Chandrasekhar, A.; Dev, S.; Abraham, W.S.; Prabhakaran, S.; Kalainathan, S.; Sridhar, R. Influence of Laser Peening on the Tensile Strength and Impact Toughness of Dissimilar Welds of Inconel 625 and UNSS32205. *Mater. Sci. Eng. A* **2016**, *676*, 88–99. [CrossRef]
19. Olden, V.; Kvaale, P.E.; Simensen, P.A.; Aaldstedt, S.; Solberg, J.K. The Effect of PWHT on the Material Properties and Microstructure in Inconel 625 and Inconel 725 Buttered Joints. In Proceedings of the 22nd International Conference on Offshore Mechanics and Artic Engineering, Cancun, Mexico, 8–13 June 2003.
20. Soares, B.A.; Schvartzman, M.M.A.M.; Campos, W.R.C. Characterization of Dissimilar Metal Welding—Austenitic Stainless Steel with Filler Metal of the Nickel Alloy. In Proceedings of the International Nuclear Atlantic Conference, Sao Paolo, Brazil, 30 September–5 October 2007.
21. Prabu, S.S.; Ramkumar, K.D.; Arivazhagan, N. Microstructural Evolution and Precipitation Behavior in Heat Affected Zone of Inconel 625 and AISI 904L Dissimilar Welds. *Mater. Sci. Eng.* **2017**, *263*, 062073. [CrossRef]
22. Naffakh, N.; Shamanian, M.; Ashrafizadeh, F. Microstructural evolutions in dissimilar welds between AISI 310 austenitic stainless steel and Inconel 657. *J. Mater. Sci.* **2010**, *45*, 2564–2573. [CrossRef]
23. Ramkumar, K.D.; Patel, S.D.; Praveen, S.S.; Choudhury, D.J.; Prabaharan, P.; Arivazhagan, N.; Xavior, M.A. Influence of Filler Metals and Welding Techniques the Structure—Property Relationships of Inconel 718 and AISI 316L Dissimilar Weldments. *Mater. Des.* **2014**, *62*, 175–188. [CrossRef]
24. Hejripour, F.; Aidun, D.K. Consumable Selection for Arc Welding between Stainless Steel 410 and Inconel 718. *J. Mater. Process. Technol.* **2017**, *245*, 287–299. [CrossRef]
25. Kumar, K.G.; Ramkumar, K.D.; Arivazhagan, N. Characterization of Metallurgical and Mechanical Properties of the multi-pass welding of Inconel 625 and AISI 316L. *J. Mech. Sci. Tech.* **2015**, *29*, 1039–1047. [CrossRef]
26. Kourdani, A.; Haghighi, R.D. Evaluating the Properties of Dissimilar Metal Welding Between Inconel 625 and 316L Stainless Steel by Applying Different Welding Methods and Consumables. *Metall. Mater. Trans. A* **2018**, *49A*, 1231–1243. [CrossRef]
27. Xie, M.X.; Shang, X.T.; Zhang, L.J.; Bai, Q.L.; Xu, T.T. Interface Characteristic of Explosive-Welded and Hot-Rolled TA1/X65 Bimetallic Plate. *Metals* **2018**, *8*, 1–14. [CrossRef]
28. Topolski, T.; Szulc, Z.; Garbacz, H. Microstructure and Properties of the Ti6Al4V/Inconel 625 Bimetal Obtained by Explosive Joining. *J. Mat. Eng. Perf* **2016**. [CrossRef]
29. Mohommed, G.H.; Ishak, M.; Ahmad, S.N.A.S.; Abdulhadi, H.A. Fiber Laser Welding of Dissimilar 2205/304 Stainless Steel Plates. *Metals* **2017**, *7*, 1–19. [CrossRef]
30. Shakil, M.; Ahmad, M.; Tariq, N.H.; Hasan, B.A.; Akhter, J.I.; Ahmed, E.; Mehmood, M.; Choudhry, M.A.; Iqbal, M. Microstructure and hardness studies of electron beam welded Inconel 625 and stainless steel 304L. *Vacuum* **2014**, *110*, 121–126. [CrossRef]
31. Kou, S. *Welding Metallurgy*, 2nd ed.; John Wiley & Sons, Inc.: New York, NY, USA, 2003; p. 151. ISBN 978-0-471-43491-7.
32. Ramkumar, K.D.; Abraham, W.S.; Viyash, V.; Arivazhagan, N.; Rabel, A.M. Investigations on the Microstructure, Tensile Strength and High Temperature Corrosion Behaviour of Inconel 625 and Inconel 718 Dissimilar Joints. *J. Manuf. Process.* **2017**, *25*, 306–322. [CrossRef]
33. Ramkumar, K.D.; Sai, R.J.; Reddy, V.S.; Gundla, S.; Mohan, T.H.; Saxena, V.; Arivazhagan, N. Effect of Filler Wires and Direct Ageing on the Microstructure and Mechanical Properties in the Multi-Pass Welding of Inconel 718. *J. Manuf. Process.* **2015**, *18*, 23–45. [CrossRef]

Article

Ageing Effects on Room-Temperature Tensile Properties and Fracture Behavior of Quenched and Tempered T92/TP316H Dissimilar Welded Joints with Ni-Based Weld Metal

Lucia Čiripová, Ladislav Falat *, Peter Ševc, Marek Vojtko and Miroslav Džupon

Institute of Materials Research, Slovak Academy of Sciences, Watsonova 47, 04001 Košice, Slovakia; lciripova@saske.sk (L.Č.); psevc@saske.sk (P.Š.); mvojtko@saske.sk (M.V.); mdzupon@saske.sk (M.D.)
* Correspondence: lfalat@saske.sk; Tel.: +421-55-792-2447

Received: 13 September 2018; Accepted: 1 October 2018; Published: 3 October 2018

Abstract: The present work is focused on the investigation of isothermal ageing effects on room-temperature tensile properties and the failure of quenched and tempered martensitic/austenitic weldments between T92 and TP316H heat-resistant steels. The dissimilar weldments were produced by gas tungsten arc welding technique using a Ni-based Thermanit Nicro 82 filler metal. The welded joints were subjected to unconventional post-welding heat treatment consisting of the welds solutionizing (1060 °C/30 min), followed by their water quenching and final stabilization tempering (760 °C/60 min). The treatment was completed by spontaneous air cooling within a tempering furnace. The welds in their initial quenched and tempered condition were subsequently aged at 620 °C for up to 2500 h. Apart from room-temperature tensile tests performed for all the welds material states, additional cross-weld hardness measurements were carried out on longitudinal sections of broken tensile specimens. The applied thermal exposure resulted in recognizable deterioration of plastic properties, whereas their effects on strength properties were rather small. The welds tensile straining and fracture evolution exhibited competitive behavior between the austenitic TP316H region and Ni-based weld metal. The observed failure locations showed significant hardness peaks due to intensive, necking-related strain hardening effects occurred during the tensile tests.

Keywords: dissimilar weld; ageing; tensile properties; hardness; failure mode

---

1. Introduction

Dissimilar welded joints between tempered martensitic and austenitic heat-resistant steels are typically used in high temperature steam circuits of ultra-supercritical (USC) power-plant boilers; e.g., for inter-linking header equipment with superheater tubes. Previous findings of several published studies (e.g., [1–4]) indicated the use of Ni-based weld metal (Ni WM) to be suitable means to suppress undesirable carbon diffusion across dissimilar metal weld interfaces and thus to retard local creation of carbon-depleted and carbon-enriched microstructural regions. It has been generally accepted that fusion welded joints of martensitic creep-resistant steels require the application of post-welding heat treatment (PWHT) in order to reduce thermally-induced and transformation-induced residual welding stresses and also to produce thermally more stable tempered martensitic microstructures with sufficient toughness [5–9]. In contrast, the welded joints of austenitic heat-resistant steels do not generally require any specific PWHT. In the case of Ni-based transition weldments between tempered martensitic and austenitic heat-resistant steels, an application of PWHT is necessary to be performed according to specifications applicable for martensitic steels welded joints [10–12]. Classical conception for PWHT processing of martensitic/austenitic weldments is based on their subcritical tempering;

i.e., the tempering below Ac1 critical transformation temperature of ferrite-to-austenite phase transformation of tempered martensitic steel base material. Conventionally established procedure for fabrication of Ni-based transition weldments between ferritic (incl. tempered martensitic) and austenitic boiler steels' pipes of greater diameters consists of an initial deposition of Ni-base alloy buttering layer onto a prepared welding edge of ferritic steel pipe [13–16]. The segment of ferritic steel base material with prepared Ni-based buttering layer is commonly subjected to conventional PWHT; i.e., subcritical tempering below the steel Ac1 critical transformation temperature. After performing this PWHT, final multi-pass welding of Ni-buttered ferritic steel segment with its austenitic steel counter-part is performed using Ni-based electrode to complete the dissimilar welded joint without the need for any further heat treatment [13–16]. However, in the case of dissimilar welds between ferritic and austenitic boiler tubes with smaller diameters, direct welding with Ni-based electrode is performed without Ni-alloy pre-buttering and then the whole weldment is subjected to PWHT [10,11,17–20].

Our previous works [18–20] were focused on the comparison between the effects of conventional "tempering PWHT" ("T PWHT") and unconventional "quenching-and-tempering PWHT" ("QT PWHT") of T92/TP316H martensitic/austenitic weldments on their high-temperature creep resistance, room-temperature hardness, impact toughness, and notch-tensile properties of thermally exposed T92 steel heat-affected zone (T92 HAZ) in relation to its long-term microstructure evolution. It has been found out that after the "QT PWHT" the microstructures and properties of all materials regions within the whole T92/TP316H weldment became homogenized in comparison to those of the weldment after the "T PWHT". The initial microstructural heterogeneity of T92 HAZ region was completely suppressed after the "QT PWHT" thanks to the weld full re-austenitization. Thus, the creep resistance of the weldments after the "QT PWHT" notably increased as a result of premature "type IV creep failure" elimination [18,19].

In contrast to our previous works on T92/TP316H martensitic/austenitic weldments with Nirod 600 Ni WM [18–20], the present study representing our continuous research work deals with isothermal ageing effects on room-temperature tensile properties and failure behavior of notch-free, quenched, and tempered T92/TP316H martensitic/austenitic weldments with Thermanit Nicro 82 Ni WM. The characterization of the overall tensile deformation and fracture behavior of the weldments without artificial stress concentrators in dependence of applied ageing exposure is presented. The correlation between thermally induced microstructural changes and mechanism of mechanical properties degradation is discussed.

2. Materials and Methods

The experimental Ni-based transition weldments between the tubes (outer diameter 38 mm, wall thickness 5.6 mm) of T92 tempered martensitic and TP316H austenitic heat-resistant steels were prepared in the company SES a. s. Tlmače, Slovakia by gas tungsten arc welding (GTAW) using "Inconel-type" filler metal Thermanit Nicro 82. The preparation of welding edges and sequence of welding passes are schematically shown in Figure 1.

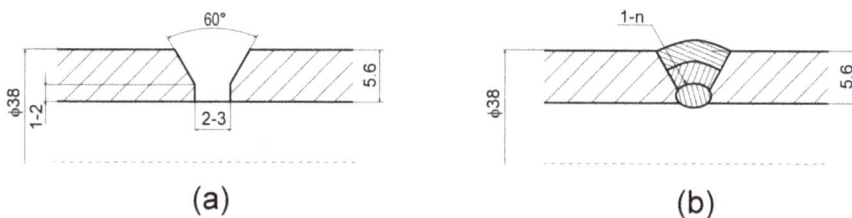

Figure 1. The preparation of welding edges (**a**) and sequence of welding passes (**b**). All dimensions are in mm. The "n" represents the number of welding passes in range from 2 to 3.

The welding conditions included welding current in the range from 70 to 110 A, voltage from 12 to 17 V, and heat input from 9 to 12 kJ/cm. The electrode diameter was 2.4 mm. The chemical composition of dissimilar base materials and Ni-based filler metal used for fabrication of the investigated dissimilar weldments is shown in Table 1.

Table 1. Chemical composition (wt.%) of individual materials used for fabrication of T92/TP316H dissimilar weldments with Ni-based weld metal.

Material	C	N	Si	Mn	Cr	Mo	W	B	Ni	Ti	V	Nb	Fe
T92	0.11	0.05	0.38	0.49	9.08	0.31	1.57	0.002	0.33	-	0.2	0.07	rest
TP316H	0.05	-	0.51	1.77	16.76	2.05	-	-	11.13	-	-	-	rest
Ni WM [#]	0.01	-	0.07	3.21	20.71	0.004	-	-	rest	0.37		2.6	0.31

[#] Thermanit Nicro 82.

The produced tubular segments of T92/TP316H weldments were subjected to unconventional QT PWHT including the welds full re-austenitization at 1060 °C for 30 min, their water quenching and subsequent subcritical tempering at 760 °C for 1 hour, completed by passive air cooling within electric resistance furnace LAC PKE 18/12R (LAC, s.r.o., Rajhrad, Czech Republic). The heating rate for obtaining individual stages of QT PWHT was maximally 150 °C/h. All further experimental work was performed using cca. 60 mm long cross-weld (c-w) samples prepared from the tubular weldments using spark erosion machine EIR-EMO 2N (Emotek s.r.o., Nove Mesto nad Vahom, Slovak Republic). Schematic c-w sampling is shown in Figure 2.

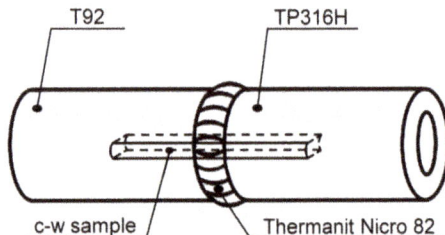

Figure 2. Schematic cross-weld (c-w) sampling.

One series of c-w samples was investigated in its initial QT PWHT state which was considered as the reference state. The second series and the third series of c-w samples were isothermally exposed at 620 °C for 500 and 2500 h, respectively. All the three material states (QT PWHT, QT PWHT + 620 °C/500 h, QT PWHT + 620 °C/2500 h) were subjected to room-temperature c-w tensile testing. Uniaxial room-temperature tensile tests of conventionally machined M6 threaded cylindrical tensile specimens (see Figure 3) were carried out by employing TIRATEST 2300 universal testing machine (TIRA GmbH, Schalkau, Germany) at a cross-head speed of 0.05 mm/min. Three individual c-w tensile test specimens were tested for each material state of the studied weldments. From recorded engineering stress-strain curves the average values of yield stress (YS) and ultimate tensile strength (UTS) were determined. The individual values of YS were graphically estimated as 0.2% proof stress. The average values of total elongation at fracture (EL) and reduction of area at fracture (RA) were obtained from micrometrical measurements performed directly on the broken tensile specimens. The evaluation of c-w tensile properties included the calculation of their standard deviations.

In order to indicate local strain hardening effects in investigated dissimilar weldments during the tensile tests, c-w Vickers hardness measurements were performed on plain surfaces of prepared longitudinal sections of broken tensile specimens after the tensile tests. The hardness measurements were performed using a Vickers 432 SVD hardness tester (Wolpert Wilson Instruments, division of INSTRON DEUTSCHLAND GmbH, Aachen, Germany) with 98 N loading for 10 s per measurement.

The reference material state was represented by the original unstrained c-w specimen (i.e., the specimen that was not subjected to tensile testing) in its initial QT PWHT condition.

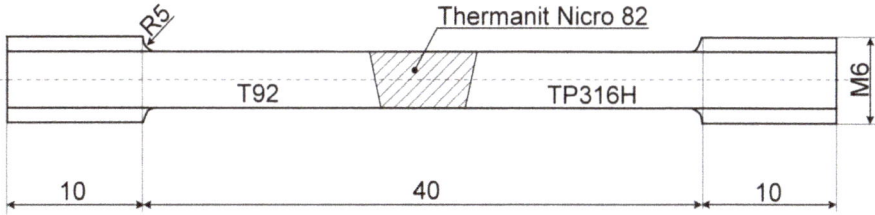

Figure 3. The tensile test specimen for cross-weld tensile testing of dissimilar weldments. All dimensions are in mm.

Microstructure, fracture path, and fracture surface analyses of the selected representative samples corresponding to individual material states were performed using light optical microscope OLYMPUS GX71 (OLYMPUS Europa Holding GmbH, Hamburg, Germany), scanning electron microscope JEOL JSM-7000F (Jeol Ltd., Tokyo, Japan) with an energy-dispersive X-ray spectroscopy (EDX) analyzer INCA X-sight model 7557 (Oxford Instruments, Abingdon, Oxfordshire, UK), and environmental scanning electron microscope EVO MA15 (Carl Zeiss Microscopy GmbH, Jena, Germany), respectively.

3. Results and Discussion

3.1. Microstructure in Initial QT PWHT State

Figure 4 shows general view on microstructural profile comprising of individual material regions of dissimilar T92/TP316H weldment with Thermanit Nicro 82 Ni-based weld metal (Ni WM) in its initial QT PWHT condition. In accordance with our previous studies about the effects of various PWHT regimes on a very similar type of dissimilar T92/TP316H weldments with Nirod 600 Ni WM [18–20], the microstructures of all material regions of the presently investigated weldment are clearly coarse-grained and homogenized without the presence of traditionally heterogeneous T92 HAZ microstructural gradient, as a result of performed re-austenitization; i.e., the solutionizing stage of performed two-step QT PWHT.

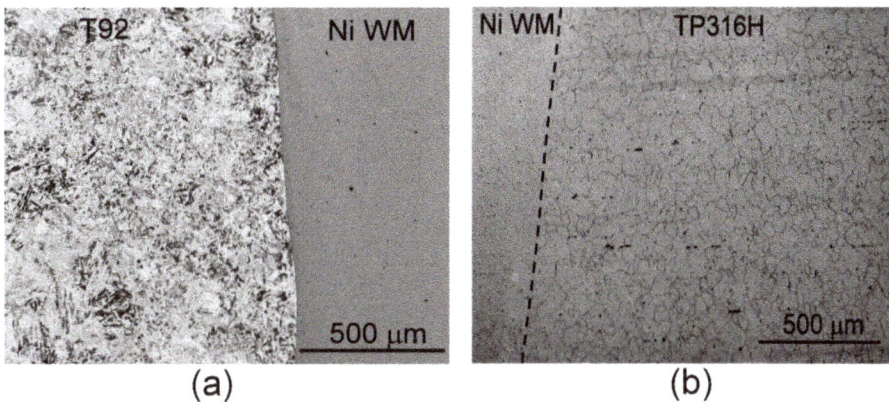

Figure 4. The general view on individual microstructural regions of T92/TP316H dissimilar weldment with Ni-based weld metal: T92/Ni WM region (**a**); Ni WM/TP316H region (**b**).

This means that the T92 steel region next to the Ni WM (Figure 4a) does not consist of several microstructural sub-regions such as coarse-grained HAZ (CGHAZ), fine-grained HAZ (FGHAZ), intercritical HAZ (ICHAZ), and subcritical HAZ (SCHAZ), which are commonly present within the HAZ regions of ferritic steels' welded joints subjected to conventional single-step "T PWHT" [17–22]. Figure 4b shows the general view on QT microstructures of NiWM and TP316H austenitic regions exhibiting polygonal grains with secondary phase precipitates. Figure 5 shows the detailed microstructure of the T92 region of the investigated weldment formed of homogenized and tempered martensite, typically consisting of tempered martensitic laths inside the blocks and packets structures within prior austenitic grains. In accordance with numerous research studies focused on normalized and tempered grade 92 steels (e.g., [6–11,23–25]), the phase composition of the T92 steel region of the investigated weldment in its initial QT PWHT condition consists of ferrite matrix and secondary phase precipitates, namely intergranular $M_{23}C_6$ (M = Cr, Fe, ...) carbides and intragranular MX (M = V, Nb; X= C, N) carbo-nitrides.

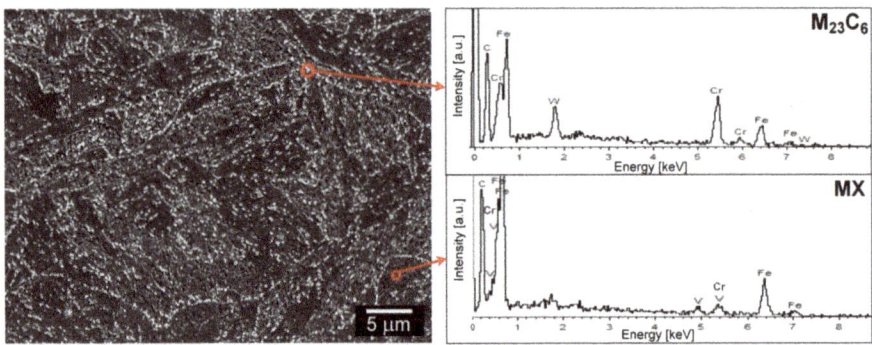

Figure 5. Scanning electron microscope (SEM)-micrograph of T92 steel region of the investigated T92/TP316H dissimilar weldment in "QT PWHT" condition and typical energy-dispersive X-ray spectroscopy (EDX)-spectra from spot-EDX analyses of $M_{23}C_6$ and MX secondary phase particles.

Figure 6 shows the initial QT PWHT microstructure of Thermanit Nicro 82 Ni WM, which consists of so-called "nickel-austenite" matrix with randomly distributed, blocky primary precipitates, namely (Ti,Nb)(C,N) or (Ti,Nb,B)(C,N) carbo-nitrides, (Nb,B)C carbides and secondary precipitates of $Cr_{23}C_6$ carbides and NbC carbides located on grain boundaries as well as in grain interiors.

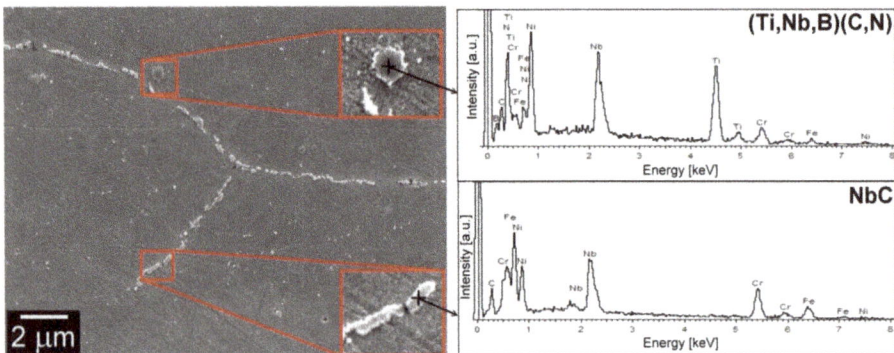

Figure 6. SEM-micrograph of Ni-based weld metal (Ni WM) region Thermanit Nicro 82 of the investigated T92/TP316H dissimilar weldment in "QT PWHT" condition and typical EDX-spectra from spot-EDX analyses of primary (Ti,Nb,B)(C,N) and secondary NbC precipitate particles.

Since Thermanit Nicro 82 filler metal does not contain any boron and nitrogen, the presence of these two elements in the primary precipitates is likely to be a result of Ni WM dilution effect due to its mixing during the welding with T92 base material, which contains both a B and N micro-alloying elements (see Table 1). In agreement with our previous works [18–20], the initial microstructure of the TP316H steel region of the studied weldment in a QT PWHT state is formed of polygonal, fully recrystallized austenitic grains with only very fine discontinuous films of $M_{23}C_6$ carbide precipitates on grain boundaries (see Figure 7).

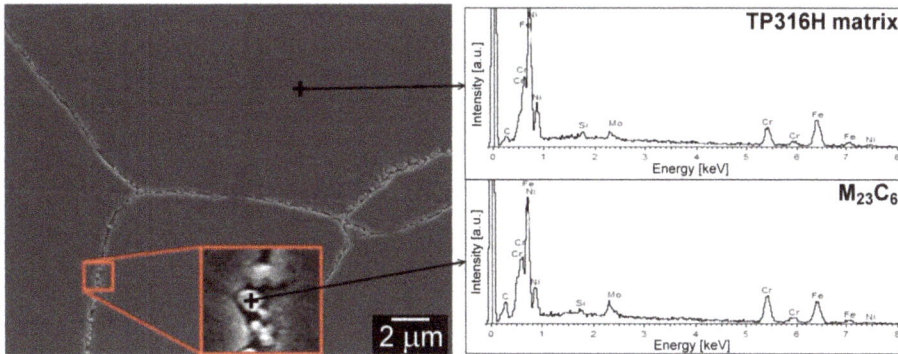

Figure 7. SEM-micrograph of TP316H steel region of the investigated T92/TP316H dissimilar weldment in "QT PWHT" condition and typical EDX-spectra from spot-EDX analyses of the steel matrix and $M_{23}C_6$ secondary phase precipitates.

This clearly indicates that the TP316H base material (BM) has not been thermally sensitized during performed QT PWHT.

3.2. Ageing Effects on Room-Temperature Tensile Properties and Hardness

The effects of isothermal ageing at 620 °C up to 2500 h of the studied weldments on their room-temperature tensile properties are shown in Figure 8.

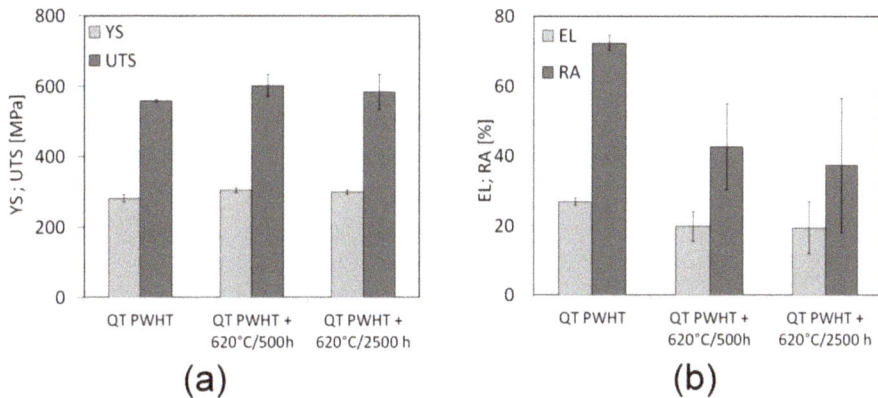

Figure 8. Ageing effects at 620 °C on room-temperature tensile properties of T92/TP316H dissimilar weldments with Ni-based weld metal.

It can be seen that applied thermal exposure of the weldments resulted only in negligible effects on their yield stress (YS) and ultimate tensile strength (UTS). In contrast, the welds ageing effects were manifested by recognizable decrease of the both measured plastic properties, namely reduction of area at fracture (RA) and total elongation (EL), whereby the RA values showed more significant deterioration at greater value scattering. Similar behavior has also been observed in our former studies [20,26–28] about ageing effects on microstructure and properties of several different types of boiler steels and their weldments. For the sake of comparison, Figure 9 shows three selected engineering stress-strain curves representing characteristics of tensile deformation behavior related to individual material states of the welds under investigation.

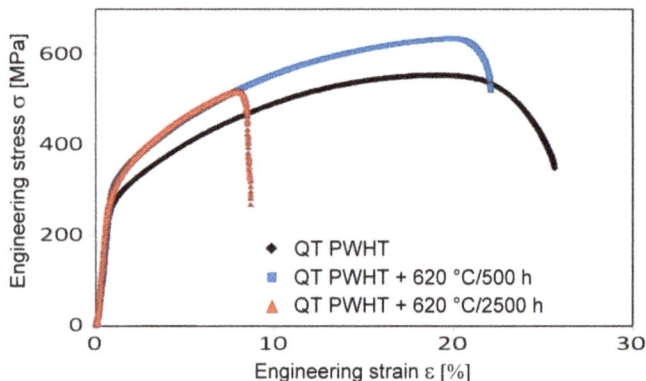

Figure 9. Ageing effects at 620 °C on tensile deformation behavior of T92/TP316H dissimilar weldments with Ni-based weld metal.

The reason for decreasing plasticity during thermal ageing is generally related to coarsening of secondary phase precipitates. Typical examples of currently observed mechanisms of the plasticity deterioration will be demonstrated and discussed in section 3.3. Figure 10 shows four selected graphs depicting the results of c-w hardness profile measurements related to individual material states of the studied weldments.

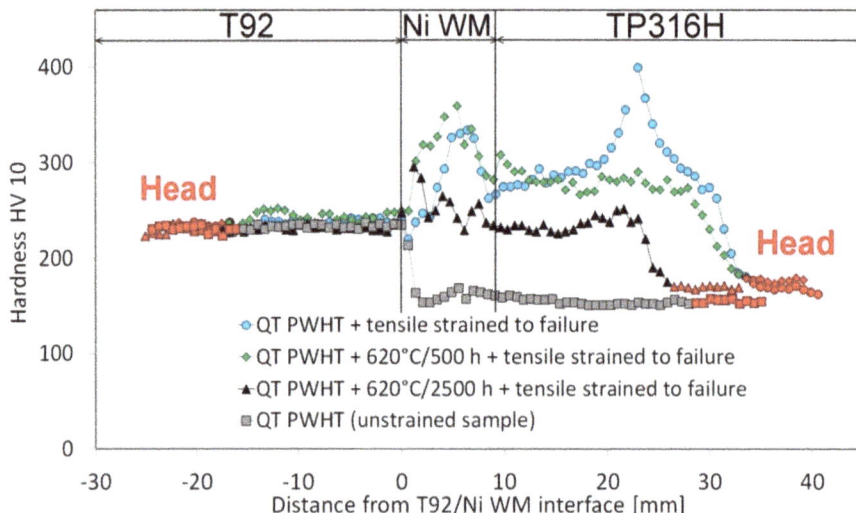

Figure 10. The effects of 620 °C isothermal ageing and subsequent room-temperature tensile testing on cross-weld hardness profiles of T92/TP316H dissimilar weldments with Ni-based weld metal.

The original undeformed c-w sample (i.e., the sample that was not subjected to tensile testing) in its initial QT PWHT state clearly shows that the T92 quenched and tempered martensitic part of the studied dissimilar welded joint exhibits the highest hardness values compared to the rest of the austenitic regions (Ni WM and TP316H). In addition, the T92 region does not exhibit a typical steep hardness gradient which is well-known in ferritic steels' weldments subjected to conventional T PWHT regime [3,13,14,18–20]. By contrast, it exhibits smooth equalized course of hardness values as a result of homogenization effect of performed QT PWHT. Visible interruptions within the next c-w hardness profiles (Figure 10) indicate the failure locations of broken test specimens after conduction of room-temperature tensile tests. It is worth noting that the austenitic regions (TP316H and Ni WM) of c-w samples after all performed tensile tests show notably increased hardness values that are even higher than those of the T92 tempered martensitic weld portions. This observation can be directly related to significantly higher strain hardening effects in the austenitic weld regions than those in the T92 tempered martensitic weld part. Figure 10 also indicates negligible deformability of the tensile samples' head portions. The complete overview of room-temperature tensile properties and all failure locations occurred after the tensile tests are summarized in Table 2.

Table 2. The complete overview of room-temperature tensile properties and failure locations of T92/TP316H dissimilar weldments with Ni-based weld metal for individual material states.

Specimen Number	Material State	YS [MPa]	UTS [MPa]	EL [%]	RA [%]	Failure Location
1		258	545	25	69	TP316H BM
2	QT PWHT	265	562	27	74	TP316H BM
3		285	561	27	74	TP316H BM
4		289	610	25	52	Ni WM
5	QT PWHT + 620 °C/500 h	304	636	21	51	Ni WM
6		292	562	14	25	Ni WM/TP316H interface
7		287	517	8	14	T92/Ni WM interface
8	QT PWHT + 620 °C/2500 h	274	602	25	61	TP316H BM
9		290	634	22	37	Ni WM/TP316H interface

According to the obtained results it can be stated that the tensile deformation and failure behavior of investigated dissimilar weldments clearly exhibited competitive manifestations between the austenitic TP316H region and Ni-based weld metal. In addition, all the fracture locations showed significant hardness peaks due to the locally most intensive, necking-related strain hardening effects occurred during the tensile tests (Figure 10). Moreover, these increased strain hardening effects were often observed not only at the final fracture location (in either TP316H or Ni WM region) but also at the location of concurrent necking area (i.e., complementarily in either Ni WM or TP316H region). Figure 11 shows photo-macrographs of selected longitudinal sections of broken tensile specimens' counterparts indicating three different tensile failure locations for the investigated dissimilar weldments in their individual material states.

Figure 11. Photo-macrographs of broken tensile specimens' counterparts showing different failure locations after room-temperature tensile tests of T92/TP316H weldments in their individual material states: "QT PWHT" (**a**); "QT PWHT + 620 °C/500 h" (**b**); "QT PWHT + 620 °C/2500 h" (**c**).

At a macroscopic scale, three different failure locations of the investigated dissimilar weldments can be distinguished, namely the failure in BM of TP316H austenitic steel (Figure 11a), the failure in central part of Ni WM Thermanit Nicro 82 (Figure 11b), and finally the interfacial failure at T92 BM/Ni

WM interface (Figure 11c). The detailed discussion on the fracture behavior of the studied weldments in relation to their microstructure will be provided in the following section.

3.3. Ageing Effects on Fracture Behavior in Relation to Microstructure

As already indicated in Table 2 and Figures 9–11, the room-temperature tensile deformation and failure behavior exhibited competitive manifestations between the austenitic TP316H steel and Thermanit Nicro 82 Ni WM regions. This observation seems to be reasonable because of generally known higher deformability of austenitic alloys compared to the ferritic (also tempered martensitic) ones. In general, the higher deformation ability of austenitic alloys at room temperature than that of the ferritic ones can be related to the existence of close packed crystallographic planes and thus higher number of active slip systems in austenitic alloys with face-centered cubic (f.c.c.) crystal structure compared to ferritic alloys with base-centered cubic (b.c.c.) crystal structure without the close packed planes [29,30]. As already discussed in previous section (Figure 10), the hardness profile of undeformed c-w sample of studied T92/TP316 welded joint before the tensile test, clearly showed lower hardness values for both the austenitic weld portions (i.e., Ni WM and TP316H) compared to T92 tempered martensitic part. However, the applied tensile loading during the tensile tests resulted in intensive strain hardening effects in both the austenitic weld regions. Thus, the preferential deformation of the austenitic weld regions (TP316H, Ni WM) during tensile straining at room temperature gave rise to their higher propensity for local plasticity exhaustion and final failure occurrence. In order to analyze mutual correlation of the weld failure location (Table 2) and the results of tensile tests (Figures 8 and 9), microstructural features just beneath fracture path and fracture surface characteristics for three different failure locations are documented in Figures 12–14. In initial QT PWHT state with a final failure location in TP316H base material, the appearance of a fracture path indicates the mixed intergranular/transgranular fracture mechanism (see Figure 12a). The austenitic grains of TP316H material beneath fracture location are significantly deformed as a result of intensive necking-related local tensile deformation. A corresponding SEM-fractograph in Figure 12b shows typical dimple tearing areas and pronounced tear ridges observed on the fracture surface within the austenitic TP316H BM region.

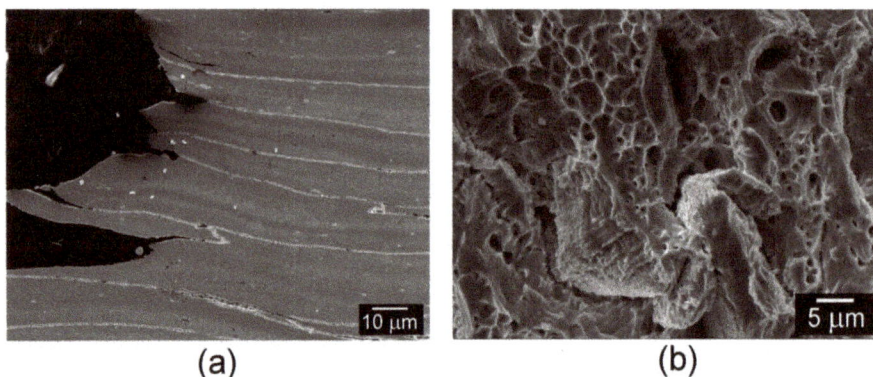

(a) (b)

Figure 12. Fracture analysis of the failure in TP316H steel region after room-temperature tensile test of T92/TP316H dissimilar weldment in "QT PWHT" condition: Fracture path and microstructure beneath the failure (**a**); Fracture surface with tear ridges and dimple areas (**b**).

Figure 13a shows a typical microstructure beneath fracture occurred in central region of Ni WM of tensile tested weldment after its isothermal ageing at 620 °C for 500 h. It preserves its original features already characterized in Section 3.1 (Figure 6); i.e., it exhibits coarse-grained blocky primary Ti-Nb or Ti-Nb-B carbo-nitrides and secondary NbC and $Cr_{23}C_6$ carbides on migrated high-angle grain boundaries. The observed fracture path in Figure 13a and corresponding SEM-fractograph in Figure 13b indicate mixed intergranular/transgranular characteristics of the fracture. Clear decohesion features can be observed at particle/matrix interfaces on grain boundaries of Ni WM (Figure 13a) that represent one example of the mechanism of the documented plasticity degradation (Figures 8 and 9). The major fracture micro-mechanism refers to the microvoid coalescence resulting in the dimpled morphology of the fracture surface (Figure 13b). However, in comparison with the fracture in TP316H region with significantly pulled-out dimples (Figure 12b), the dimpled fracture in the central region of Ni WM (Figure 13b) contains a higher amount of smaller shallow dimples indicating the material plasticity deterioration. The detailed SEM-micrograph in Figure 13c clearly shows two typical mechanisms of the plasticity deterioration, namely the cracking of the secondary phase particles and the intergranular decohesion related to the cracking at the intergranular particle/matrix interfaces.

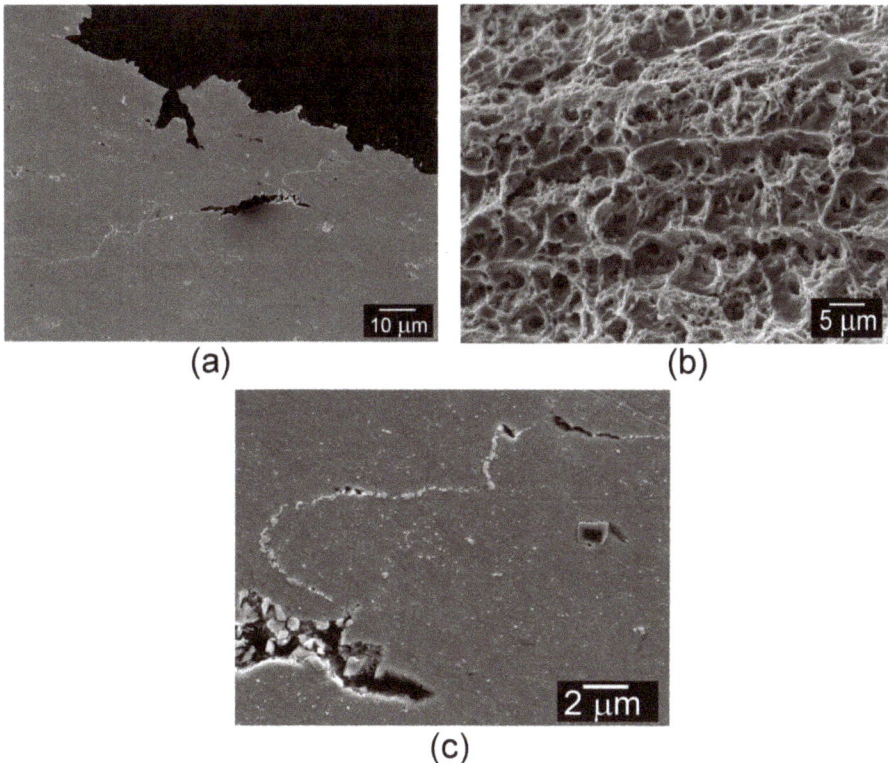

(a)   (b)

(c)

Figure 13. Fracture analysis of the failure in Ni WM central region after room-temperature tensile test of T92/TP316H dissimilar weldment in "QT PWHT + 620 °C/500 h" condition: Fracture path and microstructure beneath the failure (**a**); Fracture surface showing dimple areas (**b**); Detailed view of the mechanisms of plasticity deterioration (**c**).

The sub-fracture microstructure of tensile tested weldment thermally exposed at 620 °C for 2500 h is visualized in Figure 14a,b. The detailed observation revealed that the macroscopically classified interfacial failure (Figure 11c) did not occur directly at the T92 BM/Ni WM interface but the fracture ran partly through T92 BM and largely through precipitation-depleted zone (PDZ) of Ni WM located at a distance of about 5–8 μm from the fusion boundary (see Figure 14a,b). The occurrence of PDZ in Ni WM can presumably be related to specific welding metallurgy phenomena such as weld metal dilution, the presence of unmixed or partially melted zone, and/or type II boundaries formation [4,13,31–35].

It should be stated that all of these phenomena have been indeed noticed at several locations of Ni WM/T92 BM dissimilar weld interface of the welds investigated. However, the detailed study of Ni WM and specifically its interfacial microstructures including the characterization of densely distributed fine precipitates next to the PDZ will the subject of our separate investigation. It is worth noting that the typical mechanisms of the plasticity deterioration, namely the cracking of secondary phase particles and decohesion at particle/matrix interfaces were also clearly observed in T92 region of Figure 14b. However, since the majority of the fracture occurred in PDZ of Ni WM, a corresponding SEM-fractograph is visualized in Figure 14c. It clearly shows visible signs of thermal embrittlement represented by shallow dimples combined with indications of fine quasi-cleavage areas. This observation correlates well with the lowest plasticity determined by the tensile tests for the weld material state with the highest duration of thermal exposure (Figures 8 and 9). Complementarily to the fracture path analysis (Figure 14a,b) and fracture surface analysis (Figure 14c), an EDX analysis (Figure 14d) from the sub-fracture area confirmed that substantial portion of the fracture was located within Ni WM dilution area.

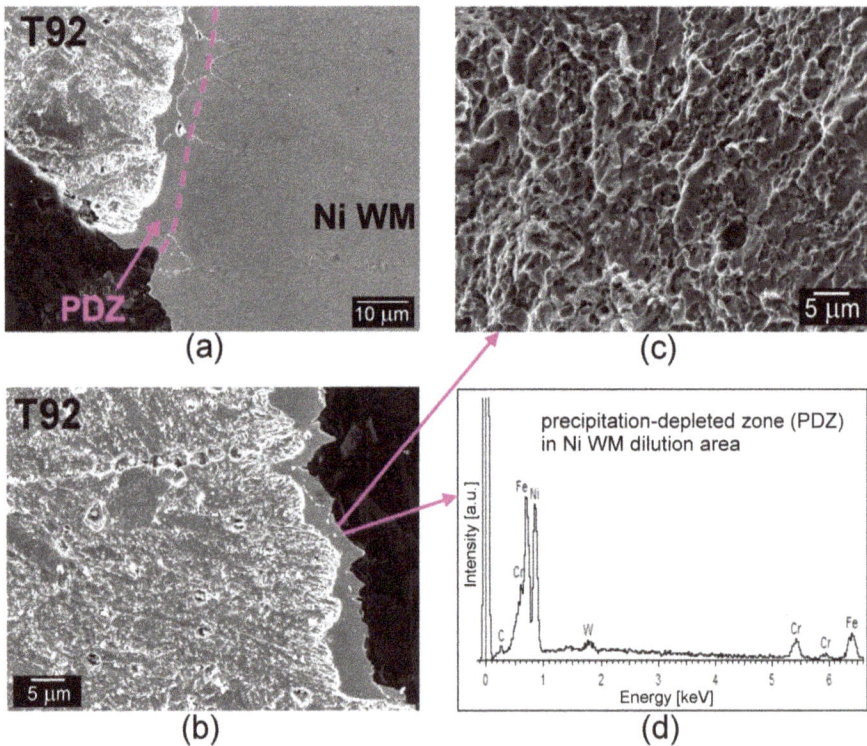

Figure 14. Fracture analysis of the failure occurred in vicinity of T92 BM/Ni WM interface after room-temperature tensile test of T92/TP316H dissimilar weldment in "QT PWHT + 620 °C/2500 h" condition: Fracture path and microstructure beneath the failure (**a,b**); Fracture surface showing dimple and intensive shear areas (**c**); Typical EDX spectrum of Ni WM dilution area at the fracture location (**d**).

The detailed SEM microstructure beneath the fracture path in Figure 15 clearly shows all indicated mechanisms of plastic properties degradation after the longest studied thermal exposure (620 °C/2500 h) of the investigated dissimilar weldment. When considering the room-temperature tensile deformation behavior of the studied dissimilar weldments, it is reasonable to expect the failure occurrence in the region with the lowest tensile strength for a given weld material state. Furthermore, it has been clearly evidenced (Figure 10) that the effects of strain hardening and plasticity exhaustion play an important role in the final failure occurrence. The observed changes of failure locations among individual weld material states (Table 2) can be directly related to gradual evolution of thermal embrittlement effects in critical weld regions during performed ageing exposure (Figures 12–15).

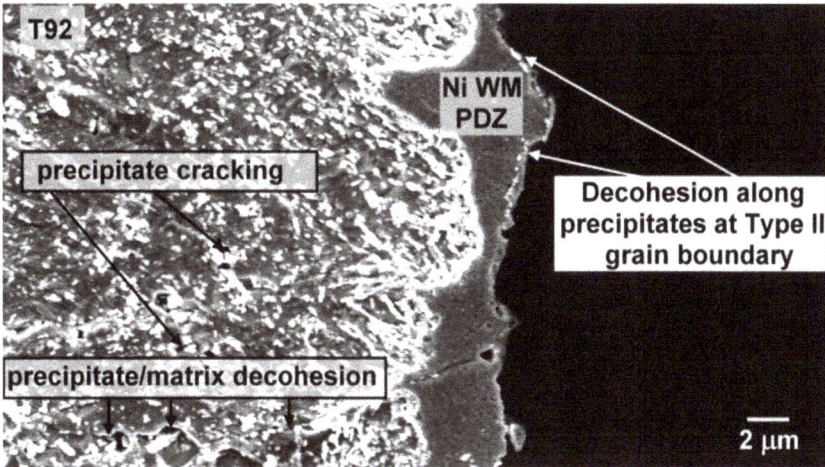

Figure 15. The mechanisms of plasticity degradation of thermally exposed (620 °C/2500 h) T92/TP316H nickel-based transition weldment subjected to tensile testing at room temperature.

The obtained results clearly indicate that the Ni WM and especially its interfacial region with T92 BM belong to the weld most critical regions with respect to failure occurrence after long-term thermal exposure.

Our future research will be focused on a continuation of the present investigation towards longer ageing experiments, creep tests, and hydrogen embrittlement studies in relation with the welds microstructure degradation characterization, aiming at a reliable prediction of their life-time performance in high-temperature steam circuits of highly-efficient environmentally friendly power plants.

4. Summary and Conclusions

Unconventionally post-weld heat treated (specifically: quenched and tempered) T92/TP316H martensitic/austenitic weldments with Ni-based weld metal (Thermanit Nicro 82) were subjected to long-term isothermal ageing at 620 °C for up to 2500 h in order to investigate the effects of thermal embrittlement evolution on room-temperature tensile properties and fracture behavior in relation with microstructure. From the obtained results the following conclusions could be drawn:

- In the initial material state i.e., in QT PWHT condition, microstructural observation and cross-weld hardness measurement of the studied weldment revealed that the use of unconventional two-step PWHT resulted in clear microstructural homogenization (i.e., no HAZ microstructural gradients) and hardness values equalization within the both weld base materials (T92 and TP316H).
- Subsequent ageing effects at 620 °C of the studied weldments resulted in gradual changes in their individual microstructures represented mainly by additional precipitation and coarsening of secondary phase precipitates. These microstructural changes resulted in significant detrimental effects on the welds plasticity (EL, RA values), whereas their effects on the welds strength (Re, Rm values) were rather small.
- The effects of thermal exposure resulted also in several changes of failure locations within the studied weldments. In the initial QT PWHT condition, the failure occurred always in TP316H steel

base material. After the ageing at 620 °C for 500 h, the weld failure location was in major cases in central part of Ni WM. The weld failures after the longest thermal exposure (620 °C/2500 h) were different for each individual tensile test. The most brittle failure was the failure in the vicinity of T92/Ni WM interface.

- By analyzing the fracture path and microstructural features beneath the fracture surfaces, the clearly observed mechanisms of the welds plasticity deterioration after thermal exposure are related to the cracking of coarse secondary phase particles and interfacial decohesion at precipitate/matrix interfaces during tensile straining. After the longest applied thermal exposure (620 °C/2500 h), interfacial decohesion occurred in Ni WM dilution area along the precipitates at Type II grain boundary.

- Fractographic analyses of fracture surfaces of broken tensile samples also revealed clear evolution of thermal embrittlement processes depending on the thermal exposure duration. In the initial QT PWHT material state, the fracture surface related to the failure in TP316H BM showed pulled-out dimples and pronounced tear ridges. Thermal exposure of the weldment at 620 °C for 500 h resulted in failure located in Ni WM. It also showed dimple morphology but without pronounced tear ridges. The most thermally embrittled material state of the weldment aged at 620 °C for 2500 h exhibited small shallow dimples combined with fine quasi-cleavage areas on the fracture surface.

- The most of fracture locations of the studied weldments in their individual material states exhibited significant hardness peaks due to the intensive, necking-related strain hardening effects that occurred during the room-temperature tensile tests.

- The tensile deformation and fracture processes of the weldments in their individual material states exhibited competitive behavior between the austenitic TP316H region and Ni-based weld metal. The observed changes in failure locations for individual material states of the weldment can be directly related to specific thermal embrittlement evolution in critical weld regions during performed isothermal ageing. The obtained results indicate that the Ni WM and T92/Ni WM interfacial region represent the most critical zones of the investigated weldment with respect to the failure occurrence after long-term thermal exposure.

Author Contributions: Conceptualization, L.Č. and L.F.; Methodology, L.Č., P.Š. and M.D.; Validation, L.Č. and M.D.; Formal Analysis, L.Č., L.F. and M.D.; Investigation, L.Č., L.F., P.Š., M.V. and M.D.; Resources, M.D.; Data Curation, L.Č., L.F. and M.D.; Writing-Original Draft Preparation, L.F.; Writing-Review & Editing, L.F.; Visualization, L.Č. and L.F.; Supervision, L.F. and P.Š.; Project Administration, P.Š. and L.F.; Funding Acquisition, P.Š.

Funding: This research was funded by "Vedecká Grantová Agentúra MŠVVaŠ SR a SAV" (project "VEGA 2/0151/16").

Acknowledgments: This work was supported by "Vedecká Grantová Agentúra MŠVVaŠ SR a SAV" under the project "VEGA 2/0151/16" and partly by "Agentúra na Podporu Výskumu a Vývoja" under the contract "APVV-16-0194". The work was also realized within the frame of the project "ITMS 26220220186".

Conflicts of Interest: The authors declare no conflict of interest.

References

1. Parker, J.D.; Stratford, G.C. The high-temperature performance of nickel-based transition joints II. Fracture behaviour. *Mater. Sci. Eng. A* **2001**, *299*, 174–184. [CrossRef]

2. Anand, R.; Sudha, C.; Karthikeyan, T.; Terrance, A.L.E.; Saroja, S.; Vijayalakshmi, M. Effectiveness of Ni-based diffusion barriers in preventing hard zone formation in ferritic steel joints. *J. Mater. Sci.* **2009**, *44*, 257–265. [CrossRef]

3. Kim, R.; Kwak, S.-C.; Choi, I.-S.; Lee, Y.-K.; Suh, J.-Y.; Fleury, E.; Jung, W.-S.; Son, T.-H. High-temperature tensile and creep deformation of cross-weld specimens of weld joint between T92 martensitic and Super304H austenitic steels. *Mater. Charact.* **2014**, *97*, 161–168. [CrossRef]

4. Mittal, R.; Sidhu, B.S. Microstructures and mechanical properties of dissimilar T91/347H steel weldments. *J. Mater. Process. Technol.* **2015**, *220*, 76–86. [CrossRef]

5. Silwal, B.; Li, L.; Deceuster, A.; Griffiths, B. Effect of postweld heat treatment on the toughness of heat-affected zone for Grade 91 steel. *Weld. J.* **2013**, *92*, 80–87.

6. Sklenička, V.; Kuchařová, K.; Svobodová, M.; Kvapilová, M.; Král, P.; Horváth, L. Creep properties in similar weld joint of a thick-walled P92 steel pipe. *Mater. Charact.* **2016**, *119*, 1–12. [CrossRef]

7. Pandey, C.; Mahapatra, M.M.; Kumar, P.; Saini, N. Dissimilar joining of CSEF steels using autogenous tungsten-inert gas welding and gas tungsten arc welding and their effect on δ-ferrite evolution and mechanical properties. *J. Manuf. Process.* **2018**, *31*, 247–259. [CrossRef]

8. Zhang, J.; Du, B.Sh.; Li, X.M.; Qin, G.L.; Zou, Y. Microstructure evolution of P92 steel weld metal after service for 8000 h. *Kovove Mater.* **2017**, *55*, 115–121. [CrossRef]

9. Wu, Q.; Lu, F.; Cui, H.; Ding, Y.; Liu, X.; Gao, Y. Microstructure characteristics and temperature-dependent high cycle fatigue behavior of advanced 9% Cr/CrMoV dissimilarly welded joint. *Mater. Sci. Eng. A* **2014**, *615*, 98–106. [CrossRef]

10. Cao, J.; Gong, Y.; Zhu, K.; Yang, Z-G.; Luo, X-M.; Gu, F-M. Microstructure and mechanical properties of dissimilar materials joints between T92 martensitic and S304H austenitic steels. *Mater. Des.* **2011**, *32*, 2763–2770. [CrossRef]

11. Chen, G.; Song, Y.; Wang, J.; Liu, J.; Yu, X.; Hua, J.; Bai, X.; Zhang, T.; Zhang, J.; Tang, W. High-temperature short-term tensile test and creep rupture strength prediction of the T92/TP347H dissimilar steel weld joints. *Eng. Fail. Anal.* **2012**, *26*, 220–229. [CrossRef]

12. Wang, W.; Wang, X.; Zhong, W.; Hu, L.; Hu, P. Failure analysis of dissimilar steel welded joints in a 3033t/h USC boiler. *Procedia Mater. Sci.* **2014**, *3*, 1706–1710. [CrossRef]

13. Karthick, K.; Malarvizhi, S.; Balasubramanian, V.; Krishnan, S.A.; Sasikala, G.; Albert, S.K. Tensile and impact toughness properties of various regions of dissimilar joints of nuclear grade steels. *Nucl. Eng. Technol.* **2018**, *50*, 116–125. [CrossRef]

14. Rathod, D.W.; Pandey, S.; Singh, P.K.; Prasad, R. Experimental analysis of dissimilar metal weld joint: Ferritic to austenitic stainless steel. *Mater. Sci. Eng. A* **2015**, *639*, 259–268. [CrossRef]

15. Wang, H.T.; Wang, G.Z.; Xuan, F.Z.; Tu, S.T. Fracture mechanism of a dissimilar metal welded joint in nuclear power plant. *Eng. Fail. Anal.* **2013**, *28*, 134–148. [CrossRef]

16. Li, Y.; Wang, J.; Han, E.-H.; Yang, C. Structural, mechanical and corrosion studies of Cr-rich inclusions in 152 cladding of dissimilar metal weld joint. *J. Nucl. Mater.* **2018**, *498*, 9–19. [CrossRef]

17. Cao, J.; Gong, Y.; Yang, Z.-G. Microstructural analysis on creep properties of dissimilar materials joints between T92 martensitic and HR3C austenitic steels. *Mater. Sci. Eng. A* **2011**, *528*, 6103–6111. [CrossRef]

18. Falat, L.; Výrostková, A.; Svoboda, M.; Milkovič, O. The influence of PWHT regime on microstructure and creep rupture behaviour of dissimilar T92/TP316H ferritic/austenitic welded joints with Ni-based filler metal. *Kovove Mater.* **2011**, *49*, 417–426. [CrossRef]

19. Falat, L.; Čiripová, L.; Kepič, J.; Buršík, J.; Podstranská, I. Correlation between microstructure and creep performance of martensitic/austenitic transition weldment in dependence of its post-weld heat treatment. *Eng. Fail. Anal.* **2014**, *40*, 141–152. [CrossRef]

20. Falat, L.; Kepič, J.; Čiripová, L.; Ševc, P.; Dlouhý, I. The effects of postweld heat treatment and isothermal aging on T92 steel heat-affected zone mechanical properties of T92/TP316H dissimilar weldments. *J. Mater. Res.* **2016**, *31*, 1532–1543. [CrossRef]

21. Nhung, L.T.; Khanh, P.M.; Hai, L.M.; Nam, N.D. The relationship between continuous cooling rate and microstructure in the heat affected zone (HAZ) of the dissimilar weld between carbon steel and austenitic stainless steel. *Acta Metall. Slovaca* **2017**, *23*, 363–370. [CrossRef]

22. Lee, S.-H.; Na, H-S.; Lee, K-W.; Lee, J-Y.; Kang, C.Y. Effect of austenite-to-ferrite phase transformation at grain boundaries on PWHT cracking susceptibility in CGHAZ of T23 steel. *Metals* **2018**, *8*, 416. [CrossRef]

23. Kral, P.; Dvorak, J.; Sklenicka, V.; Masuda, T.; Horita, Z.; Kucharova, K.; Kvapilova, M.; Svobodova, M. The effect of ultrafine-grained microstructure on creep behaviour of 9% Cr steel. *Materials* **2018**, *11*, 787. [CrossRef] [PubMed]

24. Zhao, L.; Jing, H.; Xu, L.; Han, Y.; Xiu, J. Experimental study on creep damage evolution process of Type IV cracking in 9Cr–0.5Mo–1.8W–VNb steel welded joint. *Eng. Fail. Anal.* **2012**, *19*, 22–31. [CrossRef]

25. Barbadikar, D.R.; Deshmukh, G.S.; Maddi, L.; Laha, K.; Parameswaran, P.; Ballal, A.R.; Peshwe, D.R.; Paretkar, R.K.; Nandagopal, M.; Mathew, M.D. Effect of normalizing and tempering temperatures on microstructure and mechanical properties of P92 steel. *Int. J. Press. Vessels Pip.* **2015**, *132–133*, 97–105. [CrossRef]

26. Blach, J.; Falat, L.; Ševc, P. Fracture characteristics of thermally exposed 9Cr–1Mo steel after tensile and impact testing at room temperature. *Eng. Fail. Anal.* **2009**, *16*, 1397–1403. [CrossRef]

27. Blach, J.; Falat, L. The influence of thermal exposure and hydrogen charging on the notch tensile properties and fracture behaviour of dissimilar T91/TP316H weldments. *High Temp. Mater. Process.* **2014**, *33*, 329–337. [CrossRef]

28. Falat, L.; Čiripová, L.; Homolová, V.; Kroupa, A. The influence of isothermal ageing and subsequent hydrogen charging at room temperature on local mechanical properties and fracture characteristics of martensitic-bainitic weldments for power engineering. *J. Min. Metall. Sect. B-Metall.* **2017**, *53*, 373–382. [CrossRef]

29. Cottrell, A.H. Theory of dislocations. *Prog. Met. Phys.* **1953**, *4*, 205–264. [CrossRef]

30. Hirth, J.P. Dislocations. In *Physical Metallurgy*, 4th ed.; Cahn, R.W., Haasen, P., Eds.; Elsevier Science: Amsterdam, The Netherlands, 1996; Volume III, ISBN 978-0-444-89875-3.

31. Shin, K.-Y.; Lee, J.-W.; Han, J.-M.; Lee, K.-W.; Kong, B.-O.; Hong, H.-U. Transition of creep damage region in dissimilar welds between Inconel 740H Ni-based superalloy and P92 ferritic/martensitic steel. *Mater. Charact.* **2018**, *139*, 144–152. [CrossRef]

32. Pavan, A.H.V.; Vikrant, K.S.N.; Ravibharath, R.; Singh, K. Development and evaluation of SUS304H—IN 617 welds for advanced ultrasupercritical boiler applications. *Mater. Sci. Eng. A* **2015**, *642*, 32–41. [CrossRef]

33. Choi, K.J.; Kim, T.; Yoo, S.C.; Kim, S.; Lee, J.H.; Kim, J.H. Fusion boundary precipitation in thermally aged dissimilar metal welds studied by atom probe tomography and nanoindentation. *J. Nucl. Mater.* **2016**, *471*, 8–16. [CrossRef]

34. Dong, L.; Peng, Q.; Han, E-H.; Ke, W.; Wang, L. Microstructure and intergranular stress corrosion crackingsusceptibility of a SA508-52M-316L dissimilar metal weld joint in primary water. *J. Mater. Sci. Technol.* **2018**, *34*, 1281–1292. [CrossRef]

35. Yoo, S.C.; Choi, K.J.; Bahn, C.B.; Kim, S.H.; Kim, J.Y.; Kim, J.H. Effects of thermal aging on the microstructure of Type-II boundaries in dissimilar metal weld joints. *J. Nucl. Mater.* **2015**, *459*, 5–12. [CrossRef]

![metals logo] *metals*

MDPI

Article

Optimizing the Local Strength Mismatch of a Dissimilar Metal Welded Joint in a Nuclear Power Plant

Jie Yang * and Lei Wang

School of Energy and Power Engineering, University of Shanghai for Science and Technology,
Shanghai 200093, China; wl_21th@126.com
* Correspondence: yangjie@usst.edu.cn; Tel.: +86-021-5527-2320

Received: 3 June 2018; Accepted: 25 June 2018; Published: 28 June 2018

Abstract: Local strength mismatch is one of the important factors which affects the fracture behavior of a dissimilar metal welded joint (DMWJ). The question to consider is how to improve the fracture resistance of the DMWJ effectively by optimizing the local strength mismatch. In this paper, a DMWJ in a nuclear power plant was selected, the *J*-resistance (*J*–*R*) curve and crack growth path of the DMWJ under different strength mismatches of heat affect zone (HAZ), fusion zone (FZ), and near interface zone (NIZ) were systemically studied. And then, the optimal design of the local strength mismatch was investigated. The results show that decreasing the strength of HAZ and NIZ and increasing the strength of FZ will increase the fracture resistance of a DMWJ. Increasing the strength of FZ increases the *J*–*R* curve obviously. When the $M_s(\text{HAZ}){:}M_s(\text{FZ}){:}M_s(\text{NIZ}) = 1.12{:}1.4{:}1.26$ and $M_s(\text{HAZ}){:}M_s(\text{FZ}){:}M_s(\text{NIZ}) = 1{:}1.4{:}1$, the DMWJs have the highest *J*–*R* curves. Considering that the two *J*–*R* curves are similar, it is suggested that the fracture resistance of the DMWJ can be improved only by increasing the strength of the FZ.

Keywords: local strength mismatch; dissimilar metal welded joint; fracture resistance; crack growth path; optimal design

---

1. Introduction

As an indispensable part of nuclear power plant, the dissimilar metal welded joint (DMWJ) has been widely used. The fracture behavior of a DMWJ should be considered accurately for the safe service of the nuclear power plant. Thus, the microstructural, mechanical, and fracture properties of the DMWJ have been studied extensively by experiment and finite element simulation [1–10]. Wang et al. [1] and Ding et al. [2] studied the microstructure of DMWJs by different techniques. Ming et al. [3–5] investigated the microstructure and local mechanical properties of the low-alloy-steel/nickel-based-alloy interface in DMW by experiment. Rathod et al. [6,7] studied the mechanical properties of the DMWJ by different mechanical tests comprising bend test, transverse tensile test, tensile test, Charpy impact test, and micro-hardness measurement. Hou et al. [8] characterized the microstructure and mechanical property of the fusion boundary region of DMWJ. Yang et al. [9] and Wang et al. [10] studied the fracture behavior of crack which located in the weakest location of DMWJ by experiment and finite element simulation, respectively.

Strength mismatch is also one of the important factor which affects the fracture behavior of DMWJ, and has been paid more attention to both scholars in home and abroad. Kirk et al. studied the crack-tip constraint in singe edge notched bend specimen affected by weld strength mismatch [11]. Burstow et al. investigated the crack-tip stress fields of the weld joint under different strength mismatches [12]. Cetinel et al. studied the fracture behavior of the weld nodular cast iron affected by under-matching [13,14]. An et al. studied the ductile crack initiation behavior affected by strength mismatch [15]. Rakin et al. investigated the fracture behavior of the high-strength low alloyed steel weld joint affected by strength mismatch [16].

In recent years, many scholars focused on the study of strength mismatch effect on the fracture behavior of bi-material interface. Negre et al. [17] and Samal et al. [18] investigated the local fracture resistance and crack path deviation affected by the local strength mismatch in the bi-material interface region. Zhang et al. established a material constraint parameter M to consider the effect of strength mismatch in bi-material interface crack [19]. Fan et al. studied the fracture behavior of bi-material interface region under different work hardening mismatches [20–23].

Besides, some scholars investigated the fracture behavior of DMWJ affected by strength mismatch [24–29]. Wang et al. studied the fracture behavior of DMWJ under different strength mismatches [24,25] and the fracture resistances at different crack positions [26,27]. Xue et al. [28] investigated the micro region at the crack tip of DMWJ under different mismatches of the yield stress. Zhu et al. [29] studied the fracture behavior of nuclear pressure steel A508-III DMWJ affected by strength mismatch.

This study clarified the effect of strength mismatch on the fracture behavior of welded joints, and laid the foundation for the building of accurately structure integrity assessment. In the next step, it is necessary to consider the optimal design of the local strength mismatch to improve the fracture resistance of DMWJ.

In this study, a DMWJ in nuclear power plant was selected, and the fracture behavior of the DMWJ under different strength mismatches of the heat affect zone (HAZ), fusion zone (FZ), and near interface zone (NIZ) were systemically studied. And then, the optimal design of the local strength mismatch was investigated.

2. Finite Element Numerical Calculation

2.1. Materials

The DMWJ, which was used for connecting the safe end to the pipe-nozzle in the nuclear power plants was selected. The safe-end pipe material is austenitic stainless steel (316 L), and the pipe-nozzle material is ferrite low-alloy steel (A508), as shown in Figure 1a. The DMWJ was manufactured by applying a buttering technique. The buttering material (52 Mb) and the weld material (52 Mw) are the same nickel-base alloy, but their fabrication procedures are different. The true stress-strain curves of the four materials are shown in Figure 1b.

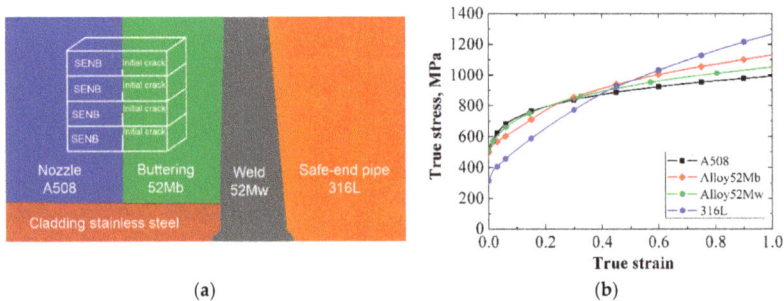

Figure 1. The dissimilar metal welded joint (DMWJ) (**a**) and the true stress-strain curves of the four materials composed of the DMWJ (**b**).

The initial crack is located in the interface of A508 and 52 Mb, as shown in Figure 1a. It is the weakest position of the DMWJ [30] and has an obvious strength mismatch. At the left side of the A508/52 Mb interface, the hardness of the material is 315 HV; at the right side of the A508/52 Mb interface, the hardness of the material is 230 HV.

To calculate the fracture behavior of the DMWJ accurately, the local heterogeneous mechanical properties are incorporated into the finite element model. The A508/52 Mb interface region is divided into three zones: HAZ, FZ, and NIZ. In addition, based on the hardness measurement and microstructure observation, the HAZ and NIZ are further divided into four-material subareas (A, B, C, and D subareas) and two-material subareas (F and G subareas), respectively. The FZ is a single-material E subarea, as shown in Figure 2a [31]. In the Figure 2a, the initial crack and the A508/52 Mb interface are located in the middle of the E subarea. The true stress-strain curves of the A, B, C, D, E, F, and G subareas are shown in Figure 2b [31].

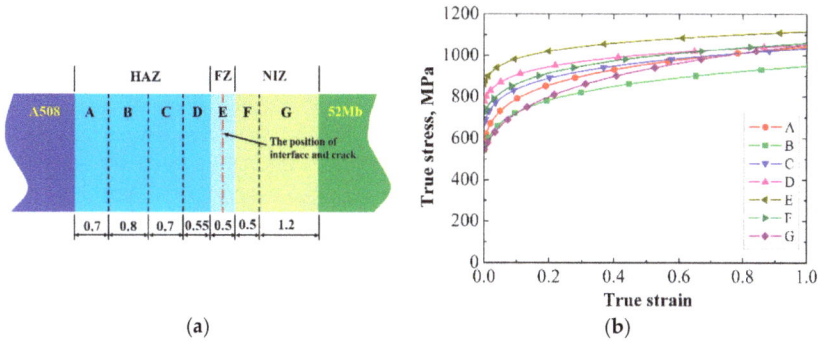

Figure 2. Different subareas in the A508/52 Mb interface region (**a**) and the true stress-strain curves of different subareas (**b**) [31].

2.2. Specimen Geometry

The SENB (singe edge notched bend) specimen was selected to study the fracture behavior of the DMWJ affected by strength mismatch. The geometry and loading configuration are shown in Figure 3. The specimen width is 14.4 mm (W = 14.4 mm), the specimen thickness is 12 mm (B = 12 mm), and the initial crack is 7.2 mm (a = 7.2 mm and a/W = 0.5), and the specimen span is 57.6 mm (L = 57.6 mm and L = 4 W).

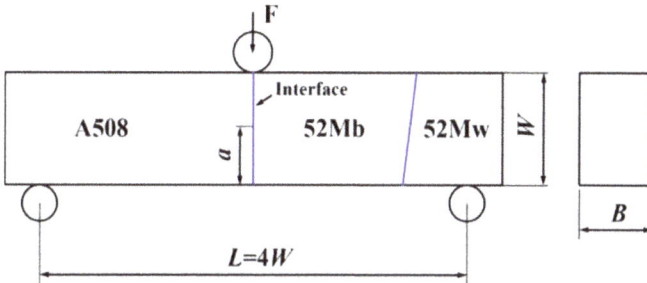

Figure 3. The geometry and loading configuration of the singe edge notched bend (SENB) specimen.

To obtain different strength mismatches, the true stress-strain curves of the HAZ, FZ, and NIZ were changed, respectively. When the true stress-strain curve of one of three zones was changed, the true stress-strain curves of subareas in this zone were changing with the same proportion, and the stress-strain curves of the other two zones remain unchanged.

Under the same true strain, the ratio of changed true stress to unchanged true stress is defined as strength mismatch coefficient M_s. The change range of M_s are 0.5, 0.6, 0.7, 0.8, 0.9, 1.0, 1.2, 1.4, 1.6, 1.8, and 2.0.

2.3. GTN Damage Model

To obtain the fracture behavior of the DMWJ under different strength mismatches, the FEM (finite element method) based on the GTN (Gurson-Tvergaard-Needleman) damage model was selected. The GTN model was firstly proposed by Gurson in the year 1977 and then modified by Tvergaard and Needleman in the year 1984. The yield function of the GTN model has the following form:

$$\phi\left(\sigma_m, \sigma_{eq}, f^*\right) = \frac{\sigma_{eq}^2}{\sigma_f^2} + 2q_1 f^* \cosh\left(\frac{3q_2\sigma_m}{2\sigma_f}\right) - 1 - q_3 f^{*2} = 0, \tag{1}$$

where σ_m is the hydrostatic stress, σ_{eq} is the von Mises equivalent stress, and σ_f is the flow stress of the "voidless" matrix material. The q_1, q_2, and q_3 are phenomenological based parameters, which are used to consider the interaction between adjacent voids. The void volume fraction f^* is the substitute of f in the original equation, which is used to take into account the gradual loss of the stress carrying capability of the material due to void coalescence. The void coalescence occurs when the void volume fraction f reaches the critical value f_c, and the fracture occurs when the f reaches the critical failure void volume fraction f_F. This GTN damage model has been implemented in ABAQUS code as a user material subroutine, and is widely used to simulate and predict the crack initiation and propagation.

Generally, the GTN model has nine parameters: the constitutive parameters q_1, q_2, and q_3, the void nucleation parameters f_N, ε_N, and S_N, the initial void volume fraction f_0, the critical void volume fraction f_C and the final failure parameter f_F, which have been listed in the Table 1 [32].

Table 1. The GTN (Gurson-Tvergaard-Needleman) parameters of different material zones [32].

Material	q_1	q_2	q_3	ε_N	S_N	f_N	f_0	f_C	f_F
A508	1.5	1	2.25	0.3	0.1	0.002	0.00008	0.04	0.25
HAZ	1.5	1	2.25	0.3	0.1	0.002	0.00015	0.04	0.25
FZ	1.5	1	2.25	0.3	0.1	0.008	0.0008	0.01	0.15
NIZ	1.5	1	2.25	0.3	0.1	0.002	0.00004	0.04	0.25
52 Mb	1.5	variable	2.25	0.3	0.1	0.002	0.000001	0.04	0.25
52 Mw	1.5	1	2.25	0.3	0.1	0.002	0.00015	0.04	0.25
316 L	1.5	variable	2.25	0.3	0.1	0.002	0.000001	0.04	0.25

The finite element code ABAQUS was selected in this study. The 3D eight-node isoperimetric elements with reduced integration (C3D8R) were used. In addition, the surface-to-surface contact (explicit) interaction type was selected. And the sliding formulation is finite sliding, the mechanical constraint formulation is kinematic contact method. Figure 4a shows the typical whole meshes of the SENB specimen, and Figure 4b shows the local meshes at the crack tip. The minimum mesh size at the crack tip is 0.1 mm × 0.1 mm [33]. The typical model contains 88,112 elements and 101,691 nodes.

A load roll is applied at the top and center of the SENB specimen, and two back-up rolls are applied at the bottom of the SENB specimen. The loading is applied at the load roll by prescribing a displacement of 6 mm, and the two back-up rolls are fixed by control displacement and rotation. The load versus load-line displacement curve can be obtained from the FEM simulation. With instantaneous crack lengths obtained at each loading point, a crack growth resistance curve can be determined, as specified in ASTM E1820 [34].

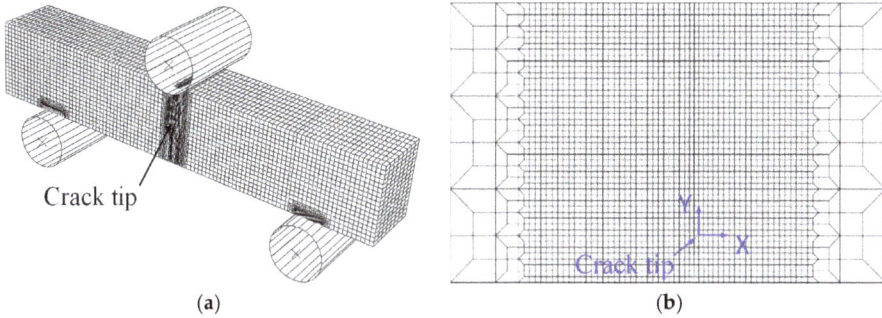

Figure 4. The whole meshes of the SENB specimen (**a**) and the local meshes along the crack growth region (**b**).

3. Results and Discussion

3.1. The Changing of Strengh Mismatch in HAZ Effect on the Fracture Behavior of DMWJ

Figure 5 shows the *J–R* curves and crack growth paths of the DMWJ when the strength of HAZ was changed individually. It can be found from Figure 5a that the *J–R* curves of the DMWJ increase firstly then decrease with increasing of the strength mismatch coefficient M_S. When the $M_S = 0.8$, the DMWJ has the highest *J–R* curve, which means that decreasing the strength of HAZ appropriately on the present basis will increase the fracture resistance of the DMWJ. When the $M_S \geq 1.6$, the *J–R* curves tend to be stable and will not change with increasing of the M_S. That is, increasing the strength of HAZ excessively is invalid for the increasing of fracture resistance.

It can also be found from Figure 5b that when the $M_S \leq 0.8$, the crack has a deviation path towards the left side, and propagates in the HAZ along the direction of 45°. When the $M_S > 0.8$, the crack has a deviation path towards the right side, and propagates along the FZ/NIZ interface. It shows that the crack growth paths deviate to the low-strength material side, and are mainly controlled by local strength mismatch. When the $M_S \leq 0.8$, the strength of D zone at the left side of crack is lower than the strength of F zone at the right side of crack, the crack has a deviation path towards the left side; when the $M_S > 0.8$, the strength of F zone at the right side of crack is lower than the strength of D zone at the left side of crack, the crack has a deviation path towards the right side. It is consistent with the results of literatures [30,35].

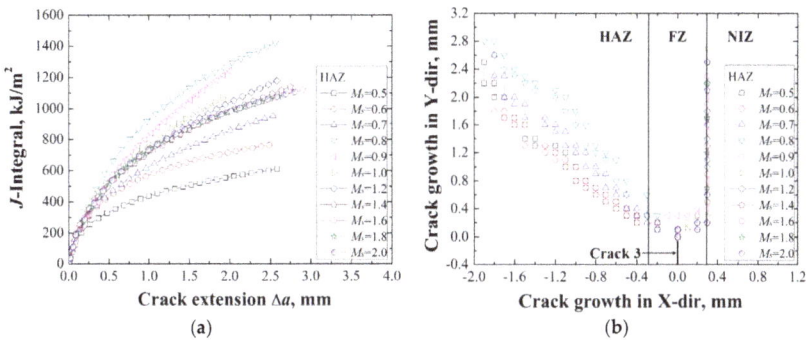

Figure 5. The *J–R* curves (**a**) and the crack growth paths (**b**) at different strengths of HAZ.

3.2. The Changing of Strength Mismatch in FZ Effect on the Fracture Behavior of DMWJ

Figure 6 shows the *J–R* curves and crack growth paths of the DMWJ when the strength of FZ was changed individually. It can be found from Figure 6a that the *J–R* curves of the DMWJ increase firstly then decrease with increasing of the strength mismatch coefficient M_s. When the M_s = 1.4, the DMWJ has the highest *J–R* curve, which means that the strength of FZ still has much room for improvement on the present basis. When the $M_s \geq$ 1.8, the *J–R* curves tend to be stable and will not change with increasing of the M_s. That is, increasing the strength of FZ excessively is invalid also for the increasing of fracture resistance.

It can also be found from Figure 6b that when the M_s < 1.4, the crack has a deviation path towards the right side, and propagates along the FZ/NIZ interface. When the M_s = 1.4, there is a crack bifurcation phenomenon. The crack has a deviation path towards the right side firstly, and then propagates to both left and right sides at the same time, as shown in Figure 7. The Figure 7 shows the VVF (void volume fraction) of the elements, which reflects the crack growth path and crack growth length of the specimen. During the loading process, the element will be regarded as losing its effectiveness when the VVF of the element reaches to the final failure parameter f_F (as shown in Table 1), and the crack will be regarded as extending an element length. For the elements in the FZ zone, the f_F is 0.15, the element color is green in the Figure 7 when the element losing its effectiveness; for the elements in the HAZ and NIZ zones, the f_F is 0.25, the element color is red in the Figure 7 when the element losing its effectiveness. Because the element length at the crack tip is known (0.1 mm), thus, the crack growth length can be calculated by counting the number of elements who lose effectiveness. When the M_s > 1.4, the crack has a deviation path towards the right side, and propagates in the NIZ along the direction of 45°.

It is also caused by the local strength mismatch. Because the crack locates in the middle of the FZ, the essence of the changing of strength mismatch in FZ is the interaction effect of D, E, and F zones in Figure 2a. Because the strength of D zone at the left side of crack is higher than the strength of F zone at the right side of crack, the crack deviated to the right side generally. When the M_s < 1.4, the strength of E zone increases with increasing of the M_s, but the strength mismatch between E and F is not enough to precipitate the crack growth path deviates from the interface, thus, the crack has a deviation path towards the right side, and propagates along the FZ/NIZ interface. When the M_s = 1.4, the mismatch between D and E and the mismatch between E and F are similar, and the crack has a bifurcation phenomenon under the competition. When the M_s > 1.4, there is a high mismatch between E and F. Dominant by the high strength mismatch, the crack penetrates the FZ/NIZ interface and propagates in the NIZ along the direction of 45°.

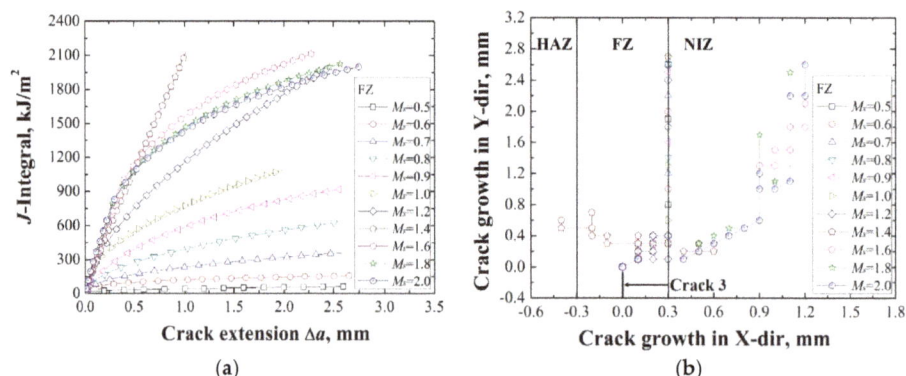

Figure 6. The *J–R* curves (**a**) and the crack growth paths (**b**) at different strengths of FZ.

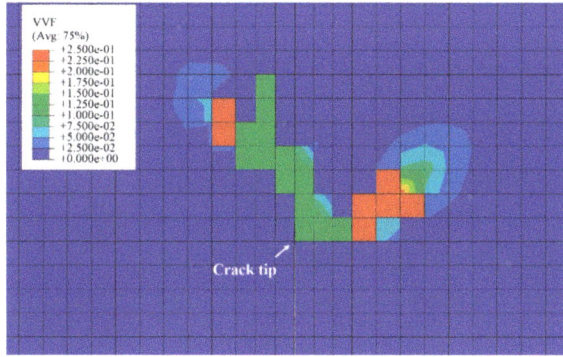

Figure 7. The crack bifurcation phenomenon when the $M_s = 1.4$.

3.3. The Changing of Strengh Mismatch in NIZ Effect on the Fracture Behavior of DMWJ

Figure 8 shows the *J–R* curves and crack growth paths of the DMWJ when the strength of NIZ was changed individually. It can be found from Figure 8a that the *J–R* curves of the DMWJ increase firstly then decrease with increasing of the strength mismatch coefficient M_s. When the $M_s = 0.9$, the DMWJ has the highest *J–R* curve, which means that decreasing the strength of NIZ appropriately on the present basis will increase the fracture resistance of DMWJ. When the $M_s = 2$, the *J–R* curve still has not to be stable and decreases with increasing of the M_s.

It can also be found from Figure 8b that when the $M_s < 0.9$, the crack has a deviation path towards the right side, and propagates in the FZ and NIZ along the direction of 45°. When $0.9 \leq M_s < 1.4$, the crack has a deviation path towards the right side, and propagates along the FZ/NIZ interface. When the $M_s \geq 1.4$, the crack has a deviation path towards the left side, and propagates along the HAZ/FZ interface.

The same as the previous analysis in Section 3.2, it is also caused by the local strength mismatch. When the $M_s < 0.9$, there is a high mismatch between E and F. Dominant by the high strength mismatch, the crack penetrates the FZ/NIZ interface and propagates in the NIZ along the direction of 45°. With increasing of the strength of NIZ, the mismatch between E and F decreases. When $0.9 \leq M_s < 1.4$, the crack does not have enough driving force (which derived from the strength mismatch) to penetrate the interface, and the crack propagates along the FZ/NIZ interface. With increasing of the strength of NIZ, when the $M_s \geq 1.4$, the strength of F zone at the right side of crack is higher than the strength of D zone at the left side of crack, the crack has a deviation path towards the left side.

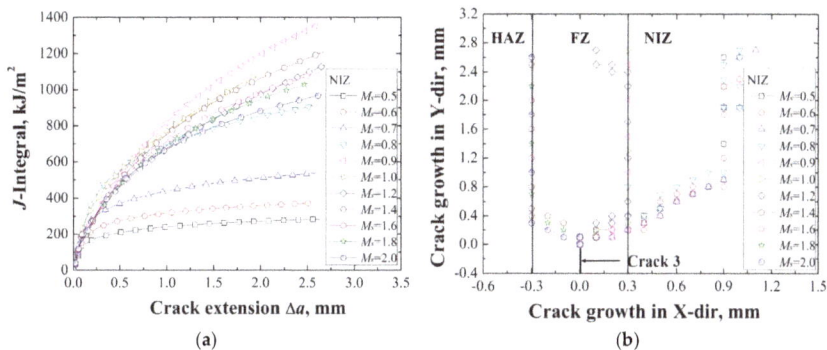

Figure 8. The *J–R* curves (**a**) and the crack growth paths (**b**) at different strengths of NIZ.

4. The Optimizing of the Local Strength Mismatch of DMWJ

Based on the results above, when the strengths of HAZ, FZ and NIZ are changed independently, the DMWJ has the highest *J*–R curve at:

M_s(HAZ):M_s(FZ):M_s(NIZ) = 0.8:1:1;
M_s(HAZ):M_s(FZ):M_s(NIZ) = 1:1.4:1;
M_s(HAZ):M_s(FZ):M_s(NIZ) = 1:1:0.9,

respectively. In order to obtain the best strength matching and the highest *J*–R curve, four sets of strength matching designs were carried out as follows:

M_s(HAZ):M_s(FZ):M_s(NIZ) = 0.8:1.12:1.008;
M_s(HAZ):M_s(FZ):M_s(NIZ) = 1.12:1.4:1.26;
M_s(HAZ):M_s(FZ):M_s(NIZ) = 1.008:1.26:0.9;
M_s(HAZ):M_s(FZ):M_s(NIZ) = 0.8:1.4:0.9.

And then, the *J*–R curves of the four sets of DMWJs were obtained and compared with other *J*–R curves, including the *J*–R curves when the strengths of HAZ, FZ, and NIZ were changed independently and the *J*–R curve of the real DMWJ, as shown in Figure 9.

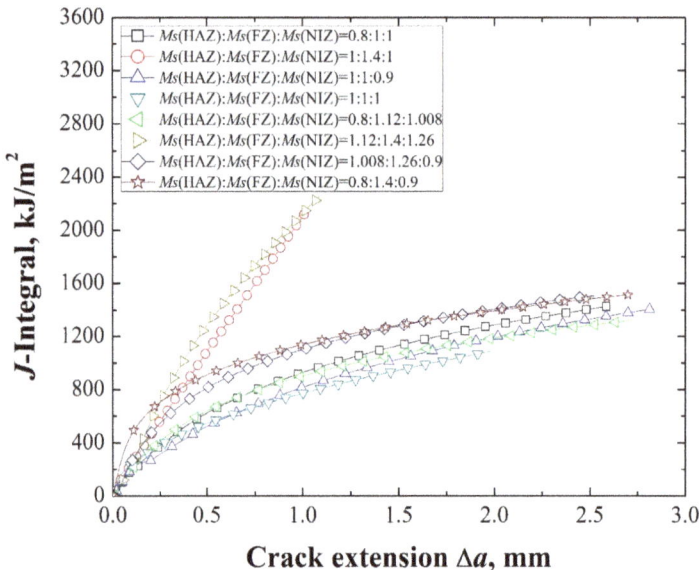

Figure 9. The *J*–R curves of the DMWJ under different strength mismatches.

It can be seen from Figure 9 that the *J*–R curves of the seven sets of optimized DMWJs are higher than the *J*–R curve of the real DMWJ. When the

M_s(HAZ):$_s$(FZ):M_s(NIZ) = 1.12:1.4:1.26

and

M_s(HAZ):M_s(FZ):M_s(NIZ) = 1:1.4:1,

the DMWJs have the highest *J*–R curves. Which means that increasing the strength of FZ will increase the *J*–R curve obviously, and will obtain the best results when the strengths of HAZ, FZ, and NIZ were wholly increased. But, the two highest *J–R* curves are similar, considering with the cost, it is suggested that the *J–R* curve of the DMWJ can be improved only by increasing the strength of the FZ.

In general, the results above show that the strength mismatch obviously effects on the *J–R* curve and crack growth path of DMWJ. As the weakest position of the nuclear power plant, the optimizing of the local strength mismatch of DMWJ is significant for the safety of the nuclear power plant. The optimizing design of the local strength mismatch of DMWJ needs to be further investigated by experiment.

5. Conclusions

(1) With increasing of the strength mismatch coefficient M_s, the *J–R* curve of DMWJ increases firstly then decreases, and tends to be stable when the M_s increases to a certain value.

(2) Decreasing the strength of HAZ and NIZ and increasing the strength of FZ increase the fracture resistance of the DMWJ. When the strengths of the HAZ, FZ, and NIZ are changed independently, the DMWJ has the highest *J–R* curve at the M_s(HAZ) = 0.8, M_s(FZ) = 1.4, and M_s(NIZ) = 0.9, respectively.

(3) Increasing the strength of FZ increases the *J–R* curve obviously. When the M_s(HAZ):M_s(FZ):M_s(NIZ) = 1.12:1.4:1.26 and M_s(HAZ):M_s(FZ):M_s(NIZ) = 1:1.4:1, the DMWJs have the highest *J–R* curves. It is suggested that the *J–R* curve of the DMWJ can be improved only by increasing the strength of the FZ.

Author Contributions: Conceptualization, J.Y.; Methodology, J.Y.; Software, L.W.; Validation, J.Y. and L.W.; Writing-Original Draft Preparation, L.W.; Writing-Review and Editing, J.Y.

Funding: This research was funded by [National Natural Science Foundation of China] grant number [51605292].

Conflicts of Interest: The authors declare no conflict of interest.

References

1. Wang, S.; Ding, J.; Ming, H.; Zhang, Z.; Wang, J. Characterization of low alloy ferritic steel–Ni base alloy dissimilar metal weld interface by SPM techniques, SEM/EDS, TEM/EDS and SVET. *Mater. Charact.* **2015**, *100*, 50–60. [CrossRef]
2. Ding, J.; Zhang, Z.; Wang, J.; Han, E.H.; Tang, W.; Zhang, M.; Sun, Z. Micro-characterization of dissimilar metal weld joint for connecting pipenozzle to safe-end in generation III nuclear power plant. *Acta Metall. Sin.* **2015**, *51*, 425–439. [CrossRef]
3. Ming, H.L.; Wang, J.Q.; Han, E.H. Comparative study of microstructure and properties of low-alloy-steel/nickel-based-alloy interfaces in dissimilar metal weld joints prepared by different GTAW methods. *Mater. Charact.* **2018**, *139*, 186–196. [CrossRef]
4. Ming, H.L.; Zhang, Z.M.; Wang, J.Q.; Han, E.H.; Wang, P.P.; Sun, Z.Y. Microstructure of a safe-end dissimilar metal weld joint (SA508-52-316L) prepared by narrow-gap GTAW. *Mater. Charact.* **2017**, *123*, 233–243. [CrossRef]
5. Ming, H.L.; Zhu, R.L.; Zhang, Z.M.; Wang, J.Q.; Han, E.H.; Ke, W.; Su, M.X. Microstructure, local mechanical properties and stress corrosion cracking susceptibility of an SA508-52M-316LN safe-end dissimilar metal weld joint by GTAW. *Mater. Sci. Eng. A-Struct.* **2016**, *669*, 279–290. [CrossRef]
6. Rathod, D.W.; Pandey, S.; Singh, P.K.; Prasad, R. Mechanical Properties Variations and Comparative Analysis of Dissimilar Metal Pipe Welds in Pressure Vessel System of Nuclear Plants. *J. Press. Vessel Technol.-Trans. ASME* **2016**, *138*, 529–540. [CrossRef]
7. Rathod, D.W.; Singh, P.K.; Pandey, S.; Aravindan, S. Effect of buffer-layered buttering on microstructure and mechanical properties of dissimilar metal weld joints for nuclear plant application. *Mater. Sci. Eng. A-Stract.* **2016**, *666*, 100–113. [CrossRef]
8. Hou, J.; Peng, Q.J.; Takeda, Y.; Kuniya, J.; Shoji, T.; Wang, J.Q.; Han, E.H.; Ke, W. Microstructure and mechanical property of the fusion boundary region in an Alloy 182-low alloy steel dissimilar weld joint. *J. Mater. Sci.* **2010**, *45*, 5332–5338. [CrossRef]
9. Yang, J.; Wang, L. Fracture mechanism of cracks in the weakest location of dissimilar metal welded joint under the interaction effect of in-plane and out-of-plane constraints. *Eng. Fract. Mech.* **2018**, *192*, 12–23. [CrossRef]

10. Wang, H.T.; Wang, G.Z.; Xuan, F.Z.; Tu, S.T. Numerical investigation of ductile crack growth behavior in a dissimilar metal welded joint. *Nucl. Eng. Des.* **2011**, *241*, 3234–3243. [CrossRef]
11. Kirk, M.T.; Dodds, R.H. The influence of weld strength mismatch on crack-tip constraint in single edge notch bend specimens. *Int. J. Fract.* **1993**, *63*, 297–316. [CrossRef]
12. Burstow, M.C.; Howard, I.C.; Ainsworth, R.A. The influence of constraint on crack tip stress fields in strength mismatched welded joints. *J. Mech. Phys. Sol.* **1998**, *46*, 845–872. [CrossRef]
13. Cetinel, H.; Uyulgan, B.; Aksoy, T. The effect of yield strength mismatch on the fracture behavior of welded nodular cast iron. *Mater. Sci. Eng. A-Struct.* **2004**, *387*, 357–360. [CrossRef]
14. Cetinel, H.; Aksoy, T. The effect of undermatching on crack tip constraint in a welded structure of nodular irons. *J. Mater. Process. Technol.* **2008**, *198*, 183–190. [CrossRef]
15. An, G.B.; Ohata, M.; Mochizuki, M.; Bang, H.S.; Toyoda, M. Effect of strength mismatch on ductile crack initiation behavior from notch root under static loading. *Adv. Fract. Strength.* **2005**, *297–300*, 756–761. [CrossRef]
16. Rakin, M.; Medjo, B.; Gubeljak, N.; Sedmak, A. Micromechanical assessment of mismatch effects on fracture of high-strength low alloyed steel welded joints. *Eng. Fract. Mech.* **2013**, *109*, 221–235. [CrossRef]
17. Negre, P.; Steglich, D.; Brocks, W. Crack extension at an interface: Prediction of fracture toughness and simulation of crack path deviation. *Int. J. Fract.* **2005**, *134*, 209–229. [CrossRef]
18. Samal, M.K.; Balani, K.; Seidenfuss, M.; Roos, E. An experimental and numerical investigation of fracture resistance behaviour of a dissimilar metal welded joint. *Proc. Inst. Mech. Eng. Part C-J. Mech. Eng. Sci.* **2009**, *223*, 1507–1523. [CrossRef]
19. Zhang, Z.L. A sensitivity analysis of material parameters for the Gurson constitutive model. *Fatigue Fract. Eng. Mater. Struct.* **1996**, *19*, 561–570. [CrossRef]
20. Fan, K.; Wang, G.; Yang, J.; Xuan, F.Z.; Tu, S.T. Numerical analysis of constraint and strength mismatch effects on local fracture resistance of bimetallic joints. *Appl. Mech. Mater.* **2015**, *750*, 24–31. [CrossRef]
21. Fan, K.; Wang, G.Z.; Tu, S.T.; Xuan, F.Z. Effects of toughness mismatch on fracture behavior of bi-material interfaces. *Procedia Eng.* **2015**, *130*, 754–762. [CrossRef]
22. Fan, K.; Wang, G.Z.; Xuan, F.Z.; Tu, S.T. Effects of work hardening mismatch on fracture resistance behavior of bi-material interface regions. *Mater. Des.* **2015**, *68*, 186–194. [CrossRef]
23. Fan, K.; Wang, G.Z.; Tu, S.T.; Xuan, F.Z. Geometry and material constraint effects on fracture resistance behavior of bi-material interfaces. *Int. J. Fract.* **2016**, *201*, 143–155. [CrossRef]
24. Wang, H.T.; Wang, G.Z.; Xuan, F.Z.; Liu, C.J. Numerical research of ductile crack growth behavior for dissimilar metal welded joint of nuclear pressure vessel. *Nucl. Power Eng.* **2012**, *33*, 36–40.
25. Wang, H. Local fracture behavior in an Alloy 52M dissimilar metal welded joint in nuclear power plants. *Nucl. Tech.* **2013**, *36*, 142–147.
26. Fan, K.; Wang, G.Z.; Xuan, F.Z.; Tu, S.T. Local failure behavior of a dissimilar metal interface region with mechanical heterogeneity. *Eng. Fail. Anal.* **2016**, *59*, 419–433. [CrossRef]
27. Fan, K.; Wang, G.Z.; Xuan, F.Z.; Tu, S.T. Local fracture resistance behavior of interface regions in a dissimilar metal welded joint. *Eng. Fract. Mech.* **2015**, *136*, 279–291. [CrossRef]
28. Xue, H.; Sun, J. Study on micro region of crack tip of welded joints under different matches of yield stress. *Hot Work. Technol.* **2016**, *45*, 239–245.
29. Zhu, Z.Q.; Jing, H.Y.; Ge, J.G.; Chen, L.G. Effects of strength mis-matching on the fracture behavior of nuclear pressure steel A508-III welded joint. *Mater. Sci. Eng. A-Struct.* **2005**, *390*, 113–117. [CrossRef]
30. Wang, H.T.; Wang, G.Z.; Xuan, F.Z.; Tu, S.T. An experimental investigation of local fracture resistance and crack growth paths in a dissimilar metal welded joint. *Mater. Des.* **2013**, *44*, 179–189. [CrossRef]
31. Wang, H.T.; Wang, G.Z.; Xuan, F.Z.; Liu, C.J.; Tu, S.T. Local mechanical properties of a dissimilar metal welded joint in nuclear power systems. *Mater. Sci. Eng.* **2013**, *568*, 108–117. [CrossRef]
32. Yang, J. Micromechanical analysis of in-plane constraint effect on local fracture behavior of cracks in the weakest locations of dissimilar metal welded joint. *Acta Metall. Sin.-Engl. Lett.* **2017**, *30*, 840–850. [CrossRef]
33. Ostby, E.; Thaulow, C.; Zhang, Z.L. Numerical simulations of specimen size and mismatch effects in ductile crack growth-Part I: Tearing resistance and crack growth paths. *Eng. Fract. Mech.* **2007**, *74*, 1770–1792. [CrossRef]

34. ASTM E1820-08a. *Standard Test Method for Measurement of Fracture Toughness*; American Society for Testing and Materials: Philadelphia, PA, USA, 2008.

35. Yang, J.; Wang, G.Z.; Xuan, F.Z.; Tu, S.T.; Liu, C.J. An experimental investigation of in-plane constraint effect on local fracture resistance of a dissimilar metal welded joint. *Mater. Des.* **2014**, *53*, 611–619. [CrossRef]

MDPI

St. Alban-Anlage 66

4052 Basel

Switzerland

Tel. +41 61 683 77 34

Fax +41 61 302 89 18

www.mdpi.com

Metals Editorial Office

E-mail: metals@mdpi.com

www.mdpi.com/journal/metals